高等学校工程创新型"十二五"规划计算机教材

Android 程序设计教程

方　欣　赵红岩　主编
郭龙源　张登奇　冯战申　副主编

电子工业出版社
Publishing House of Electronics Industry
北京·BEIJING

内 容 简 介

本教材从初学者的角度出发，通过通俗易懂的语言、丰富多彩的实例、关键代码的分析，详细介绍 Android 平台基础知识以及进行项目开发应该掌握的基本应用技术。全书共分 9 章，内容包括 Android 操作系统基础知识，开发环境搭建，Android 项目的组成及开发流程，常用基本组件的使用，事件处理机制，常用高级组件，组件之间的通信技术，多媒体技术，数据存储技术和网络通信技术等。本教材注重应用实例开发，由浅入深、循序渐进地将理论知识和实例紧密结合进行介绍、剖析和实现，以加深读者对 Android 系统基础知识和基本应用的理解，帮助读者系统全面地掌握 Android 程序设计的基本思想和基本应用技术，快速提高开发技能，为进一步深入学习 Android 应用开发打下坚实的基础。

本教材可作为本科计算机科学与技术、计算机网络、信息工程、电子信息等专业的程序设计课程的教材，也作为 Android 程序设计技术的培训教材，同时可供自学者及从事计算机应用的工程技术人员参考。

未经许可，不得以任何方式复制或抄袭本书之部分或全部内容。
版权所有，侵权必究。

图书在版编目 (CIP) 数据

Android 程序设计教程 / 方欣，赵红岩主编. —北京：电子工业出版社，2014.5
高等学校工程创新型"十二五"规划计算机教材
ISBN 978-7-121-22716-5

Ⅰ. ①A… Ⅱ. ①方… ②赵… Ⅲ. ①移动终端－应用程序－程序设计－高等学校－教材 Ⅳ. ①TN929.53

中国版本图书馆 CIP 数据核字（2014）第 056214 号

策划编辑：袁　玺
责任编辑：郝黎明
印　　刷：涿州市京南印刷厂
装　　订：涿州市京南印刷厂
出版发行：电子工业出版社
　　　　　北京市海淀区万寿路 173 信箱　邮编：100036
开　　本：787×1 092　1/16　印张：23　字数：588.8 千字
版　　次：2014 年 5 月第 1 版
印　　次：2016 年 1 月第 2 次印刷
定　　价：46.50 元

凡所购买电子工业出版社图书有缺损问题，请向购买书店调换。若书店售缺，请与本社发行部联系，联系及邮购电话：(010)88254888。
质量投诉请发邮件至 zlts@phei.com.cn，盗版侵权举报请发邮件至 dbqq@phei.com.cn。
服务热线：(010)88258888。

前 言

随着移动通信与 Internet 向移动终端的普及，网络和用户对移动终端的要求越来越高，Google 于 2007 年 11 月推出专为移动终端设计的软件平台——Android。由于该平台开源以及使用 Java 作为开发语言的特点，受到越来越多程序设计人员的青睐，支持的硬件厂商也在不断增加。

与此同时，企业对 Android 项目开发方面的应用型人才需求越来越大。"全面贴近企业需求，无缝打造专业实用人才"是目前普通高等院校计算机专业教育的改革方向。

本教材于 2012 年 10 月份开始筹备，到 2013 年 8 月份编写完成。在这期间，编者不断和外界公司、企业联系，了解他们的人才需求，整个教材的编写充分结合软件企业的用人需求，经过了充分的调研和论证，具有系统性、实用性等特点。目标是为了让尽量多的初学者少走弯路，尽快掌握基础知识，创造出更多、更好的基于 Android 平台的应用程序，满足用人单位的需要。

本教材从初学者的角度出发，通过通俗易懂的语言、丰富多彩的实例、关键代码的分析，详细介绍 Android 平台基础知识以及进行项目开发应该掌握的基本应用技术。全书共分 9 章，内容包括 Android 操作系统基础知识，Android 系统开发环境的搭建，Android 项目的组成，项目开发流程，常用基本组件的使用（例如：文本框、按钮、文本编辑框、单选框等），事件处理机制，常用高级组件的使用（例如：列表显示组件、对话框组件、画廊组件、选项卡组件等），组件之间的通信技术，多媒体技术，数据存储技术和网络通信技术等。本教材注重应用实例开发，整个教材由浅入深、循序渐进地将理论知识和实例紧密结合进行介绍、剖析和实现，加深读者对 Android 系统基础知识和基本应用的理解，帮助读者系统全面地掌握 Android 程序设计的基本思想和基本应用技术，快速提高开发技能，为进一步深入学习 Android 应用开发打下坚实的基础。

本教材是编者多年来教学和软件开发经验的总结，编者对书中的内容进行了精心设计和安排，力求达到内容丰富，结构清晰。教材中给出的实例，简单实用，易于教学和读者自学；阅读本教材，并结合上机实验，能在较短的时间内掌握 Android 项目开发的基本技能。本教材随书配有教学课件，并附带了书中给出的 118 个实例源代码，所有源代码都经过反复调试，在 Android 开发平台能导入直接运行。上述教学资源均可以在华信教育资源网（www.hxedu.com.cn）免费注册后下载。

本教材可作为本科计算机科学与技术、计算机网络、信息工程、电子信息等专业的程序设计课程的教材，也可作为 Android 程序设计技术的培训教材，同时可供自学者及从事计算机应用的工程技术人员参考。本教材要求读者最好具有一定的 Java 语言基础，具有面向对象基础和其他 GUI 设计经验的人员也可以学习本书。

全书由方欣、赵红岩担任主编，郭龙源、张登奇、冯战申担任副主编，另外，廖艳等人对本书做了一些图表的绘制、校对和纠错等工作。

本教材的编写得到了"受复杂系统优化与控制湖南省普通高等学校重点实验室"的资助。

本教材的作者大都来自教学一线，我们的目标是编写一本优秀实用的教材，但是由于作者水平有限，虽然经过多次审校，可能依然存在一些不足之处，敬请读者和同行专家批评指正。

本教材教学网站：http://61.187.92.238:8100/android

编 者
2014 年 4 月

目 录

第1章 Android 概述 ··············· 1
1.1 智能手机的发展 ············· 1
1.2 智能手机操作系统简介 ······· 2
1.3 Android 操作系统简介 ······· 3
1.3.1 Android 操作系统的发展 ······· 3
1.3.2 Android 操作系统的特点 ······· 5
1.4 搭建 Android 系统开发环境 ····· 5
1.4.1 安装 JDK ····················· 6
1.4.2 安装 Eclipse ·················· 8
1.4.3 安装 Android SDK ············ 9
1.5 开发第一个 Android 项目 ········ 13
1.6 封装第一个 Android 项目 ········ 14
本章小结 ······························ 16
习题 ································· 16

第2章 Android 中的项目 ··········· 17
2.1 Android 项目的组成 ············ 17
2.1.1 几个关键文件夹和文件 ······· 18
2.2 扩充 firstDemo 项目 ············ 22
2.3 Activity 简介 ···················· 27
2.4 Android 中的常用包 ············· 28
2.5 Android 项目的大致开发流程 ····· 29
2.6 Android 中常见文件介绍 ········· 29
本章小结 ···························· 30
习题 ································· 30

第3章 Android 开发常用组件 ······· 31
3.1 Android 平台中的 View 类 ······· 31
3.2 文本显示组件 TextView ········· 33
3.2.1 TextView 组件常见的属性和方法 ···· 33
3.2.2 TextView 组件的使用实例 ···· 34
3.3 按钮组件 Button ················ 37
3.3.1 Button 组件常见的属性和方法 ····· 38
3.3.2 Button 组件使用实例 ········· 38
3.4 编辑框组件 EditText ············ 39
3.4.1 EditText 组件常见的属性和方法 ···· 39
3.4.2 EditText 组件使用实例 ······· 40

3.5 图片视图组件 ImageView ······· 42
3.5.1 ImageView 组件常用的属性和方法 ···· 42
3.5.2 ImageView 组件使用实例 ···· 43
3.6 图片按钮组件 ImageButton ····· 44
3.6.1 ImageButton 组件常用的属性和方法 ···· 44
3.6.2 ImageButton 组件使用实例 ···· 44
3.7 单选按钮组件 RadioGroup ······ 45
3.7.1 RadioGroup 组件常见的属性 ···· 46
3.7.2 RadioGroup 组件使用实例 ···· 46
3.8 复选框组件 CheckBox ·········· 47
3.8.1 CheckBox 组件常见的属性和方法 ···· 48
3.8.2 CheckBox 组件使用实例 ····· 48
3.9 下拉列表框组件 Spinner ········ 49
3.9.1 Spinner 组件常见的属性和方法 ···· 49
3.9.2 Spinner 组件使用实例 ········ 50
3.10 信息提示框组件 Toast ········· 54
3.10.1 Toast 组件常见的属性和方法 ····· 54
3.10.2 Toast 组件使用实例 ········· 54
3.11 相对布局管理器组件 RelativeLayout ···· 55
3.11.1 RelativeLayout 组件常用的属性和方法 ···· 56
3.11.2 RelativeLayout 组件使用实例 ···· 56
3.12 线性布局管理器组件 LinearLayout ···· 58
3.12.1 LinearLayout 组件常用的属性和方法 ···· 59
3.12.2 LinearLayout 组件使用实例 ···· 59
3.13 框架布局管理器组件 FrameLayout ···· 61
3.13.1 FrameLayout 组件常用的属性和方法 ···· 61
3.13.2 FrameLayout 的使用举例 ···· 61
3.14 表格布局管理器组件 TableLayout ···· 62
3.14.1 TableLayout 组件常用的属性和方法 ···· 62
3.14.2 TableLayout 的使用举例 ···· 64

3.15 布局管理器的嵌套 ………………… 65
本章小结 ………………………………………… 67
习题 ……………………………………………… 67

第 4 章 Android 中的事件处理 ………… 68
4.1 Android 中的事件处理基础 …………… 68
  4.1.1 事件处理的过程 …………………… 68
  4.1.2 事件处理模型 ……………………… 68
4.2 单击事件 OnClickListener ……………… 69
  4.2.1 单击事件基础 ……………………… 69
  4.2.2 单击事件实例 ……………………… 70
4.3 长按事件 OnLongClickListener ………… 71
  4.3.1 长按事件基础 ……………………… 71
  4.3.2 长按事件实例 ……………………… 72
4.4 焦点改变事件 OnFocusChange
  Listener ………………………………… 73
  4.4.1 焦点改变事件基础 ………………… 73
  4.4.2 焦点改变事件实例 ………………… 74
4.5 键盘事件 OnKeyListener ……………… 76
  4.5.1 键盘事件基础 ……………………… 76
  4.5.2 键盘事件实例 ……………………… 76
4.6 触摸事件 onTouchEvent ………………… 80
  4.6.1 触摸事件基础 ……………………… 80
  4.6.2 触摸事件实例 ……………………… 81
4.7 选择改变事件 OnCheckedChange ……… 82
  4.7.1 选择改变事件基础 ………………… 82
  4.7.2 RadioGroup 选择改变事件实例 …… 83
  4.7.3 CheckBox 选择改变事件实例 …… 84
4.8 选项选中事件 OnItemSelected ………… 88
  4.8.1 选项选中事件基础 ………………… 88
  4.8.2 OnItemSelected 选项选中事件实例 … 88
4.9 日期和时间监听事件 …………………… 92
  4.9.1 日期和时间选择器组件 …………… 92
  4.9.2 DatePicker 和 TimePicker 组件使用
     实例 ………………………………… 93
  4.9.3 日期和时间的设置 ………………… 95
  4.9.4 日期和时间监听事件 ……………… 97
4.10 菜单事件 ……………………………… 99
  4.10.1 菜单事件基础 …………………… 99
  4.10.2 选项菜单 OptionsMenu ………… 100
  4.10.3 上下文菜单 ContextMenu ……… 104

  4.10.4 子菜单 SubMenu ………………… 107
本章小结 ………………………………………… 109
习题 ……………………………………………… 109

第 5 章 Android 常用高级组件 ………… 110
5.1 滚动视图组件 ScrollView ……………… 110
  5.1.1 ScrollView 组件常见的属性
     和方法 ……………………………… 110
  5.1.2 ScrollView 组件使用实例 ………… 111
5.2 列表显示组件 ListView ………………… 112
  5.2.1 ListView 组件常见的属性和方法 … 112
  5.2.2 SimpleAdapter 类 …………………… 114
5.3 可展开的列表组件
  ExpandableListView …………………… 118
  5.3.1 ExpandableListView 组件基础 …… 118
  5.3.2 ExpandableListView 组件实例 …… 120
5.4 进度条组件 ProgressBar ………………… 124
  5.4.1 ProgressBar 组件基础知识 ………… 124
  5.4.2 ProgressBar 组件实例 ……………… 126
5.5 拖动条组件 SeekBar …………………… 127
  5.5.1 SeekBar 组件基础知识 …………… 128
  5.5.2 SeekBar 组件实例 ………………… 128
5.6 星级评分条组件 RatingBar ……………… 129
  5.6.1 RatingBar 组件基础 ………………… 130
  5.6.2 RatingBar 组件实例 ………………… 130
5.7 自动完成文本框
  AutoCompleteTextView ………………… 131
  5.7.1 AutoCompleteTextView 组件基础 … 132
  5.7.2 AutoCompleteTextView 组件实例 … 132
5.8 对话框组件 Dialog ……………………… 133
  5.8.1 警告对话框：AlertDialog ………… 134
  5.8.2 AlertDialog 组件实例 ……………… 136
  5.8.3 自定义对话框 ……………………… 139
  5.8.4 带进度条的对话框 ProgressDialog … 141
5.9 图片切换组件 ImageSwitcher …………… 144
5.10 画廊组件 Gallery ……………………… 147
5.11 选项卡组件 TabHost …………………… 152
  5.11.1 TabHost 组件基础 ………………… 153
  5.11.2 TabHost 组件实例 ………………… 156
本章小结 ………………………………………… 159
习题 ……………………………………………… 159

第6章 Android 组件之间的通信 160
6.1 Android 四大组件简介 160
6.2 Intent 简介 161
6.2.1 利用 Intent 启动 Activity 161
6.2.2 利用 Intent 在 Activity 之间传递数据 162
6.2.3 Intent 组件传递数据实例 163
6.3 深入了解 Intent 167
6.3.1 Intent 的构成 167
6.3.2 Intent 常用用法示例 169
6.3.3 Intent 操作实例 174
6.4 Activity 的生命周期 176
6.5 Android 中的消息处理机制 179
6.5.1 消息处理机制基础 179
6.5.2 一个简单的消息处理实例 181
6.5.3 线程基础知识 183
6.5.4 异步处理工具类：AsyncTask 187
6.6 Service 192
6.6.1 Service 基础 192
6.6.2 Service 的启动和停止 193
6.6.3 绑定 Service 194
6.6.4 Service 的生命周期 198
6.6.5 跨进程调用 Service（AIDL 服务） 199
6.6.6 Service 系统服务 204
6.7 BroadcastReceiver 207
6.7.1 BroadcastReceiver 基础 207
6.7.2 BroadcastReceiver 组件操作实例 208
6.7.3 通过 BroadCast 启动 Service 210
本章小结 212
习题 212

第7章 Android 多媒体技术 213
7.1 Android 中图形的绘制 213
7.1.1 图形绘制基础 213
7.1.2 图形绘制实例 215
7.2 Android 中图像的处理 217
7.2.1 图像的获取 217
7.2.2 对获取的图像进行处理 218
7.2.3 图像处理实例 219
7.3 Android 中的动画 221
7.3.1 Tween 动画 221
7.3.2 创建动画实例 222
7.3.3 通过 XML 文件来创建动画 226
7.3.4 Frame 动画 227
7.3.5 动画监听器：AnimationListener 230
7.3.6 动画操作组件：LayoutAnimationController 232
7.4 Android 中的媒体播放 232
7.4.1 Android 中音频播放 234
7.4.2 Android 中视频播放 240
7.5 Android 中的照相机 244
7.6 Android 中的媒体录制 249
7.6.1 Android 中的录音 250
7.6.2 Android 中的录像 253
本章小结 256
习题 256

第8章 Android 数据存储技术 257
8.1 使用 SharedPreferences 存储数据 257
8.1.1 使用 SharedPreferences 存储数据 258
8.1.2 使用 SharedPreferences 读取数据 260
8.2 使用文件存储数据 261
8.2.1 手机内存中的文件存储和读取 262
8.2.2 SD 卡中的文件存储和读取 264
8.2.3 读取资源文件 268
8.3 使用数据库存储数据 269
8.3.1 创建数据库及表 272
8.3.2 操作数据库 275
8.3.3 数据查询操作 278
8.4 使用 ContentProvider 存储数据 280
8.4.1 ContentProvider 基础 280
8.4.2 创建自己的 ContentProvider 283
8.4.3 操作联系人的 ContentProvider 284
8.4.4 多媒体信息的 ContentProvider 291
本章小结 296
习题 296

第9章 Android 网络通信技术 297
9.1 Android 网络通信技术基础 297
9.1.1 Android 中的 HTTP 协议基础 297
9.1.2 Android 中的 Socket 基础 299
9.1.3 Android 中的 Web Service 基础 300

        9.1.4 Android 中的蓝牙基础 ……………301
        9.1.5 Android 中的 Wi-Fi 基础 …………301
9.2 WebView 组件介绍…………………………302
        9.2.1 WebView 组件基础知识……………302
        9.2.2 使用 WebView 加载网页 …………304
        9.2.3 使用 WebView 加载 HTML 文件…307
        9.2.4 使用 WebView 加载 JSP 文件……309
        9.2.5 JavaScript 调用 WebView 中
              的数据 ……………………………311
        9.2.6 WebView 调用中 JavaScript
              的数据 ……………………………313
9.3 利用 HttpURLConnection 开发 HTTP
    程序 ………………………………………315
        9.3.1 HttpURLConnection 基础…………315
        9.3.2 HttpURLConnection 通信：GET
              方式 ………………………………316
        9.3.3 HttpURLConnection 通信：POST
              方式 ………………………………321
9.4 利用 HttpClient 开发 HTTP 程序 ……325

        9.4.1 HttpClient 通信基础………………325
        9.4.2 HttpClient 通信：GET 方式………327
        9.4.3 HttpClient 通信：POST 方式……330
        9.4.4 数据的实时更新……………………332
9.5 利用 Socket 交换数据……………………334
        9.5.1 基于 TCP 协议的 Socket 通信……335
        9.5.2 基于 UDP 协议的 Socket 通信……338
        9.5.3 利用 Socket 实现简易的聊天室…340
9.6 Web Service 通信 …………………………344
9.7 蓝牙通信 …………………………………350
        9.7.1 蓝牙通信基础………………………350
        9.7.2 蓝牙通信实现………………………352
        9.7.3 蓝牙通信实例………………………356
9.8 WiFi 通信 …………………………………357
本章小结 …………………………………………359
习题 ………………………………………………359

参考文献 ………………………………………360

第 1 章　Android 概述

学习目标：
- 了解智能手机的发展史及常见的手机操作系统
- 了解 Android 操作系统的发展及其特点
- 搭建 Android 系统开发环境
- 开发第一个 Android 程序
- 了解 Android 程序封装过程
- 了解 Android Market

随着人们生活水平的提高，手机已经逐渐从奢侈品发展成为十分普及的电子产品。经过了一次又一次的技术变革，手机已成为具有独立操作系统的智能设备，而不再仅仅是一个语音通信工具。

1.1　智能手机的发展

1. 智能手机的定义

智能手机（Smartphone），是指像个人电脑一样，具有独立的操作系统，可以由用户自行安装软件、游戏等第三方服务商提供的程序，通过此类程序对手机的功能进行扩充，并可以通过移动通信网络实现无线网络接入的手机的总称。

2. 智能手机的发展

1973 年 4 月 3 日，摩托罗拉公司前高管马蒂·库珀在曼哈顿的实验网络上测试了他的一台电话，他把电话打给了贝尔实验室的一名科学家，这是世界上公认的第一台手机，马蒂·库珀也被称为"现代手机之父"。

随着时间的推移，手机功能也在不断扩充，除了打电话之外，同时还具备了 PC 机的功能，例如：玩游戏、收发电子邮件及网页浏览等功能，这就是所说的智能手机。

全球首款智能手机是美国 IBM 公司在 1994 年投放市场的"IBM Simon"。这款手机配备了使用手写笔的触摸屏，除了通话功能之外，还具备 PDA 及游戏功能，操作系统采用的是夏普 PDA 的"Zaurus OS"。

1996 年，芬兰诺基亚公司推出了名为"Nokia 9000 Communicator"的折叠式智能手机。Nokia 9000 Communicator 受到了商务人士的青睐，后来逐步演变为 1998 年上市的"诺基亚 9110"和"诺基亚 9110i"，又推出了采用 Symbian 系统的机型。1997 年，瑞典爱立信公司推出了与 Nokia 9000 Communicator 相似的"GS88"手机。该手机的说明书中首次出现了"智能手机"一词。

进入 2000 年以后，市场上出现了很多采用面向 PDA 及嵌入设备的通用操作系统的智能手机。这些手机使用 Symbian、Palm OS 及 Windows CE 等操作系统。

首次采用 Symbian 操作系统的智能手机是爱立信"Ericsson R380 Smartphone"。之后，诺基亚公司也于 2000 年投放了采用 Symbian 操作系统的智能手机，（后来诺基亚的智能手机便一直使用 Symbian 操作系统），Symbian 操作系统一度成为占主导地位的手机操作系统。

2001 年 2 月配备 Palm 操作系统的手机"Kyocera 6035"上市。

美国微软公司于 2002 年发布了"Microsoft Windows Powered Smartphone 2002",该手机配备的是 Windows CE 智能手机系统,后来更名为"Windows Mobile",韩国三星电子及夏普等公司向市场投放了多款采用这种操作系统的智能手机。

加拿大 RIM(Research In Motion)公司于 2003 年推出了首款"黑莓"(BlackBerry)手机。该手机融合了电子邮件、SMS 及 Web 浏览等功能。

以上这些手机均以企业用户为目标,以嵌入商务软件的形式提供,基本未向普通消费者推广。让普通消费者购买并使用智能手机,掀起这股潮流的是美国苹果公司于 2007 年 6 月投放市场的 iPhone。这款手机配备有以触摸屏完成的用户界面(UI)、基本与个人电脑等同的 Web 浏览器和电子邮件功能,以及与 iTunes 软件联动的音乐播放软件等,从而将智能手机提高到了任何人都能使用的水平。

随后,美国谷歌公司于2007年11月发布了智能手机软件平台 Android 系统。2008年,美国 T-Mobile USA 公司推出了首款配备 Android 系统的智能手机"T-Mobile G1"。此后,美国摩托罗拉移动公司、三星电子以及日本与瑞典的合资公司索尼爱立信移动通信等公司都相继推出了基于 Android 系统的智能手机。

微软公司在 iPhone 与 Android 成功之后也转变了市场方针,于 2009 年 2 月宣布开发面向普通消费者的"Windows Mobile 6.5"及"Windows Phone 7"。采用 Windows Mobile 6.5 系统的手机于 2009 年 10 月投放市场,Windows Phone 7 手机则于 2010 年 10 月问世。

2011 年后,"双核"智能手机推出。摩托罗拉公司、LG 公司以及三星公司发布了采用双核处理器的智能手机产品,而 HTC 公司发布的双核处理器智能手机主频更是已经高达 1.2GHz。智能手机的硬件发展进入了一个新的阶段。

未来的手机将偏重于安全和数据通信。一方面加强个人隐私的保护,另一方面加强数据业务的研发,各种多媒体功能被引入进来,手机将会具有更加强劲的运算能力,成为个人的信息终端,而不是仅仅具有通话和文字消息的功能。

3. 智能手机与 3G

3G(Third Generation)指的是第三代移动通信技术。

相对于第一代模拟制式手机(1G)和第二代 GSM、TDMA 等数字手机(2G),第三代手机是指将无线通信与国际互联网等多媒体通信结合的新一代移动通信系统。它能够处理图像、音乐、视频流等多种媒体形式,提供包括网页浏览、电话会议、电子商务等多种信息服务。

国际电信联盟(ITU)在 2000 年 5 月确定三大主流无线接口标准:
- WCDMA　　　　支持者主要是以 GSM 系统为主的欧洲厂商
- CDMA2000　　　美国高通北美公司为主导
- TDS-CDMA　　　中国大陆独自制定的 3G 标准

目前,中国移动采用的是 TDS-CDMA 标准,中国电信采用的是 CDMA2000 标准,中国联通采用的是 WCDMA2000 标准。

1.2 智能手机操作系统简介

智能手机就是安装了某个操作系统的手机,能够安装在手机上的操作系统有:Android、iOS、Windows Mobile、Symbian、BlackBerry、Palm 等。

1. Android

Android（中文名：安卓）系统是由 Google 公司推出的基于 Linux 平台的开源手机操作系统，由于开源以及使用 Java 作为开发语言的特点，越来越受到广大用户的青睐，支持的硬件厂商也越来越多。目前市面上几大操作系统中，Android 系统的市场占有率最高，上升速度最快。

2．iOS（iPhone OS 的简称）

iOS 是由苹果公司为 iPhone 开发的基于 Mac 环境的操作系统，采用 Objective-C 为主要开发语言，主要用于 iPhone、iPad Touch 以及 iPad 等终端设备。iOS 支持多点触控，能给用户提供全新的体验但是目前只能应用于苹果公司的设备上。

3．Windows Phone 7

Windows Phone 7（前身为 Windows Mobile）是 Microsoft 公司为移动设备推出的 Windows 操作系统，该系统有很多先天的优势，有庞大的用户群，但是由于硬件要求极高，导致硬件设备价格也高，在一定程度上限制了它的发展。

4．Symbian

Symbian（中文名：塞班）是一个实时、多任务的 32 位操作系统，具有功耗低、内存占用少等特点，非常适合手机等移动设备使用。Symbian 操作系统曾经是市场占有率最高的手机操作系统，随着越来越多手机操作系统的出现，尤其是 Android 系统的出现，Symbian 系统的发展遇到了瓶颈，被迫于 2010 年 2 月进行开源。

5．BlackBerry

BlackBerry（中文名：黑莓）是 RIM 公司开发的手机操作系统，以前这个系统曾经显赫一时，现在由于面临着 Android 和 iOS 两大阵营的冲击，其用户群在逐渐减少。

6．Palm

Palm 操作系统是 Palm 公司推出的 32 位嵌入式操作系统，早期主要应用于掌上电脑，该公司 2010 年被惠普收购，惠普公司在 Palm 系统的基础上推出了 Web OS，现在成为惠普平板电脑上的操作系统。

7．Bada

Bada 是韩国三星公司自主研发的智能手机平台，支持 Flash 界面，对于 SNS 应用有着很好的支持，于 2009 年 11 月 10 日发布。

1.3 Android 操作系统简介

1.3.1 Android 操作系统的发展

Android 一词最早出现于法国作家利尔亚当在 1886 年发表的科幻小说《未来的夏娃》中。他将聪明美丽的机器人女孩起名为 Android。

美国 Google（谷歌）公司早在 2002 年就进入了移动领域，可是由于手机操作系统企业和手机企业相对封闭，提高了行业的进入门槛，谷歌的目标是将传统互联网和移动互联网进行融合，但没有合适的手机系统合作伙伴。

Android 公司由安迪·鲁宾创办，谷歌公司在 2005 年收购了这个公司，安迪·鲁宾继续负责 Android 项目的研发工作。

2007 年 11 月 5 日，谷歌公司正式向外展示了 Android 1.0 操作系统，提供了基础的智能手机功能：闹钟、API 示例、浏览器、计算器、摄像头、联系人、开发工具包、拨号应用、电子邮件、地图（包含街景）、信息服务、音乐、图片、设置等。

该系统发布之后不久就有一款装有 Android 1.0 系统的手机 T-Mobile G1 问世，手机由运营商 T-Mobile 定制，台湾 HTC 公司代工制造。T-Mobile G1 是世界上第一款使用 Android 操作系统的手机，手机的全名为 HTC Dream。

2009 年 4 月，谷歌正式推出了基于 Android 1.5 系统的手机，加入了输入法框架支持，视频录像等功能。9 月份，谷歌发布了 Android 1.6 系统，并且推出了装载 Android 1.6 正式版的手机 HTC Hero G3，凭借出色的外观设计以及全新的 Android 1.6 操作系统，HTC Hero G3 成为当时全球最受欢迎的手机。

2009 年 10 月份，谷歌发布了 Android 2.0 操作系统，改进了桌面主题，联系人管理，完善了蓝牙通信，以及 OpenGL ES 2.0 的支持，新增了多点触控的支持。Android 2.0 版本的代表机型为 NEXUS One，这款手机为谷歌旗下第一款自主品牌手机，由 HTC 代工生产，NEXUS One 这款手机在 2010 年 1 月正式发售。

在 2010 年 5 月份，谷歌正式发布了 Android 2.2 操作系统，支持应用安装到 SD 卡上，运行效率有了大幅的提升，支持更大内存，开始支持 Flash 播放器和 FLV 视频媒体解码。采用 Android 2.2 操作系统的手机比较出众的有 HTC Desire HD，除了 HTC，三星的 GALAXY S 也是一款 Android 2.2 操作系统的手机。

在 2010 年 12 月，谷歌正式发布了 Android 2.3 操作系统，在多媒体库方面有了大幅的改变，同时引入了近距离数据通信协议的支持。Android 2.3 代表机型 GALAXY S II、HTC Sensation 等。

2011 年 2 月 3 日谷歌在发布了专用于平板电脑的 Android 3.0 系统，对于大屏幕高分辨率的平板电脑进行了界面的优化，同时支持多核 CPU，高性能 2D 和 3D 图形性能，在娱乐方面有了大幅的增强，同时全新的开发附件协议，将使其在 USB 外设有了大幅的支持，这是首个基于 Android 的平板电脑专用操作系统。

2011 年 5 月 11 日 Google 发布 Android 3.1，部分功能做了小幅改进，在虚拟键盘等方面有了小幅的变化。新版本最大的改变是将 Android 手机系统与平板系统再次合并，方便开发者。

2011 年 7 月 13 日 Google 发布了 Android 3.2 操作系统，对于 7 英寸的屏幕在 1024×600 分辨率的设备进行了界面的优化，解决了早期系统仅支持 10.1 英寸大平板的问题。

2011 年 10 月 19 日在香港发布 Android 4.0，最明显的是 Android 4.0 界面 UI 做了重新设计。在系统性能方面也做了大幅改进。同时适用于手机和平板。

2012 年 6 月 28 日发布 Android 4.1，它使系统变得更快更流畅，优化了系统操作体验，增加了包括 Google Now 和更丰富的通知中心在内的很多新功能。

2012 年 10 月 30 日发布 Android 4.2，增强了 Google Now 功能，增加了对航班信息查询、酒店和餐厅预订、电影和音乐推荐的支持，并且平板用户还能自由切换账户。即将推出的版本是 Android 5.0。

现在，Android 系统不但应用于智能手机，已经延伸到其他便携式和嵌入式设备（平板电脑、电子书、上网本、高清电视等）。支持 Android 系统主要厂商包括 HTC、三星、摩托罗拉、华为、中兴、联想、小米、LG、戴尔、宏基、华硕、海信等公司。

开放手机联盟（Open Handset Alliance）是 Google 公司于 2007 年 11 月 5 日宣布组建的一个全球性的联盟组织。这一联盟将支持 Google 发布的 Android 手机操作系统或者应用软件，共同开发名为 Android 的开放源代码的移动系统。开放手机联盟包括手机制造商、手机芯片厂商和移动运营商几类。目前，联盟成员数量众多，这也是 Android 迅猛发展的一个原因。

1.3.2 Android 操作系统的特点

Android 系统是基于 Linux 开放性内核的操作系统,具有如下特点:

1)开放性,Android 平台允许任何移动终端厂商加入到 Android 联盟中来。开放性可以使其拥有更多的开发者,专业人士可以利用开放的源代码来进行二次开发,打造出个性化的 Android 系统。而且开放性可以缩短开发周期,降低开发成本,也有利于 Android 的发展。

2)应用程序无界限,Android 系统上的应用程序可以通过标准 API 接口访问核心移动设备功能。

3)应用程序是在平等条件下创建的,移动设备上的应用程序可以被替换或扩展。

4)应用程序可以轻松地嵌入网络,应用程序可以轻松地嵌入 HTML、JavaScript 和样式表,还可以通过 Web View 控件显示网络内容。

5)应用程序可以并行运行,Android 系统是多任务环境,应用程序可以并行运行。

Android 操作系统的缺点:

1)安全问题

由于 Android 系统的开源和快速发展以及应用程序审核机制的不完善等原因,导致 Android 程序应用方面出现一些恶意软件。2009 年 11 月 10 日 Android 平台出现了第一个恶意间谍软件:Mobile Spy。2010 年 8 月 12 日,出现了第一个木马病毒:Trojan-SMS.Android OS.FakePlayer.a。在这些恶意软件的影响下,用户在不经意间就可能泄露自己的隐私。因此,2011 年 11 月 20 日,Google 公司宣布启动 Android 应用审核、取缔、清扫行为,定期对电子市场中不合格、低质量、违法、恶意程序进行清理。

2)稳定性问题

由于 Android 系统的开源,各个厂商都能对代码进行二次开发,由于开发水平的原因,一些厂商开发出来的应用程序可能会导致系统崩溃等后果。

3)必须用高配置弥补系统上的缺陷

Android 的 UI 渲染遵循传统电脑模式的主线程普通优先级,当触摸 Android 手机屏幕的时候,系统后台的程序并没有停止,仍然在继续运行之中,这就是 Android 系统不流畅的原因之一。Android 系统缺乏有效的硬件加速也是一个原因,在不同的 Android 手机上的硬件加速存在巨大差异。

1.4 搭建 Android 系统开发环境

Android 系统的开发环境可以搭建在 Windows XP 及以上的操作系统中,在 WIN7 下的安装方法与 Windows XP 下的安装方式大致相同,要注意的是,Windows 7 下要 64 位的文件,而 Windows XP 下要 32 位的文件,下面在 Windows 7 旗舰版环境下进行 Android 环境的搭建。

在搭建环境之前,需要准备如图 1.1 所示的三个文件。

图 1.1 要准备的三个文件

下载地址分别如下:

Android SDK 下载地址:http://developer.Android.com/sdk/index.html

Eclipse 下载地址:http://www.eclipse.org/downloads/

Java JDK 下载地址:http://java.sun.com/javase/downloads/index.jsp

注意：Android SDK 两种下载版本，一种是包含具体的 SDK 版本，另一种是只有升级工具，而不包含具体的 SDK 版本，可以在线升级，建议采用这种形式。

1.4.1 安装 JDK

1. 安装 jdk 程序

1）双击 jdk-7u21-windows-x64 文件，运行该程序，弹出如图 1.2 所示的安装向导界面。

2）单击"下一步"按钮，弹出如图 1.3 所示界面，单击"更改"按钮，可以更改 JDK 的安装路径，这里更改为"D:\Android\java\jdk1.7.0_21"，如图 1.3 所示，然后单击"下一步"按钮。

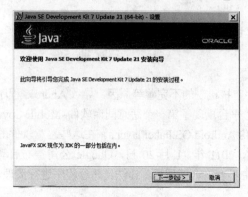

图 1.2 安装向导界面

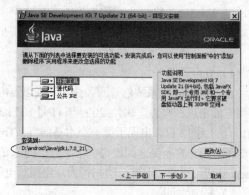

图 1.3 更改 JDK 安装目录

3）弹出正在复制新文件的界面，如图 1.4 所示。

4）文件复制完成后，提示安装 JRE，建议和 JDK 安装在同一个盘符下，更改目录为"D:\Android\Jre7"，如图 1.5 所示。

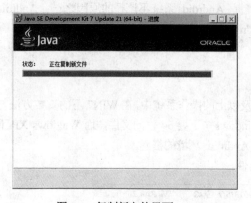

图 1.4 复制新文件界面

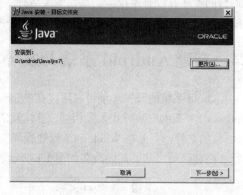

图 1.5 更改 JRE 安装目录

5）单击"下一步"按钮，开始复制文件并安装 JRE，文件复制完成后，弹出成功安装界面，如图 1.6 所示。

2. 设置环境变量

1）右键单击"我的电脑"，选择"属性"选项，在弹出的窗口中选择"高级系统设置"选项，如图 1.7 所示。

2）在弹出的窗口中选择"高级"选项，再单击"环境变量（N）按钮"，如图 1.8 所示。

3）在弹出的窗口中，单击"新建（N）"按钮，如图 1.9 所示。设置 JAVA_HOME 变量的值为："d:\Android\java\jdk1.7.0_21"，如图 1.10 所示。类似地，新建 classpath 变量，其值为：".;%JAVA_HOME%\lib\tools.jar;%JAVA_HOME%\lib\dt.jar;%JAVA_HOME%\bin;"。

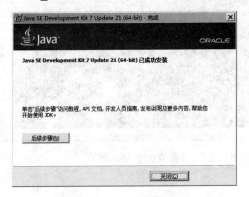

图 1.6　JDK 安装成功

图 1.7　高级系统设置

图 1.8　设置环境变量

图 1.9　新建环境变量

4）双击"系统变量"中的"Path"选项，如图 1.11 所示，打开 Path 变量修改窗口，在最后添加"%JAVA_HOME%\bin;"，（或者 d:\Android\java\jdk1.7.0_21\bin），如图 1.12 所示。

图 1.10　新建环境变量 JAVA_HOME

图 1.11　修改系统变量 Path

3. 检查 JDK 是否安装成功

打开 cmd 窗口，输入 "java –version" 命令，查看 JDK 的版本信息，如图 1.13 所示，如能正常显示版本信息，表示 JDK 已经安装成功。

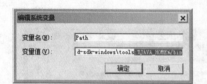

图 1.12 在最后添加值%JAVA_HOME%\bin

图 1.13 查看 JDK 的版本信息

1.4.2 安装 Eclipse

1）解压缩 Eclipse-SDK-4.2.2-win32-x86_64.zip 到指定目录，例如：D:\Android\Eclipse，如图 1.14 所示。

2）运行 eclipse.exe，设置 Workspace，也就是指定一个开发目录，如图 1.15 所示。

图 1.14 Eclipse 文件目录

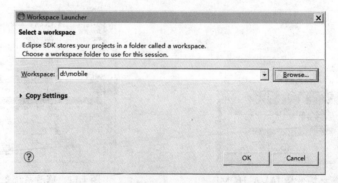

图 1.15 指定一个开发目录 d:\mobile

1.4.3 安装 Android SDK

1. 解压缩 Android -sdk-r18-windows 文件

解压缩 Android -sdk-r18-windows.zip 文件到指定文件夹（如 D:\Android \Android -sdk-windows），将 Android SDK 中的 tools 文件夹的绝对路径添加到系统变量 Path 中，方法与上面讲述的相同，添加的值为 "D:\Android \Android -sdk-windows\tools"，单击"确定"按钮。

2. 查看 Path 设置是否生效

打开 cmd 窗口，输入 "android –h" 命令，弹出如图 1.16 所示信息，表示设置已经生效。

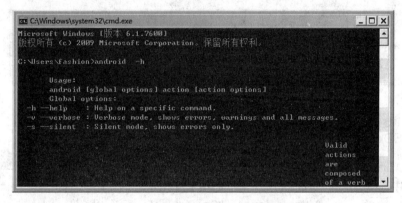

图 1.16　查看 Android-h 输出信息

3. 安装 Android Development Tools（ADT）

1）打开 Eclipse 程序，选择菜单中的 "Help→Install New Software" 选项，如图 1.17 所示。

2）在界面的 "Work with:" 对话框中输入：http://dl-ssl.google.com/Android/eclipse，然后，勾选 "Android DDMS" 和 "Android Development Tools" 等选项，如图 1.18 所示。

3）单击 "Next" 按钮，勾选需要安装的 Android SDK 包，如图 1.19 所示。注意：安装过程中有个别地方需要接受许可，然后等待安装完成，这可能需要比较长的时间。

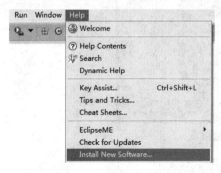

图 1.17　安装新的软件

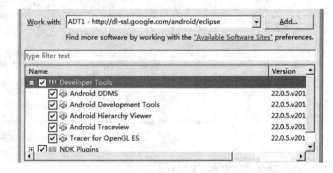

图 1.18　设置 Work with：的值，选择要安装的软件

安装完毕，重启 Eclipse 软件，完成安装。如果在 Eclipse 软件的工具栏中弹出如图 1.20 所示的图标，表示 SDK Manager 安装成功。

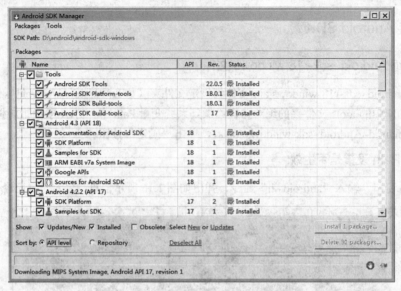

图 1.19　勾选需要安装的 Android SDK 包

4. 设定 SDK Location

1）打开 Eclipse 程序，选择菜单中的"Window→Preferences"选项，在界面中选中其中的"Android"节点，如图 1.21 所示。

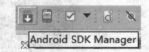

图 1.20　工具栏上的 SDK Manager

2）设定"SDK Location"为 D:\android\android-sdk-windows（Android SDK 解压的目录），单击"OK"按钮后，再次打开这个窗口，可以看到 SDK 列表，如图 1.21 所示。同时，在 D:\android\android-sdk-windows\platforms 目录下应该有这些 SDK 的文件夹，如图 1.22 所示。

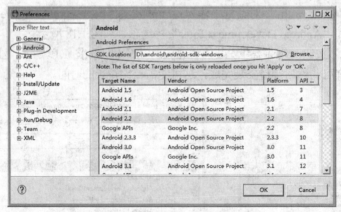

图 1.21　设置 SDK Location

5. 创建 Android 虚拟设备 AVD

1）打开 Eclipse 程序，选择菜单中的"Run→Run Configurations"选项，如图 1.23 所示。

2）在弹出的窗口中，选择"Target"标签，单击"Manager"按钮，如图 1.24 所示。

3）在弹出的界面中单击"Device Definitions"标签，选择列表中的一个设备，单击"Create AVD"按钮，如图 1.25 所示。

第 1 章　Android 概述

图 1.22　SDK 目录列表

图 1.23　选择 Run Configurations 选项

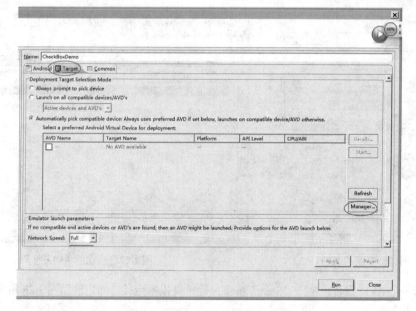

图 1.24　进入 AVD Manager 界面

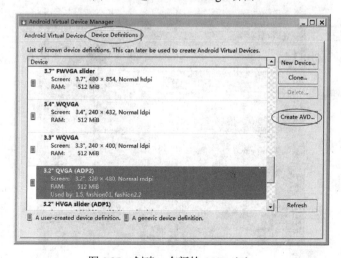

图 1.25　创建一个新的 AVD（1）

4）在如图 1.26 所示的界面中，对 AVD 进行相应的参数设置，例如：名称、设备的分辨率、使用的 API 级别、模拟机上的软键盘、是否有照相功能、SD 卡的大小等，可以根据需要自行设置，设置完毕后，单击"OK"按钮。

5）类似地，可以设置不同 Android 系统的模拟机，在 AVD 列表中已经添加了刚才设置的 AVD 设备，如图 1.27 所示。

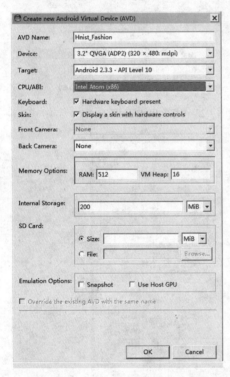

图 1.26　创建一个新的 AVD（2）

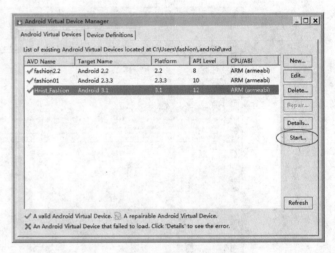

图 1.27　建立的 AVD 列表

6）选择一个 AVD 设备，单击"Start"按钮，等待模拟器运行，如出现如图 1.28 所示的界面，表示安装 AVD 设备成功。

图 1.28　运行 AVD 设备界面

1.5 开发第一个 Android 项目

开发环境配置完毕后，就可以进行 Android 项目开发了。打开 Eclipse 程序，弹出如图 1.29 所示的欢迎界面，按照后续的步骤建立一个 Android 项目。

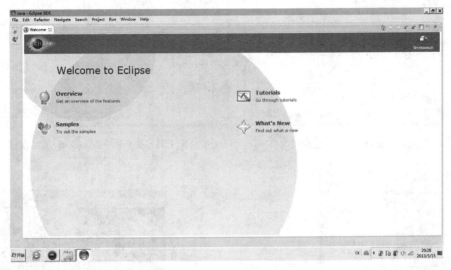

图 1.29　Eclipse 软件欢迎界面

1）选择菜单中的"File→New→Android Application Project"选项（如果没有这个选项，选择Other），如图 1.30 所示。

2）在弹出的窗口中，选择"Android→Android Application Project"选项，如图 1.31 所示，单击"Next"按钮。

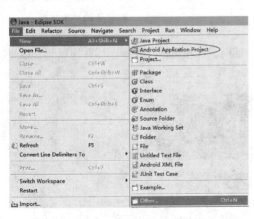

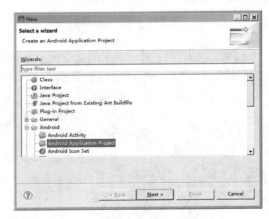

图 1.30　建立一个 Android 项目（1）　　　图 1.31　建立一个 Android 项目（2）

3）在弹出的窗口中，输入项目的名称，例如 firstDemo，包名称可以自己输入（例如"org.hnist.firstdemo"），也可以使用默认值，其他选项使用取默认值，如图 1.32 所示，单击"Next"按钮。

4）在后续弹出的窗口中，所有选项均使用默认值，单击"Finish"按钮完成配置，如图 1.33 所示。

5）返回到 Eclipse 界面，左侧项目栏中已经有了 firstDemo 项目，里面有许多的文件夹和文件，右键单击 firstDemo 项目，选择"Run as → 1.Android Application"选项，如图 1.34 所示。

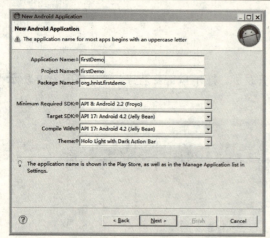

图 1.32　给 Android 项目、包命名　　　　图 1.33　完成新的 Android 项目创建

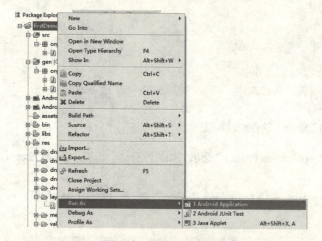

图 1.34　运行 Android 项目　　　　图 1.35　模拟机的开机界面

6）等待一段时间，将弹出如图 1.35 所示的界面，使用鼠标左键按住开锁图标并拖曳到右边。

7）弹出 firstDemo 的运行结果，显示了"Hello world!"，如图 1.36 所示。

图 1.36　第一个程序的运行结果

1.6　封装第一个 Android 项目

　　Android 程序开发完成后，为了方便用户使用，还需要将程序封装后上传至 Google Play 应用程序商店，经过商店审核通过后才能向用户提供下载和安装服务。接下来讲解如何进行 Android 程序的封装。

1）在 Eclipse 中，如图 1.37 所示，选择"File→Export"选项，出现如图 1.38 所示界面，选择"Android→Export Android Application"选项，然后单击"Next"按钮。

2）在弹出的窗口中输入需要封装的项目名称，如图 1.39 所示，在 Project 输入框中输入项目名，例如：刚才建立的"firstDemo"，然后单击"Next"按钮。

3）要导出项目需要先建立一个证书，在图 1.40 所示的界面中，输入证书文件保存的位置，例如：d:\MyfirstDemo，然后单击"Next"按钮。

4）填写完整的证书信息，如图 1.41 所示，然后单击"Next"按钮。

图 1.37　输出 Android 项目（1）

图 1.38　输出 Android 项目（2）

图 1.39　输入要封装的 Android 项目的名称

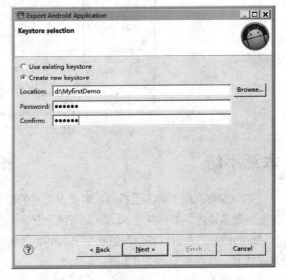
图 1.40　输入要封装的 Android 项目的输出位置

5）在弹出的界面中，输入要导出的 apk 文件存放的位置，例如：d:\MyfirstDemo\first.apk，如图 1.42 所示，然后单击"finish"按钮。

6）打开"d:\MyfirstDemo"文件夹，可以发现"first.apk"文件已经建立，如图 1.43 所示。这个 apk 文件就可以安装在手机上运行，也可以在 Android Market 注册后发布。

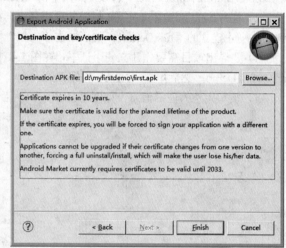

图 1.41 填写完整的证书信息　　　　图 1.42 输入要导出的 apk 文件存放的位置

图 1.43 文件夹中的 apk 文件

本章小结

本章简要介绍了智能手机的发展史及常见的手机操作系统，Android 操作系统的发展及其特点和缺点，着重讲述了如何搭建 Android 系统开发环境，如何建立一个 Android 项目的基本过程，如何运行 Android 项目，以及如何封装发布 Android 项目等。目的是为了让用户对 Android 的运行环境和项目的建立、发布有个基本的了解。

习题

1. 简要描述 Android 操作系统的特点和缺点。
2. 在 Windows 7 的环境下配置 Android 系统开发环境。
3. 根据 1.5 小节的步骤说明，建立一个 Android 项目，命名为：HelloDemo，包名称为：org.hnist.hello，封装形成 apk 文件，然后安装到手机上看是否能运行？

第 2 章　Android 中的项目

学习目标：
- 了解 Android 的项目的结构
- 掌握 Android 项目各个常用部分的相互关系
- 了解 Activity 的基础知识
- 了解 Android 项目中的开发包
- 了解 Android 项目中的大致开发流程
- 了解 Android 项目中常见的文件

第 1 章中在没有编写一条代码的情况下就建立了一个 Android 项目，并能够在屏幕上显示"Hello world!"，项目建立后在 Eclipse 界面的左边导航栏出现了很多的文件和文件夹，它们都是用来干什么的？显示的"Hello world!"能不能换成别的字符？本章对此会有详细的介绍。

2.1　Android 项目的组成

第 1 章中我们建立了一个 Android 项目：firstDemo，在 Eclipse 软件中，看到该项目下有一些文件和文件夹，如图 2.1 所示。打开其中一个文件夹，里面存在一些文件，如图 2.2 所示。

图 2.1　firstDemo 项目

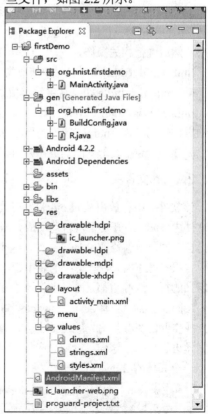

图 2.2　展开的 firstDemo 项目

2.1.1 几个关键文件夹和文件

按照图 2.2 所列出的文件和文件夹顺序，逐个介绍这些文件夹和文件的作用。

1. src 文件夹

该文件夹是存放项目源代码的，单击文件夹前面的"+"，可逐层展开，文件夹中有个 MainActivity.java 文件，是 Activity 程序，类似于 Java 中的主类。可以理解为它是一个 UI 的容器，是直接面向用户的类。双击打开 MainActivity.java 文件会看到如下代码：

```
package org.hnist.firstdemo;             //程序所在包为 org.hnist.firstdemo
import android.app.Activity;             //导入类 Android.os.Bundle
import android.os.Bundle;                //导入类 Android.app.Activity
import android.view.Menu;                //导入类 Android.view.Menu
public class MainActivity extends Activity {  //定义 Activity 程序
@Override
    protected void onCreate(Bundle savedInstanceState) {//覆写 onCreate 方法
        super.onCreate(savedInstanceState);     //调用父类的 onCreate()方法
        setContentView(R.layout.activity_main); //调用布局文件 activity_main.xml
...
```

android.app.Activity 类：因为几乎所有的活动都是与用户交互的，所以 Activity 类关注创建和显示控件，可以用方法 setContentView(View)将自己的用户界面（UI）放在其中。

有两个方法是几乎所有的 Activity 子类都可以实现的。

onCreate(Bundle)：初始化 Activity 程序。

setContentView(int)：指定由哪个文件指定布局（比如 activity_main.xml），可以将这个界面显示出来，然后进行相关操作。

findViewById(int)：在 UI 中获取指定的控件，后面会有详细的介绍。

android.os.Bundle 类：从字符串值映射各种可封装的（Parcelable）类型。

public class MainActivity extends Activity {}：定义 Activity 程序，里面可以包含控件和对这些控件的操作代码，是 Android 项目的核心部分。

@Override：表示下面的语句是可以改写的，它是一个标记，没有实质性作用，可以省略。

2. gen 文件夹

该文件夹中存放由 Android 开发工具自动生成的所有文件。目录中最重要的就是 R.java 文件，这个文件由 Android 开发工具自动生成。Android 开发工具会根据 res 目录的 xml 界面文件、图标以及常量，自动同步更新修改 R.java 文件。这个文件是只读文件，一般不要去修改它。打开 R.java 文件，其中包含很多静态类，且静态类的名字都与 res 中的一个名字对应，即 R 类定义该项目所有资源的索引，它定义的每个资源值都是唯一的，不会和系统冲突。

R.java 里面一般有 attr、drawable、id、raw、layout、string 以及 xml 标签等，通过 R.java 可以很快地查找需要的资源，另外编译器也会检查 R.java 列表中的资源是否被使用到，没有被使用到的资源不会编译进软件中，这样可以减少应用程序在手机中占用的空间。

一个具体的 R.java 代码如下：

```
package org.hnist.firstdemo;
public final class R {
```

```
        public static final class attr {
        }
        public static final class dimen {
            public static final int activity_horizontal_margin=0x7f040000;
            public static final int activity_vertical_margin=0x7f040001;   }
        public static final class drawable {
            public static final int ic_launcher=0x7f020000;    }
        public static final class id {
            public static final int action_settings=0x7f080000;    }
        public static final class layout {
            public static final int activity_main=0x7f030000;    }
        public static final class menu {
            public static final int main=0x7f070000;    }
        public static final class string {
            public static final int action_settings=0x7f050001;
            public static final int app_name=0x7f050000;
            public static final int hello_world=0x7f050002;    }
        public static final class style {
            public static final int AppBaseTheme=0x7f060000;
            public static final int AppTheme=0x7f060001;       }}
```

3. Android 4.2.2

表示现在使用的 Android SDK 的版本是 4.2.2。

4. assets

该文件夹中包含应用系统需要使用到的资源文件，比如音频、视频等较大的文件。注意：这些文件不会在 R.java 中自动生成 ID。

5. res 文件夹

该文件夹为资源目录，包含项目中较小的资源文件，并将编译进应用程序。向此目录添加资源时，会被 R.java 自动记录。新建一个项目 res 目录下会有三个自动生成的子目录：drawabel、layout、values。

1）drawabel：包含应用程序要用到图标文件（*.png、*.jpg），hdpi 表示存放大约 240dpi 的高分辨率图片资料，ldpi 表示存放大约 120dpi 的低分辨率图片资料，mdpi 表示存放大约 160dpi 的中等分辨率图片资料，xhdpi 表示存放大约 320dpi 的超高分辨率图片资料。

2）layout：存放界面布局文件（activity_main.xml），界面布局文件主要用于放置不同的显示组件。在 MainActivity.java 中通过 setContentView(R.layout.activity_main)语句来调用布局文件 activity_main.xml。

一个具体的 activity_main.xml 的代码如下，**注意：后面的汉字是对本行代码的解释，实际的程序中不能有这些解释的汉字。**

```
<RelativeLayout                                  //采用的相对布局模式
    xmlns:android=http://schemas.android.com/apk/res/android
    xmlns:tools="http://schemas.android.com/tools"
    android:layout_width="match_parent"        //布局管理器的宽度为屏幕宽度
    android:layout_height="match_parent"       //布局管理器的高度为屏幕高度
    android:paddingBottom="@dimen/activity_vertical_margin"   //是指控件中内容
        距离控件底边距离为 dimen.xml 中变量 activity_vertical_margin 设定的值
    android:paddingLeft="@dimen/activity_horizontal_margin"   //是指控件中内容
        距离控件左边距离为 dimen.xml 中变量 activity_horizontal_margin 设定的值
```

```
    android:paddingRight="@dimen/activity_horizontal_margin"    //是指控件中
内容距离控件右边距离为 dimen.xml 中变量 activity_horizontal_margin 设定的值
    android:paddingTop="@dimen/activity_vertical_margin"        //是指控件中
内容距离控件上边距离为 dimen.xml 中变量 activity_vertical_margin 设定的值
    tools:context=".MainActivity" >           //说明你当前的布局文件所在的对
    象是 MainActivity 对应的那个 activity。
    <TextView                                     //设置一个文本显示组件
        android:layout_width="wrap_content"       //组件的宽度为文字的宽度
        android:layout_height="wrap_content"      //组件的高度为文字的高度
        android:text="@string/hello_world" />     //文本显示的内容为 string.xml
                                                    中的 hello_world 所定义的值。
</RelativeLayout>                                 //相对布局模式结束
```

3）values：该文件夹里面可以有多个 XML 文件，以便存放不同类型的数据。例如字符串（string.xml）、颜色文件（colors.xml）、数组文件（arrays.xml）、尺寸文件（dimens.xml）和类型文件（styles.xml）等，例如：一个 dimen.xml 文件的代码：

```
<resources>
    <dimen name="activity_horizontal_margin">16dp</dimen> 定义 activity_horizontal_
            margin 的值为 16dp
    <dimen name="activity_vertical_margin">16dp</dimen>   定义 activity_vertical_
            margin 的值为 16dp
</resources>
```

Android 系统中定义的单位有以下几种类型：

px（Pixels 像素）：对应屏幕上的实际像素点。例如，320*480 的屏幕在横向有 320 个像素，在纵向有 480 个像素。

in（Inches 英寸）：屏幕物理长度单位，1 英寸等于 2.54 厘米。例如，形容手机屏幕大小，可以用 3.2（英）寸、3.5（英）寸、4（英）寸来表示。这些尺寸是指屏幕的对角线长度。

mm（Millimeters 毫米）：屏幕物理长度单位。

pt（Points 磅）：屏幕物理长度单位，大小为 1 英寸的 1/72。

dp（与密度无关的像素）：逻辑长度单位，在 160 dpi 屏幕上，1dp=1px=1/160 英寸。随着密度变化，对应的像素数量也变化，但并没有直接的变化比例。

dip：与 dp 相同，多用于 Google 示例中。

sp（与密度和字体缩放度无关的像素）：与 dp 类似，但是可以根据用户的字体大小首选项进行缩放。

注意：在项目中尽量使用 dp 作为空间大小单位，sp 作为和文字相关大小的单位。

例如：一个 strings.xml 文件的代码：

```
<?xml version="1.0" encoding="utf-8"?>
<resources>
    <string name="app_name">firstDemo</string>//定义 app_name 的值为 firstDemo
    <string name="action_settings">Settings</string>  //定义 action_settings
                                                        的值为 Settings
    <string name="hello_world">Hello world!</string>  //定义 hello_world 的值为
                                                        hello world!
</resources>
```

4）menu 目录：主要放置设计的 OptionsMenu 和 ContextMneu 的菜单项。

6. AndroidManifest.xml

每个应用程序都有一个 AndroidManifest.xml 文件在项目根目录里。这个文件提供了关于这个应用程序的基本信息,记录了应用程序中所使用的各种组件。程序在编译之前必须知道这些信息。要开发 Activity、Broadcast、Service 等应用都要在 AndroidManifest.xml 中进行定义。另外如果要使用系统自带的服务,如拨号服务、应用安装服务、GPRS 服务等都必须在 AndroidManifest.xml 中声明权限。当新添加一个 Activity 的时候,也需要在这个文件中进行相应配置,只有配置好后,才能调用此 Activity。

AndroidManifest.xml 是用来存储一些关于 Android 项目的配置数据,主要包含以下功能:

1)命名应用程序的 Java 应用包,这个包名用来唯一标识应用程序;

2)描述应用程序的组件——活动、服务、广播接收者、内容提供者;对实现每个组件和公布其功能的类进行命名。这些声明使得 Android 系统了解这些组件以及它们在什么条件下可以被启动;

3)决定应用程序组件运行在哪个进程里面;

4)声明应用程序所必须具备的权限,用以访问受保护的部分 API 以及和其他应用程序的交互;

5)声明应用程序其他的必备权限,用以组件之间的交互;

6)列举测试设备 Instrumentation 类,用来提供应用程序运行时所需的环境配置及其他信息,这些声明只在程序开发和测试阶段存在,发布前将被删除。

一般 AndroidManifest.xml 包含如下设置:application、permissions、Activities、intent filters 等。

例如,一个 AndroidManifest.xml 文件的代码如下:

```xml
<?xml version="1.0" encoding="utf-8"?>
<manifest xmlns:android="http://schemas.android.com/apk/res/android"
    package="org.hnist.firstdemo"           //应用程序的包名
    android:versionCode="1"                 //开发者内部的版本号
    android:versionName="1.0" >             //发给用户的版本号
  <uses-sdk
    android:minSdkVersion="8"               //应用程序的最低支持的 sdk
    android:targetSdkVersion="17" />        //应用程序的最高支持的 sdk
  <application
    android:allowBackup="true"              //允许备份文件
    android:icon="@drawable/ic_launcher"    //索引的图标在系统的位置
    android:label="@string/app_name"        //应用程序的名字 就是在安装到手机
                                              上的名字
    android:theme="@style/AppTheme" >
    <activity
      android:name="org.hnist.firstdemo.MainActivity"   默认启动的Activity
      android:label="@string/app_name" >  //这是 activity 的名字
      <intent-filter>
        <action android:name="android.intent.action.MAIN" /> //主程序
        <category android:name="android.intent.category.LAUNCHER" />
                                              //表示放到手机应用程序的列表里
      </intent-filter>
    </activity>
  </application>
</manifest>
```

本章仅对 AndroidManifest.xml 文件进行了简单介绍,在后续的章节还会进行详细介绍。

7. default.properties

项目环境信息文件,一般不需要修改此文件。

上面的文件和文件夹都和项目有着直接的关系,不了解对应关系的话不要随便修改。

2.2 扩充 firstDemo 项目

下面对第一个项目 firstDemo 用四种不同方法进行扩充，在现有的基础上增加了一个文本显示组件和按钮组件，通过这些不同的实例，来说明这些文件之间的一些关系。

1）修改布局文件 Activity_main.xml，增加一个文本显示组件和按钮组件。

在原有代码基础上，直接在 Activity_main.xml 文件中增加如下代码：

```xml
<RelativeLayout                                       //采用的是相对布局模式
    xmlns:android="http://schemas.android.com/apk/res/android"
    xmlns:tools="http://schemas.android.com/tools"
    android:layout_width="match_parent"               //布局管理器的宽度为屏幕宽度
    android:layout_height="match_parent"              //布局管理器的高度为屏幕高度
    android:paddingBottom="@dimen/activity_vertical_margin"//设置于底部的距离
    android:paddingLeft="@dimen/activity_horizontal_margin"//设置于左边的距离
    android:paddingRight="@dimen/activity_horizontal_margin"//设置于右边的距离
    android:paddingTop="@dimen/activity_vertical_margin"   //设置于顶部的距离
    tools:context=".MainActivity" >
    <TextView                                         //定义一个文本显示组件
        android:id="@+id/textView1"                   //组件的名字叫 textView1
        android:layout_width="wrap_content"           //定义组件的宽度
        android:layout_height="wrap_content"          //定义组件的高度
        android:text="@string/hello_world" />         //从资源文件中读取要显示的内容
    <TextView                                         //定义一个文本显示组件
        android:id="@+id/textView2"                   //组件的名字叫 textView2
        android:layout_width="wrap_content"           //定义组件的宽度
        android:layout_height="wrap_content"          //定义组件的高度
        android:layout_below="@+id/textView1"         //位于 textView1 组件的下方
        android:layout_centerHorizontal="true"        //垂直居中
        android:layout_marginTop="20dp"               //与顶部边界的距离为 20dp
        android:text="欢迎您使用本系统！" />           //显示的文字为"欢迎您使用本系统"
    <Button                                           //定义一个按钮组件
        android:id="@+id/mybutton"                    //定义组件名字为 mybutton
        android:layout_width="fill_parent"            //组件的宽度为屏幕的宽度
        android:layout_height="wrap_content"          //定义组件的高度
        android:layout_alignLeft="@+id/textView1"     //位于 textView1 组件的左边
        android:layout_alignParentBottom="true"       //贴紧父元素的下边缘
        android:layout_marginBottom="260dp"           //与底部边界的距离为 260dp
        android:text="我是按钮！" />                   //按钮上的文字为"我是按钮！"
</RelativeLayout>
```

其他的不做任何修改，保存并运行该项目，出现如图 2.3 所示的结果。

图 2.3 firstDemo 项目修改后的运行结果

布局管理文件除了像上面那样通过输入代码的形式增加组件外，还可以使用如下的方式增加组件。

1）双击打开 Activity_main.xml 文件；

2）单击 "Graphical Layout" 选项，弹出如图 2.4 所示界面；

3）在左侧选择要加入的组件，拖曳到右侧的手机屏幕上；

4）在屏幕上移动组件，将它放在合适的地方，保存后，再单击 "activity_main.xml" 进入到该文件，发现组件相应的代码已经添加。

这个时候打开 R.java 文件，可以发现里面多了一些代码，如图 2.5 所示。

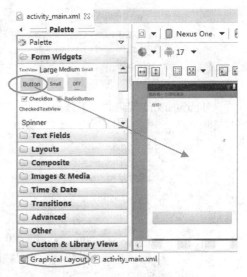

```
public static final class id {
    public static final int action_settings=0x7f080003;
    public static final int mybutton=0x7f080002;
    public static final int textView1=0x7f080000;
    public static final int textView2=0x7f080001;
```

图 2.5　firstDemo 修改后项目 R.java 文件 id 项的内容

图 2.4　用拖曳方式添加组件

注意：在 R.java 中有了定义的 id 才能在程序中通过 findViewById() 方来调用，获得相应的组件。

2）修改布局文件和 string.xml 文件，增加一个文本显示组件和按钮组件。

在 Activity_main.xml 文件中增加代码（粗体部分），如下所示：

```
<RelativeLayout                                          //采用的是相对布局模式
    xmlns:android="http://schemas.android.com/apk/res/android"
    xmlns:tools="http://schemas.android.com/tools"
    android:layout_width="match_parent"                  //布局管理器的宽度为屏幕宽度
    android:layout_height="match_parent"                 //布局管理器的高度为屏幕高度
    android:paddingBottom="@dimen/activity_vertical_margin"//设置于底部的距离
    android:paddingLeft="@dimen/activity_horizontal_margin"//设置于左边的距离
    android:paddingRight="@dimen/activity_horizontal_margin"//设置于右边的距离
    android:paddingTop="@dimen/activity_vertical_margin"    //设置于顶部的距离
    tools:context=".MainActivity" >
    <TextView                                            //定义一个文本显示组件
        android:id="@+id/textView1"                      //组件的名字叫 textView1
        android:layout_width="wrap_content"              //定义组件的宽度
        android:layout_height="wrap_content"             //定义组件的高度
        android:text="@string/hello_world" />            //从资源文件中读取要显示的内容
    <TextView                                            //增加一个文本显示组件
        android:id="@+id/textView2"                      //组件的名字叫 textView2
        android:layout_width="wrap_content"              //定义组件的宽度
        android:layout_height="wrap_content"             //定义组件的高度
        android:layout_below="@+id/textView1"            //位于 textView1 组件的下方
        android:layout_centerHorizontal="true"           //垂直居中
        android:layout_marginTop="20dp"                  //与顶部边界的距离为 20dp
        android:text="@string/txt"/> //显示的文字从文件 string.xml 中 txt 变量获得
    <Button                                              //增加一个按钮组件
        android:id="@+id/mybutton"                       //定义组件名字为 mybutton
        android:layout_width="fill_parent"               //组件的宽度为屏幕的宽度
        android:layout_height="wrap_content"             //定义组件的高度
```

```
        android:layout_alignLeft="@+id/textView1"    //位于 textView1 组件的左边
        android:layout_alignParentBottom="true"      //贴紧父元素的下边缘
        android:layout_marginBottom="260dp"          //与底部边界的距离为 260dp
        android:text=""@string/but"" />  //显示的文字从文件 string.xml 中 but 变量获得
</RelativeLayout>
```

对 strings.xml 代码做如下修改：

```
<?xml version="1.0" encoding="utf-8"?>
<resources>
    <string name="app_name">firstDemo</string>   //定义 app_name 的值为 firstDemo
    <string name="action_settings">Settings</string>   //定义 action_settings
                                                           的值为 Settings
    <string name="hello_world">Hello world!</string>   //定义 hello_world 的值
                                                           为 hello world!
    <string name="txt">欢迎您使用本系统!</string>  //定义 txt 的值为"欢迎您使用本系统!"
    <string name="but">我是按钮!</string>          //定义 but 的值为"我是按钮!"
</resources>
```

保存并运行该项目，也会出现如图 2.3 所示的结果。再对 strings.xml 代码做如下修改：

```
<?xml version="1.0" encoding="utf-8"?>
<resources>
    <string name="app_name">我的第一个项目演示</string>   //定义 app_name 的值
    <string name="action_settings">Settings</string>   //定义 action_settings 的值
    <string name="hello_world">你好!</string>           //定义 hello_world 的值
    <string name="txt">欢迎您使用本系统!</string>        //定义 txt 的值
    <string name="but">我是按钮!</string>                //定义 but 的值
</resources>
```

其他的文件不做修改，保存并运行该项目，会弹出如图 2.6 所示的结果。

图 2.6　修改后的运行结果

3）修改布局文件和 MainActivity.java 文件，增加一个文本显示组件和按钮组件。

在上面的修改基础上对 Activity_main.xml 文件进行修改，代码如下：

```
<RelativeLayout                                           //采用的是相对布局模式
    xmlns:android="http://schemas.android.com/apk/res/android"
    xmlns:tools="http://schemas.android.com/tools"
    android:layout_width="match_parent"      //布局管理器的宽度为屏幕宽度
    android:layout_height="match_parent"     //布局管理器的高度为屏幕高度
    android:paddingBottom="@dimen/activity_vertical_margin"//设置于底部的距离
    android:paddingLeft="@dimen/activity_horizontal_margin"//设置于左边的距离
```

```xml
        android:paddingRight="@dimen/activity_horizontal_margin"//设置于右边的距离
        android:paddingTop="@dimen/activity_vertical_margin"//设置于顶部的距离
        tools:context=".MainActivity" >
    <TextView                                           //定义一个文本显示组件
        android:id="@+id/textView1"                     //组件的名字叫 textView1
        android:layout_width="wrap_content"             //定义组件的宽度
        android:layout_height="wrap_content"            //定义组件的高度
        android:text="@string/hello_world" />           //从资源文件中读取要显示的内容
    <TextView                                           //定义一个文本显示组件
        android:id="@+id/textView2"                     //组件的名字叫 textView2
        android:layout_width="wrap_content"             //定义组件的宽度
        android:layout_height="wrap_content"            //定义组件的高度
        android:layout_below="@+id/textView1"           //位于 textView1 组件的下方
        android:layout_centerHorizontal="true"          //垂直居中
        android:layout_marginTop="20dp"                 //与顶部边界的距离为 20dp
    <Button                                             //增加一个按钮组件
        android:id="@+id/mybutton"                      //定义组件名字为 mybutton
        android:layout_width="fill_parent"              //组件的宽度为屏幕的宽度
        android:layout_height="wrap_content"            //定义组件的高度
        android:layout_alignLeft="@+id/textView1"       //位于 textView1 组件的左边
        android:layout_alignParentBottom="true"         //贴紧父元素的下边缘
        android:layout_marginBottom="260dp"             //与底部边界的距离为 260dp
</RelativeLayout>
```

注意：这里没有对文本显示组件和按钮组件设置 text 的值，需要在 MainActivity.java 程序中设置，对 MainActivity.java 文件进行修改，代码如下：

```java
package org.hnist.firstdemo;
import android.os.Bundle;                   //导入 os.Bundle 类
import android.app.Activity;                //导入 app.Activity 类
import android.widget.TextView;             //导入 widget.TextView 类
import android.widget.Button;               //导入 widget.Button 类
public class MainActivity extends Activity { //定义主程序入口
    private TextView txt=null;              //定义 txt 文本组件
    private Button but=null;                //定义 but 按钮组件
    @Override
     protected void onCreate(Bundle savedInstanceState) {//覆写 onCreadte 方法
        super.onCreate(savedInstanceState);       //调用父类的 onCreate()方法
        setContentView(R.layout.activity_main);   //调用布局文件 activity_main.xml
        this.txt=(TextView)super.findViewById(R.id.textView2);
                                                  //取得 txt 文本显示组件
        txt.setText("欢迎您使用本系统！");         //设置 txt 组件显示的内容
        this.but=(Button)super.findViewById(R.id.mybutton);//取得 but 按钮组件
        but.setText(super.getString(R.string.but)); } }    //设置 but 按钮上显示的
                     内容来源于 string.xml 中 but 定义的值
```

其他的文件不做修改，保存并运行该项目，会出现如图 2.6 所示的结果。

这种方法适合初学者，比较简单、直观。首先在布局管理器中定义要使用的组件，然后在 Activity 程序中通过 setContentView()方法调用这个布局管理文件，布局管理文件里的组件则通过 findViewById()方法来获得，然后对组件进行相应的操作，例如通过 setText()方法设置在组件上要显示的内容。

事实上，也可以不使用布局管理文件，直接在 MainActivity.java 上添加组件，设置参数。

4）修改 MainActivity.java 文件，增加一个文本显示组件。

对 MainActivity.java 文件做如下修改：

```java
package org.hnist.firstdemo;
import android.os.Bundle;                    //导入 os.Bundle 类
import android.app.Activity;                 //导入 app.Activity 类
import android.widget.TextView;              //导入 widget.TextView 类
public class MainActivity extends Activity {//定义主程序入口
    @Override
    public void onCreate(Bundle savedInstanceState) {  //覆写 onCreadte 方法
        super.onCreate(savedInstanceState);            //调用父类的 onCreate()方法
        //setContentView(R.layout.activity_main);      //注意这里没有调用布局文件
        TextView txt=new TextView(this);               //定义一个文本显示组件
        txt.setText(super.getString(R.string.txt));    //设置文本显示组件显示的文字
        super.setContentView(txt); } }                 //设置显示该组件
```

这里直接通过程序的方式生成组件，因此不需要加载布局管理文件，要注意的是组件生成之后要利用 setContentView()方法将组件显示出来。

上面代码中的语句 txt.setText(super.getString(R.string.txt))是通过调用 string.xml 文件中的变量 txt 的值来设置文本显示组件上显示的文字，也可以直接在这里使用 txt.setText("欢迎您使用本系统!")，运行效果一样。程序运行结果如图 2.7 所示。

图 2.7　修改后的运行结果

这种方法比较灵活，但是利用这种方式所生成的组件，每次只能显示一个组件，虽然布局管理文件中存在按钮组件，但是没有在 Activity 程序中调用布局文件，因此没有显示按钮。在后续的课程中可以定义一个布局管理器对象，然后在其中添加多个组件。

在模拟机上运行速度可能会有些慢，Android 程序也可以在真实手机中调试运行。将手机通过 USB 接口连接电脑，在电脑上安装相应手机的驱动程序，在 Eclipse 程序中选择"Window→show view→other →Android →devices"选项，单击"OK"按钮，就可以看到手机的名字，如图 2.8 所示。

图 2.8　在 Eclipse 中发现真实手机

右键单击项目，选择在真实的手机上运行，如图 2.9 所示，就可以在手机上显示程序的运行界面。

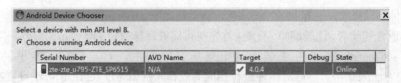

图 2.9　设置程序在真实手机上运行

2.3 Activity 简介

Android 项目共包括四个应用程序组件（Component），一个 Android 应用程序是一个包（Package），包中可能包含下面的一个或者多个 Android 组件。

1）活动（Activity）：Activity 是最基本的 Android 应用程序组件，在应用程序中，一个 Activity 通常就是一个单独的用户界面。每一个 Activity 都可以视为一个独立的类，并且从 Activity 基类中继承而来，Activity 类将会显示由视图（View）控件组成的用户接口，并对事件（Event）做出响应。大多数的应用程序都会有多个用户界面，因此便会有多个相应的 Activity 程序。

一个活动一般对应界面中的一个屏幕显示，可以理解成一个界面，每一个 Activity 在界面上可以包含按钮、文本框等多种可视的 UI 元素。

2）广播接收器（Broadcast Receiver）：广播接收器用于让应用程序对一个外部事件做出响应。例如：电话呼入事件、数据网络可用通知或者指定时间进行通知等。

3）服务（Service）：服务是具有一段较长生命周期但没有用户界面的程序。例如：一个正在从播放列表中播放歌曲的媒体播放器在后台运行。

4）内容提供者（Content Provider）：应用程序能够将它们的数据保存到文件或 SQLite 数据库中，甚至是任何有效的设备中。当需要将数据与其他的应用共享时，内容提供者组件将会很有用。一个内容提供者类实现了一组标准的方法，从而能够让其他应用程序保存或读取此内容提供者处理的各种数据类型。

简单地说 Activity 就是一个用户所能看到的屏幕，是应用程序的界面容器，可以放置各种各样的控件、设置处理事件（例如按键事件、触摸屏事件等）、为用户显示指定的 View，启动其他 Activity 等。Activity 中所有操作都与用户密切相关，是负责与用户交互的组件，所有应用的 Activity 都继承于 android.app.Activity 类。

创建一个 Activity，要注意以下的四个方面：

1）一个 Activity 就是一个类，并且这个类要继承 Activity；

```
import android.app.Activity;
```

2）需要覆写 onCreate 方法（应用程序启动后第一个运行的函数，由 Android 框架决定）；

```
public void onCreate(Bundle savedInstanceState) {//覆写 onCreate 方法
    super.onCreate(savedInstanceState);        //调用父类的 onCreate()方法
    setContentView(R.layout.activity_main); }  //调用布局文件 activity_main.xml
```

3）每一个 Activty 都需要在 AndroidManifest.xml 文件当中进行配置；

```
<activity
    android:name="org.hnist.firstdemo.MainActivity"//设置默认启动的 Activity
    android:label="@string/app_name" >             //定义 activity 的名字
    <intent-filter>
        <action android:name="android.intent.action.MAIN" />    //主程序
        <category android:name="android.intent.category.LAUNCHER" />//表示放
                到手机应用程序的列表里
    </intent-filter>
</activity>
```

4）根据需求为 Activity 添加必要的控件。

通过 findViewById（控件 id）方法可以得到所要显示的控件。例如：

```
this.txt=(TextView)super.findViewById(R.id.textView2);  //取得 txt 文本显示组件
```

Activity 的启动可以通过 Launcher 组件也可以通过 Activity 内部调用 startActivity 接口来启动。Activity 之间的数据传递，主要是通过一个 Action 来完成的，如果需要传递参数可通过 Intent 来进行参数的传递，传递数据的过程可以是双向的，这些将在本书的第 6 章进行详细的介绍。

2.4 Android 中的常用包

在 Android 应用程序开发中，使用的是 Java 语言，除了需要熟悉 Java 语言的基础知识之外，还需要了解 Android 提供的扩展的 Java 功能，Android 提供了一些扩展的 Java 类库，类库又分为若干个包，每个包中包含若干个类。Android Java API 包含 40 多个包和 700 多个类，这些为编写 Android 应用程序提供了一个功能丰富的平台。

下面对一些常用的重要包进行简要的描述，如表 2-1 所示。

表 2-1 重要包描述

包名称	功能描述
android.app	实现 Android 的应用程序模型，提供基本的运行环境
android.bluetooth	提供一些类来处理蓝牙功能
android.content	包含各种的对设备上的数据进行访问和发布的类
android.database	通过内容提供者浏览和操作数据库
android.database.sqlite	实现 android.database 包，将 SQLite 用作物理数据库
android.graphics	底层的图形库，包含画布、颜色过滤、点、矩形，可以将它们直接绘制到屏幕上
android.graphics.drawable	实现绘制协议和背景图像，支持可绘制对象动画
android.graphics.drawable.shapes	实现各种形状
android.hardware	实现与物理照相机相关的类
android.location	定位和相关服务的类
android.media	提供一些类管理多种音频、视频的媒体接口
android.net	提供帮助网络访问的类，超过通常的 java.net.*接口
android.net.wifi	管理 WiFi 连接
android.opengl	提供 OpenGL 的工具
android.os	表示可通过 Java 编程语言访问的操作系统服务
android.provider	提供类访问 Android 的内容提供者
android.speech	包含用于语音识别的常量。这个包只在 1.5 版和更新版本中提供
android.speech.tts	提供从文本到语音转换的支持
android.telephony	提供与拨打电话相关的 API 交互
android.telephony.gsm	可用于根据基站来收集手机位置
android.text	包含文本处理类
android.view	提供基础的用户界面接口框架
android.util	涉及工具性的方法，例如时间日期的操作
android.webkit	默认浏览器操作接口
android.widget	包含通常派生自 View 类的所有 UI 控件。主要的部件包括 Button、Checkbox、Chronometer、AnalogClock、DatePicker、DigitalClock、EditText、ListView、FrameLayout、GridView、ImageButton、MediaController、ProgressBar、RadioButton、RadioGroup、RatingButton、Scroller、ScrollView、Spinner、TabWidget、TextView、TimePicker、VideoView 和 ZoomButton

例如：希望在 Android 程序中用 Color.BLACK 来表示黑色，就要调用包含颜色的包 android.

graphics.Color 包，通常使用 import android.graphics.Color；语句进行引用，引用后就可以直接使用 Color.BLACK 来表示黑色，Color.RED 来表示红色，等等，如果事先不进行引用，直接使用 Color.BLACK 来表示黑色的话，程序将会报错。

2.5 Android 项目的大致开发流程

在开发一个 Android 项目前，先要对这个项目进行一些最基本的分析，规划好项目的基本开发步骤，确保项目顺利开发。一般来说，Android 项目开发步骤包括以下几个方面。

1）对项目进行分析：了解项目的主要功能、有哪些必需的界面及界面之间的跳转关系、需要的数据及数据的来源、是否需要服务器的支持、是否需要后台服务等。

2）架构设计：将整个项目进行分解，确定在 Activity 里设计的项目有哪些、哪个是进行网络连接、哪个是数据库处理等。

3）界面设计：确定程序的主界面、各模块界面、列表、查看、编辑界面、菜单、按钮、对话框、提示信息的设计、界面总体颜色设计，使项目更加美观统一。

4）数据操作和存储：确定项目需要的数据来源、如何存储和读取等。

5）代码的编写：对分解的各个子模块进行代码编写，包括控件、事件、菜单、页面跳转等。

6）程序调试。

当然，这里只是一个简单的介绍，如果真正开发一个较大的 Android 项目，可能还会考虑得更多。例如：可行性分析、用户需求分析、项目进度设计、项目的总体设计、详细设计等，读者可以参考软件工程的相关书籍进行了解。

2.6 Android 中常见文件介绍

1. Java 文件——应用程序源文件

Android 的应用程序使用 Java 来开发，开发完成后的 Java 程序放在 src 文件夹下，一个 Android 项目可以有一个或多个 Java 程序。

2. xml 文件——Android 上的资源文件

Android 中的资源是指非代码部分，主要是外部文件，例如布局管理文件、activity_main.xml 文件等。在代码中访问资源文件，是通过 R 类中定义的资源文件类型和资源文件名称来访问的，具体格式为：R.资源文件类型.资源文件名称。

3. Class 文件——Java 编译后的目标文件

在 J2SE 环境中 Java 程序编译成 class 就可以直接运行了，但是在 Android 平台上 class 文件不能直接运行，要使用 Dalvik 来运行应用程序，Android 平台上的 class 文件实际上只是编译过程中的中间文件，需要链接成 dex 文件后才能在 Dalvik 上运行。

4. Dex 文件——Android 平台上的可执行文件

Google 公司在 Android 平台上使用了自己的 Dalvik 虚拟机来运行程序，这种虚拟机执行的并非 Java 代码，而是另一种代码：dex 格式的代码。在编译 Java 代码之后，通过 Android 平台上的工具可以将 Java 代码转换成 Dex 代码。

5. apk 文件——Android 上的安装文件

apk 是 Android 程序安装包的扩展名，一个 Android 安装包包含了与某个 Android 应用程序相关的所有文件。apk 文件将 AndroidManifest.xml 文件、应用程序代码（.dex 文件）、资源文件和其他文件封装为一个压缩包。

本章小结

本章着重介绍了 Android 项目的结构，主要文件夹和文件的用途，通过四种不同的方法对第一个项目 firstDemo 进行扩充，加深对项目各组成部分之间关系的理解，简单介绍了 Activity 及建立 Activity 要注意的几个方面，简单介绍了 Android 项目中的开发包，项目开发的大致流程等。目的是为了让用户深入了解 Android 项目各组成部分之间的关系，对 Android 项目开发有一个大致了解。

习题

1. 简要描述 Android 项目开发的大致开发流程。
2. 在网络上下载 Android API 文档，了解其用途。
3. 新建一个项目，命名为：HelloDemo，包名称为：org.hnist.hello，运行项目，在屏幕上显示一行文字"信息提交！"，显示两个按钮，一个为"确定"按钮，一个为"取消"按钮。要求用两种方式实现：调用布局管理文件和不调用布局管理文件。
4. 如果要修改标题的提示文字为"欢迎使用成绩管理系统"，应该在哪个文件里修改？
5. 如果要修改运行项目的最小版本号为 8，应该在哪个文件里修改？
6. 要在 res\layout 文件夹下建立一个 xml 文件，应该如何操作？
7. 要在 src 文件夹下建立一个 Java 文件，应该如何操作？

第 3 章 Android 开发常用组件

学习目标：
- 了解 Android 中的 View 类；
- 掌握 Android 中的 TextView 等常用基本组件及操作方法；
- 掌握 Android 中的 4 种布局管理器的使用。

前面的章节对 firstDemo 项目进行了扩充，添加了按钮，其实按钮和文字显示都是通过 Android 中的组件实现的，Android 平台提供了大量的控件来简化 Android 项目的开发，本章主要将介绍 Android 平台提供的 View、TextView 等常用基本组件和布局管理器组件及其基本操作。

3.1 Android 平台中的 View 类

前面章节介绍的按钮和文本显示组件都是 View 类的子类，Android 平台中的 android.view.View 类包含了大多数的图形显示组件，其层次关系如下：

```
java.lang.Object
    android.view.View
```

要在 Android 程序中使用 View 类的话，必须在程序中使用下面的语句，否则会报错。

```
import android.android.view.View;          //导入 android.view.View 类
```

除了按钮和文本显示组件外，android.view.View 类中还定义了许多图形组件，如表 3-1 所示，这些组件都在 android.widge 包中定义。

表 3-1 部分常见图形组件列表

组件名称	类名称	描述
TextView	android.widget. TextView	文本的显示组件
Button	android.widget. Button	普通按钮组件
EditText	android.widget. EditText	编辑文本框组件
CheckBox	android.widget. CheckBox	表示复选框组件
RadioGroup	android.widget. RadioGroup	表示单选按钮组件
Spinner	android.widget. Spinner	下拉列表框
DatePicker	android.widget. DatePicker	日期选择组件
TimePicker	android.widget. TimePicker	时间选择组件
ImageView	android.widget. ImageView	图片显示组件
ImageButton	android.widget. ImageButton	图片按钮组件
Toast	android.widget. TextView	信息提示框组件

通过前面的学习可知，将组件显示在屏幕上可以调用布局管理文件中设置的组件，也可以在 Activity 程序中通过代码来设置。View 组件有相应的属性和方法，参数的值可以在布局管理中设置，也可以在 Activity 程序中通过代码来设置，表 3-2 列出了 View 组件常用的属性及对应方法。

表 3-2 View 组件常用属性及对应方法

属性名称	方法名称	描述
android:background	public void setBackgroundResource (int resid)	设置组件背景
android:clickable	public void setClickable (boolean clickable)	是否可以产生单击事件
android:contentDescription	public void setContentDescription (CharSequence contentDescription)	定义视图的内容描述
android:drawingCacheQuality	public void setDrawingCacheQuality (int quality)	设置绘图时所需要的缓冲区大小
android:focusable	public void setFocusable (boolean focusable)	设置是否可以获得焦点
android:focusableInTouchMode	public void setFocusableInTouchMode (boolean focusableInTouchMode)	在触摸模式下配置是否可以获得焦点
android:id	public void setId (int id)	设置组件 ID
android:longClickable	public void setLongClickable (boolean longClickable)	设置长按事件是否可用
android:minHeight		定义视图的最小高度
android:minWidth		定义视图的最小宽度
android:padding	public void setPadding (int left, int top, int right, int bottom)	填充所有的边缘
android:paddingBottom	public void setPadding (int left, int top, int right, int bottom)	填充下边缘
android:paddingLeft	public void setPadding (int left, int top, int right, int bottom)	填充左边缘
android:paddingRight	public void setPadding (int left, int top, int right, int bottom)	填充右边缘
android:paddingTop	public void setPadding (int left, int top, int right, int bottom)	填充上边缘
android:scaleX	public void setScaleX (float scaleX)	设置 X 轴缩放
android:scaleY	public void setScaleY (float scaleY)	设置 Y 轴缩放
android:scrollbarSize		设置滚动条大小
android:scrollbarStyle	public void setScrollBarStyle (int style)	设置滚动条样式
android:visibility	public void setVisibility (int visibility)	设置是否显示组件
android:layout_width		定义组件显示的宽度
android:layout_height		定义组件显示的长度
layout_toRightOf		位于组件的右边
layout_above		位于组件的上方
layout_below		位于组件的下方
layout_toLeftOf		位于组件的左边
android:layout_gravity		组件文字的对齐位置
android:layout_margin		设置文字的边距
android:layout_marginTop		上边距
android:layout_marginBottom		下边距
android:layout_marginLeft		左边距
android:layout_marginRight		右边距
android:background		设置背景颜色

在 Android 平台中 View 类是一个包含子类最多的一个类,常见的组件有以下几种。
- 文本类组件:TextView、EditText、……
- 按钮类组件:Button、ImageButton、……
- 选择类组件:RadioButton、CheckBox、……
- 列表类组件:Spinner、ListView、……
- 图像类组件:ImageView、Gallery、……
- 时间类组件:DatePicker、TimePicker、……

- 布局类组件：RelativeLayout、LinearLayout、……
- 提示类组件：Toast、……
- 菜单类组件：Menu、……

……

下面介绍 View 类中几个常见的组件，以加深对 Activity 操作的认识。

3.2 文本显示组件 TextView

文本显示组件的主要作用是在屏幕上显示文字，在 Android 程序中用 TextView 组件来实现，其层次关系如下：

```
java.lang.Object
    android.view.View
        android.widget.TextView
```

直接子类：Button, CheckedTextView, Chronometer, DigitalClock, EditText。

间接子类：AutoCompleteTextView，ExtractEditText, MultiAutoCompleteTextView，RadioButton，ToggleButton，CheckBox，CompoundButton。

要在 Android 程序中使用 TextView 组件必须要在程序中使用下面的语句。

```
import android.widget.TextView;                //导入widget.TextView类
```

3.2.1 TextView 组件常见的属性和方法

TextView 组件继承了 View 类，所以前面介绍的 View 类的属性它都具备，除此之外，该组件还有其他属性，见表 3-3。表 3-4 列出了该组件的常用方法。

表 3-3 TextView 组件常用属性列表

属性名称	描述
android:autoLink	设置是否当文本为 URL 链接/email/电话号码/map 时，文本显示为可点击的链接。可选值（none/web/email/phone/map/all）
android:ellipsize	设置当文字过长时，该控件该如何显示。有如下值设置： start——省略号显示在开头； end——省略号显示在结尾； middle——省略号显示在中间； marquee——以跑马灯的方式显示（动画横向移动）
android:gravity	设置文本位置，如设置成"center"，文本将居中显示
android:marqueeRepeatLimit	在 ellipsize 指定 marquee 的情况下，设置重复滚动的次数，当设置为 marquee_forever 时表示无限次
android:maxLength	限制显示的文本长度，超出部分不显示
android:lines	设置文本的行数，设置两行就显示两行，即使第二行没有数据
android:password	以小点"."显示文本
android:selectAllOnFocus	如果文本是可选择的，让它获取焦点而不是将光标移动为文本的开始位置或者末尾位置。TextView 中设置后无效果
android:shadowColor	指定文本阴影的颜色，需要与 shadowRadius 一起使用
android:shadowRadius	设置阴影的半径。设置为 0.1 就变成字体的颜色了，一般设置为 3.0 的效果比较好
android:text	设置显示文本
android:textColor	设置文本颜色
android:textColorLink	文字链接的颜色
android:textSize	设置文字大小，推荐度量单位"sp"，如"15sp"
android:height	设置文本区域的高度，支持度量单位：px(像素)/dp/sp/in/mm
android:width	设置文本区域的宽度，支持度量单位：px(像素)/dp/sp/in/mm

表 3-4 TextView 组件常用方法列表

方法	描述
public void setText(CharSquence str)	设置组件显示文字
public String setText getText()	获得 TextView 对象的文本
public int length()	获得 TextView 中的文本长度
public void getEditableText()	取得文本的可编辑对象，通过这个对象可对 TextView 的文本进行操作，如在光标之后插入字符
public void getAutoLinkMask()	返回自动连接的掩码
public void setTextColor()	设置文本显示的颜色
public void setHintTextColor()	设置提示文字的颜色
public void setLinkTextColor()	设置链接文字的颜色
public void setGravity()	设置当 TextView 超出了文本本身时横向以及垂直对齐

3.2.2 TextView 组件的使用实例

TextView 组件要显示在屏幕上可以调用布局管理文件中设置的组件，例如：

```
<TextView                                       //定义一个文本显示组件
    android:id="@+id/mytxt"                     //组件的名字叫 mytxt
    android:layout_width="wrap_content"         //定义组件的宽度
    android:layout_height="wrap_content"        //定义组件的高度
    android:text="欢迎您使用本系统！"            //显示的文字为"欢迎您使用本系统"
    …… />
```

代码中"android:"后面的都是 TextView 组件的属性，这些属性可以参照表 3-3 进行设置，也可以在 Activity 程序中通过调用 TextView 相应的方法来实现，例如：

```
TextView txt=new TextView(this);                //定义一个文本显示组件
txt.setText("欢迎您使用本系统！");               //设置文本显示组件显示的文字
super.setContentView(txt);                      //设置显示该组件
```

实例 3-1：各种属性的运用

按照如下步骤新建一个 Android 项目：

1）打开 Eclipse 程序，选择菜单中的"File→New→Other"选项。

2）在弹出的窗口中，选择"Android→Android Application Project"项目，单击"Next"按钮。

3）在弹出的窗口中，输入项目的名称：exam3_1，包名称：org.hnist.demo，其他的选项取默认值，单击"Next"按钮。

4）在后续弹出的窗口中，所有选项采用默认值，直到单击"Finish"按钮完成。

5）在 res 文件夹下选择 layout，双击 Activity_main.xml，打开布局管理文件，作如下的修改：

```
<RelativeLayout                                                    //相对布局开始
    xmlns:android="http://schemas.android.com/apk/res/android"
    xmlns:tools="http://schemas.android.com/tools"
    android:layout_width="match_parent"         //布局管理器的宽度为屏幕宽度
    android:layout_height="match_parent"        //布局管理器的高度为屏幕高度
    android:background="#000000"                //设置布局管理器背景颜色为黑色
    tools:context=".MainActivity" >
    <TextView                                   //定义一个文本显示组件
        android:id="@+id/mytxt1"                //组件的名字叫 mytxt1
```

```
        android:layout_width="wrap_content"         //组件的宽度为文字的宽度
        android:layout_height="wrap_content"        //组件的高度为文字的高度
        android:layout_centerHorizontal="true"      //设置文字水平居中显示
        android:background="#FF00FF00"              //设置文字背景颜色
        android:text="www.hnist.cn" />              //设置显示的文字为 www.hnist.cn
<TextView                                           //定义一个文本显示组件
        android:id="@+id/mytxt2"                    //组件的名字叫 mytxt2
        android:layout_width="wrap_content"         //组件的宽度为文字的宽度
        android:layout_height="wrap_content"        //组件的高度为文字的高度
        android:layout_below="@+id/mytxt1"          //该组件位于组件 mytxt1 的下方
        android:layout_centerHorizontal="true"      //设置文字水平居中显示
        android:layout_marginTop="25dp"             //距离上面组件 25dp
        android:textColor="#FF00FF00"               //设置文字的颜色"#FF00FF00"绿色
        android:textSize="20sp"                     //设置文字的大小为 20sp
        android:textStyle="bold"                    //设置文字加粗显示
        android:text="www.hnist.cn" />              //设置显示的文字为 www.hnist.cn
<TextView                                           //定义一个文本显示组件
        android:id="@+id/mytxt3"                    //组件的名字叫 mytxt3
        android:layout_width="wrap_content"         //组件的宽度为文字的宽度
        android:layout_height="wrap_content"        //组件的高度为文字的高度
        android:layout_below="@+id/mytxt2"          //该组件位于组件 mytxt2 的下方
        android:layout_marginTop="35dp"             //距离上面组件 35dp
        android:autoLink="all"                      //设置超级链接
        android:textSize="20sp"                     //设置文字的大小为 20sp
        android:text="www.hnist.cn" />              //设置显示的文字为 www.hnist.cn
<TextView                                           //定义一个文本显示组件
        android:id="@+id/mytxt4"                    //组件的名字叫 mytxt4
        android:layout_width="wrap_content"         //组件的宽度为文字的宽度
        android:layout_height="wrap_content"        //组件的高度为文字的高度
        android:layout_alignParentLeft="true"       //设置文字左对齐显示
        android:layout_below="@+id/mytxt3"          //该组件位于组件 mytxt3 的下方
        android:layout_marginTop="20dp"             //距离上面组件 35dp
        android:password="true"                     //设置文字以密文方式显示
        android:textSize="15sp"                     //设置文字的大小为 15sp
        android:text="www.hnist.cn"/>               //设置显示的文字为 www.hnist.cn
<TextView                                           //定义一个文本显示组件
        android:id="@+id/mytxt5"                    //组件的名字叫 mytxt5
        android:layout_width="wrap_content"         //组件的宽度为文字的宽度
        android:layout_height="wrap_content"        //组件的高度为文字的高度
        android:layout_above="@+id/mytxt6"          //该组件位于组件 mytxt6 的上方
        android:layout_marginBottom="16dp"          //距离下面组件 16dp
        android:layout_centerHorizontal="true"      //组件水平居中
        android:shadowColor="#FFFF0000"             //设置文本阴影的颜色
        android:shadowRadius="3.0"                  //设置阴影的半径为 3.0
        android:textSize="20sp"                     //设置文字的大小为 20sp
        android:text="www.hnist.cn"/>               //设置显示的文字为 www.hnist.cn
<TextView                                           //定义一个文本显示组件
        android:id="@+id/mytxt6"                    //组件的名字叫 mytxt6
        android:layout_width="80dp"                 //组件的宽度为 80dp
        android:layout_height="wrap_content"        //组件的高度为文字的高度
        android:layout_centerHorizontal="true"      //组件水平居中
```

```
            android:layout_centerVertical="true"          //组件垂直居中
            android:ellipsize="marquee"                    //设置为滚动的文字
            android:focusable="true"                       //设置可以获得焦点
            android:focusableInTouchMode="true"            //设置触摸模式下可以获得焦点
            android:gravity="center"                       //设置文本居中显示
            android:marqueeRepeatLimit="marquee_forever"//设置滚动的次数为无限次
            android:singleLine="true"                      //设置为单行显示文字内容
            android:text="www.hnist.cn" />                 //设置显示的文字为 www.hnist.cn
</RelativeLayout>                                          //相对布局结束
```

保存文件，程序运行结果如图 3.1 所示。

单击带有链接的 www.hnist.cn 能够链接到www.hnist.cn网页，最下面的文字实现了"跑马灯"的效果。

实例 3-2：使用样式文件简化 xml 代码

程序中的文字如果使用某种指定格式的效果，可以建立样式文件，以便于维护和使用。样式文件格式与 strings.xml 文件类似，一般是在 res\values 文件夹中的 styles.xml 文件中建立的。

按照前面的步骤建立一个项目，项目的命名为：exam3_2，包名称为：org.hnist.demo。

图 3.1 实例 3-1 运行结果

在 res\values 文件夹中双击 styles.xml 文件，在文件中输入如下代码：

```
<resources>
    <style name="AppBaseTheme" parent="android:Theme.Light"></style>
    <style name="AppTheme" parent="AppBaseTheme">    </style>
<style name="my_styles" >                             //定义 my_styles 样式
    <item name="android:layout_width">wrap_content</item> //组件的宽度为文字的宽度
    <item name="android:layout_height">wrap_content</item>//组件的高度为文字的高度
    <item name="android:layout_centerHorizontal">true</item>//组件水平居中
    <item name="android:textSize">20sp</item>          //设置文字的大小为 20sp
    <item name="android:textColor">#FF00FF00</item>    //设置文字颜色
    <item name="android:textStyle">bold</item>         //设置文字加粗显示
    <item name="android:shadowColor">#FFFF0000</item>  //设置文字阴影颜色
    <item name="android:shadowRadius">3.0</item>       //设置文字阴影半径
</style>                                               //结束样式定义
</resources>
```

一旦样式文件建立好了，就可以被调用，显示出来的文字就是前面定义好的样式。例如：

```
<RelativeLayout
    xmlns:android="http://schemas.android.com/apk/res/android"
    xmlns:tools="http://schemas.android.com/tools"
    android:layout_width="match_parent"                //布局管理器的宽度为屏幕宽度
    android:layout_height="match_parent"               //布局管理器的高度为屏幕高度
    tools:context=".MainActivity" >
    <TextView                                          //定义一个文本显示组件
        android:id="@+id/mytxt1"                       //组件的名字叫 mytxt1
        style="@style/my_styles"                       //文字的显示样式调用 my_styles 样式
        android:background="@drawable/a1"              //设置背景图像为 a1.jpg（注意这里将
        a1.jpg 文件先复制到 drawable 某个文件夹下，程序否则会报错）
        android:gravity="center"                       //文字居中显示
```

```
        android:autoLink="phone"                     //设置电话号码链接
        android:text="拨打手机: 13207304569" />        //设置显示文字
    <TextView                                        //定义一个文本显示组件
        android:id="@+id/mytxt2"                     //组件的名字叫 mytxt2
        style="@style/my_styles"                     //文字的显示样式调用 my_styles 样式
        android:layout_below="@+id/mytxt1"           //组件位于 mytxt1 组件的下方
        android:background="#FFFF0000"               //设置文字背景颜色
        android:text="湖南理工学院" />                 //设置显示的文字为湖南理工学院
</RelativeLayout>
```

保存文件，程序运行结果如图 3.2 所示。虽然没有在布局文件里对文字进行颜色字号等属性进行设置，但是因为调用了样式文件，所以显示的文字在颜色、阴影等方面都有变化。

单击号码，可以进入电话拨打界面，如图 3.3 所示。

图 3.2 实例 3-2 运行结果

图 3.3 拨打电话界面

实例 3-3：在 Activity 程序中动态创建 TextView 组件

新建一个项目，项目的命名为：exam3_3，包名称为：org.hnist.demo，在项目中对 MainActivity.java 文件做如下修改：

```
package org.hnist.demo;                              //包名
import android.os.Bundle;                            //导入 os.Bundle 类
import android.app.Activity;                         //导入 app.Activity 类
import android.graphics.Color;                       //导入 graphics.Color 类，就能识别 RED 等
import android.widget.TextView;                      //导入 widget.TextView 类
public class MainActivity extends Activity {
    protected void onCreate(Bundle savedInstanceState) {//覆写 onCreate()方法
        super.onCreate(savedInstanceState);          //调用父类的 onCreate()方法
        //setContentView(R.layout.activity_main);    注意这里没有调用布局文件
        TextView txt=new TextView(this);             //定义一个文本显示组件
        txt.setText("www.hnist.cn");                 //设置组件文字为 www.hnist.cn
        txt.setBackgroundColor(Color.GREEN);         //设置背景颜色为绿色
        txt.setTextColor(Color.RED);                 //设置文字颜色为红色
        txt.setTextSize(30);                         //设置文字大小为 30dp
        super.setContentView(txt); } }               //设置 txt 组件显示
```

保存文件，程序运行结果如图 3.4 所示。

3.3 按钮组件 Button

按钮组件是在人机交互时使用较多的组件，当用户进行某些选择的时候，就可以通过按钮的操作来接收用户的选择。Button 组件的层次关系如下：

图 3.4 实例 3-3 运行界面

```
java.lang.Object
    android.view.View
        android.widget.TextView
            android.widget.Button
```

直接子类：CompoundButton

间接子类：CheckBox, RadioButton, ToggleButton

要在 Android 程序中使用 Button 组件必须在程序中使用下面的语句：

```
import android.widget.Button;            //导入 widget.Button 类
```

3.3.1 Button 组件常见的属性和方法

由于 Button 组件继承了 TextView 类，TextView 类的属性，Botton 组件都可以使用，常见属性见表 3-3 所示。

Button 组件一般用来做事件处理，用于接收用户的选择并执行相应的程序或操作，本章节会有详细介绍，Button 组件常用的方法如表 3-5 所示。

表 3-5 Button 组件常用的方法列表

方法	描述
public Boolean onKeyDown()	当用户按键时，该方法被调用
public Boolean onKeyUp()	当用户按键弹起后，该方法被调用
public Boolean onKeyLongPress()	当用户保持按键时，该方法被调用
public Boolean onKeyMultiple()	当用户多次调用时，该方法被调用
public void invalidateDrawable()	刷新 Drawable 对象
public void scheduleDrawable()	定义动画方案的下一帧
public void unscheduleDrawable()	取消 scheduleDrawable 定义的动画方案
public Boolean onPreDraw()	设置视图显示
public void sendAccessibilityEvent()	发送事件类型指定的 AccessibilityEvent。发送请求之前，需要检查 Accessibility 是否打开
public void sendAccessibilityEventUnchecked()	发送事件类型指定的 AccessibilityEvent。发送请求之前，不需要检查 Accessibility 是否打开
public void setOnKeyListener()	设置按键监听

3.3.2 Button 组件使用实例

因为 Button 组件是 TextView 的子类，如果将上面例子中的"TextView"换成"Button"也可以顺利执行。

实例 3-4：各种属性的运用

新建一个项目，项目的命名为：exam3_4，包名称为：org.hnist.demo，在 res 文件夹下选择 layout，双击 activity_main.xml，将代码逐行输入。

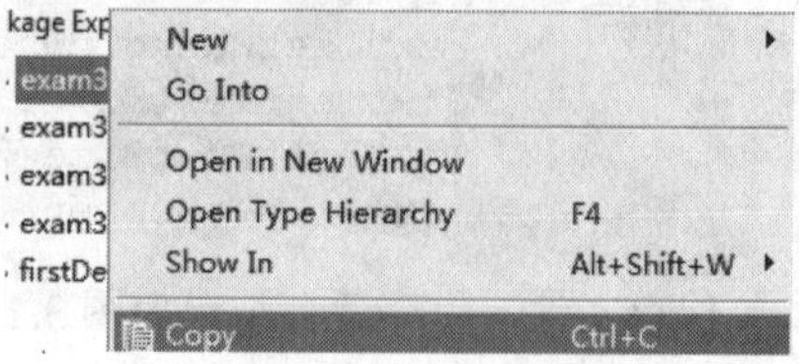

也可以按照如下步骤进行：

1）右键单击 exam3_1，选择"Copy"命令，如图 3.5 所示；

图 3.5 复制文件

2）在左边空白处单击鼠标右键，选择"Paste"命令，如图 3.6 所示，在弹出的对话框中输入 exam3_4，如图 3.7 所示。

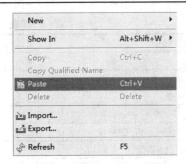

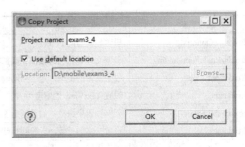

图 3.6　粘贴文件　　　　　　　　　　　图 3.7　名字改为 exam3_4

在 res 文件夹下选择 layout，双击 activity_main.xml，将其中的 TextView 替换为 Button（这里没有给出代码和注释，可以参照实例 3-1。）程序运行结果如图 3.8 所示。类似的，按钮也可以调用定义好的 style.xml 文件，也能够在 Activity_main.java 中动态实现，这里就不再赘述，用户可参考前面的实例自己完成。

需要说明的是，按钮一般不用来显示漂亮的文字效果，而是主要用来做事件处理，接收用户的选择后执行其他的程序或操作，通俗地说，就是当单击或双击该按钮后要进行的操作。

3.4　编辑框组件 EditText

文本显示组件（TextView）的功能只是显示一些文字信息，如果要想在屏幕上显示可以输入文本的组件，Android 平台提供了编辑框组件 EditText，EditText 组件也是 TextView 的一个子类，其层次关系如下：

图 3.8　exam3_4 运行结果

```
java.lang.Object
    android.view.View
        android.widget.TextView
            android.widget.EditText
```

直接子类：AutoCompleteTextView, ExtractEditText
间接子类：MultiAutoCompleteTextView
要在 Android 程序中使用 EditText 组件必须在程序中使用下面的语句：

```
import android.widget.EditText;                //导入 widget.EditText 类
```

3.4.1　EditText 组件常见的属性和方法

EditText 组件继承了 TextView 类，除了具备 TextView 类的属性外，其他常用属性如表 3-6 所示。

表 3-6　EditText 组件常用的属性列表

属性名称	描述
android:capitalize	设置英文字母大写类型
android:cursorVisible	设定光标为显示/隐藏，默认显示。如果设置 false，即使选中了也不显示光标栏
android:digits	设置允许输入哪些字符。如"1234567890.+-*/%\n()"
android:editable	设置是否可编辑。仍然可以获取光标，但是无法输入
android:hint	Text 为空时显示的文字提示信息，可通过 textColorHint 设置提示信息的颜色

续表

属性名称	描述
android:imeOptions	设置软键盘的 Enter 键
android:includeFontPadding	设置文本是否包含顶部和底部额外空白，默认为 true
android:inputType	设置文本的类型，用于帮助输入法显示合适的键盘类型
android:ems	设置 TextView 的宽度为 N 个字符的宽度
android:maxLength	限制输入字符数。如设置为 5，那么仅可以输入 5 个汉字/数字/英文字母
android:lines	设置文本的行数，设置两行就显示两行，即使第二行没有数据
android:numeric	如果被设置，该 TextView 有一个数字输入法。有如下值设置：integer 正整数、signed 带符号整数、decimal 带小数点浮点数
android:password	以小点"."显示文本
android:phoneNumber	设置为电话号码的输入方式
android:selectAllOnFocus	如果文本是可选择的，让它获取焦点而不是将光标移动为文本的开始位置或者末尾位置。TextView 中设置后无效果
android:textAppearance	设置文字外观

EditText 组件的属性可以在布局文件里通过上面的参数进行设置，也可以在 Activity 文件中通过调用方法来实现，表 3-7 给出了 EditText 组件的常用方法。

表 3-7 EditText 组件常用的方法列表

方法	描述
public void setImeOptions()	设置软键盘的 Enter 键
public Charsequence getImeActionLable()	设置 IME 动作标签
public Boolean getDefaultEditable()	获取是否默认可编辑
public void setEllipse()	设置文件过长时控件的显示方式
public void setFreeezesText()	设置保存文本内容及光标位置
public Boolean getFreeezesText()	获取保存文本内容及光标位置
public void setGravity()	设置文本框在布局中的位置
public int getGravity()	获取文本框在布局中的位置
public void setHint()	设置文本框为空时，文本框默认显示的字符
public Charsequence getHint()	获取文本框为空时，文本框默认显示的字符
public void setIncludeFontPadding()	设置文本框是否包含底部和顶端的额外空白
public void setMarqueeRepeatLimit()	在 ellipsize 指定 marquee 的情况下，设置重复滚动的次数，当设置为 marquee_forever 时表示无限次

3.4.2 EditText 组件使用实例

实例 3-5：各种属性的运用

新建一个项目，项目的命名为：exam3_5，包名称为：org.hnist.demo，在 res 文件夹下选择 layout 文件夹，双击 activity_main.xml 文件，将代码做如下修改：

```
<RelativeLayout
    xmlns:android="http://schemas.android.com/apk/res/android"
    xmlns:tools="http://schemas.android.com/tools"
    android:layout_width="match_parent"
    android:layout_height="match_parent">
    <EditText
        android:id="@+id/edit1"
        android:layout_width="fill_parent"
```

```
            android:layout_height="wrap_content"
            android:text="请输入您所在的单位："
            android:textSize="20dp" />
    <!-- 用于不可编辑的文本框 -->
    <EditText
            android:id="@+id/edit2"
            android:layout_width="fill_parent"
            android:layout_height="wrap_content"
            android:layout_below="@+id/edit1"     //设置组件位于edit1下方
            android:editable="false"              //设置文本框不可编辑
            android:text="www.hnist.cn"
            android:textSize="20dp" />
    <!-- 用于输入数字的文本框 -->
      <EditText
            android:id="@+id/edit3"
            android:layout_width="fill_parent"
            android:layout_height="wrap_content"
            android:layout_below="@+id/edit2"
            android:text="请输入电话号码"
            android:phoneNumber="true"            //设置文本框只能输入电话号码
            android:textColorHint="#FFFF0000"     //设置文字颜色
            android:selectAllOnFocus="true"/>     //设置选中全部内容
      <!-- 用于输入密码的文本框 -->
    <EditText
        android:id="@+id/edit4"
        android:layout_width="fill_parent"
        android:layout_height="wrap_content"
        android:layout_below="@+id/edit3"
        android:hint="请输入密码"
        android:password="true"                    //设置文本框输入为密文显示
        android:textColorHint="#FF000000" />
</RelativeLayout>
```

保存文件，程序运行结果如图3.9所示。

可以在编辑框中输入数据，默认状态是输入英文，如果是要输入中文，首先进入手机的设置"Setting"，在列表中选择"language&keyboard"，并将"谷歌拼音输入法"选中。然后，只要在需要输入中文的地方，长按输入框，就会弹出选择框，进入输入法选项里面找到谷歌输入法并选中它，回到程序中，这时就可以输入中文了。

1. 在Activity文件中动态创建EditText组件

1) 在activity_main.xml文件中添加如下代码：

图3.9 exam3_5运行结果

```
<EditText
    android:id="@+id/edit5"
    android:layout_width="fill_parent"
    android:layout_height="wrap_content"
    android:layout_below="@+id/edit4"/>
```

保存文件，程序运行结果如图3.10所示，增加了一个没有内容的编辑框。

2）对 MainActivity.java 文件做如下修改：

```
package org.hnist.demo;
import android.os.Bundle;
import android.app.Activity;
import android.widget.EditText;                        //导入widget.EditText类
public class MainActivity extends Activity {
    private EditText myedit=null;                      //定义一个编辑框组件
    @Override
    protected void onCreate(Bundle savedInstanceState) {
      super.onCreate(savedInstanceState);
      setContentView(R.layout.activity_main);          //调用布局文件
      this.myedit=(EditText)super.findViewById(R.id.edit5);//获取编辑框组件
      this.myedit.setText("请输入您的姓名"); }}         //设置编辑框里的内容
```

保存文件，程序运行结果如图 3.11 所示，增加的编辑框里面有了提示内容。

图 3.10 添加了新编辑框的 exam3_5 运行结果

图 3.11 添加了新编辑框内容的 exam3_5 运行结果

3.5 图片视图组件 ImageView

ImageView 组件的主要功能是为图片展示提供一个容器，它可以加载各种来源（如资源或图片库）的图片文件，并提供例如缩放和着色等各种显示选项。其层次关系如下：

```
java.lang.Object
    android.view.View
        android.widget.ImageView
```

直接子类：ImageButton, QuickContactBadge
间接子类：ZoomButton
要在 Android 程序中使用 ImageView 组件必须在程序中使用下面的语句：

```
import android.widget.ImageView;                       //导入widget.ImageView类
```

3.5.1 ImageView 组件常用的属性和方法

ImageView 组件是 View 的子类，所以 View 的属性和方法它都具备，另外还有以下常见的属性和方法，如表 3-8 和表 3-9 所示。

表 3-8 ImageView 常见属性列表

属性名称	描述
android:adjustViewBounds	设置该属性为真可以在 ImageView 调整边界时保持图片的纵横比例（需要与 MaxWidth、MaxHeight 一起使用）
android:baseline	视图内基线的偏移量
android:baselineAlignBottom	如果为 true，图像视图将基线与父控件底部边缘对齐
android:cropToPadding	如果为真，会剪切图片以适应内边距的大小（单独设置无效，需要与 scrollY 一起使用）
android:maxHeight	为视图提供最大高度的可选参数（单独使用无效，需要与 setAdjustViewBounds 一起使用）
android:maxWidth	为视图提供最大宽度的可选参数
android:scaleType	控制为了使图片适合 ImageView 的大小，相应地变更图片大小或移动图片
android:src	设置可绘制对象作为 ImageView 显示的内容
android:tint	为图片设置着色颜色

表 3-9 ImageView 常见方法列表

方法	描述
public void setMaxHeight (int maxHeight)	定义图片的最大高度
public void setMaxWidth (int maxWidth)	定义图片的最大宽度
public void setImageResource (int resId)	定义显示图片的 ID
public void setImageBitmap (Bitmap bm)	定义显示图片
public void setAlpha (int alpha)	设置透明度
public void setAdjustViewBounds (boolean adjustViewBounds)	调整边框时是否保持可绘制对象的比例
public void setBaselineAlignBottom (boolean aligned)	设置是否设置视图底部的视图基线
public final void setColorFilter (int color)	为图片设置着色选项
public void setImageDrawable	设置可绘制对象为该 ImageView 显示的内容
public void setImageResource (int resId)	通过资源 ID 设置可绘制对象为该 ImageView 显示的内容
ublic void setImageURI (Uri uri)	设置指定的 URI 为该 ImageView 显示的内容
public void setScaleType (ImageView.ScaleType scaleType)	控制图像应该如何缩放和移动，以使图像与 ImageView 一致
public void setSelected (boolean selected)	改变视图的选中状态

3.5.2 ImageView 组件使用实例

实例 3-6：ImageView 的使用举例

新建一个项目，项目的命名为：exam3_6，包名称为：org.hnist.demo，将已有图片 a1.jpg 和 aa.jpg 文件复制到\res\drawable-hdpi 文件夹下，修改布局管理器文件使之加载 a1.jpg 图片文件，然后修改 Main_Activity.java 文件加载 aa.jpg 图片文件。

1）修改 activity_main.xml 文件：

```xml
<RelativeLayout
    xmlns:android="http://schemas.android.com/apk/res/android"
    xmlns:tools="http://schemas.android.com/tools"
    android:layout_width="match_parent"
    android:layout_height="match_parent"
    tools:context=".MainActivity" >
    <ImageView                                       //定义一个 ImageView
        android:layout_width="wrap_content"
        android:layout_height="wrap_content"
        android:src="@drawable/a1" />                //加载 a1.jpg 文件
    <ImageView                                       //定义一个 ImageView
        android:id="@+id/myimg"                      //定义一个 ImageView
        android:layout_width="wrap_content"
        android:layout_height="wrap_content"
        android:layout_centerHorizontal="true"       //图片水平居中显示
```

```
        android:layout_marginTop="100dp"/>
    </RelativeLayout>
```

2）修改 MainActivity.java 文件：

```
    package org.hnist.demo;
    import android.os.Bundle;
    import android.app.Activity;
    import android.widget.ImageView;                    //导入 widget.ImageView 类
    public class MainActivity extends Activity {
    private ImageView img= null;                        //定义一个 ImageView 对象 img
      @Override
      protected void onCreate(Bundle savedInstanceState) {
          super.onCreate(savedInstanceState);
          setContentView(R.layout.activity_main);
          this.img=(ImageView) super.findViewById(R.id.myimg); //获得 ImageView 组件
          this.img.setImageResource(R.drawable.aa); }} //设置显示的图片为 aa.jpg
```

保存文件，程序运行结果如图 3.12 所示，程序界面中显示了 2 张图片，其中 a1.jpg 是在布局管理器中加载的，aa.jpg 是在 Main_Activity 中加载的。

3.6 图片按钮组件 ImageButton

ImageButton 是带有图标的按钮，与 Button 组件类似，可以直接使用 ImageButton 定义，其层次关系如下：

图 3.12 exam3_6 运行结果

```
    java.lang.Object
    android.view.View
        android.widget.ImageView
            android.widget.ImageButton
```

要在 Android 程序中使用 ImageButton 组件，必须在程序中使用下面的语句：

```
    import android.widget.ImageButton;                 //导入 widget.ImageButton 类
```

3.6.1 ImageButton 组件常用的属性和方法

ImageButton 是 ImageView 的子类，因此 ImageView 的属性和方法它都具备，和 Button 组件类似，它的主要目的不是用来显示图片，而是用作事件处理，接收用户的选择后执行其他的程序或操作。

3.6.2 ImageButton 组件使用实例

实例 3-7：ImageButton 的使用举例

新建一个项目，项目的命名为：exam3_7，包名称为：org.hnist.demo，将图片 a1.jpg 和 aa.jpg 文件复制到\res\drawable-hdpi 文件夹下，修改布局管理器文件加载 a1.jpg 图片文件到按钮上，然后修改 MainActivity.java 文件加载 aa.jpg 图片文件到按钮上。

1）修改 activity_main.xml 文件：

```
    <RelativeLayout xmlns:android="http://schemas.android.com/apk/res/android"
        xmlns:tools="http://schemas.android.com/tools"
        android:layout_width="match_parent"
        android:layout_height="match_parent"
```

```xml
    tools:context=".MainActivity" >
  <ImageButton                                              //定义一个 ImageButton 组件
      android:layout_width="fill_parent"
      android:layout_height="wrap_content"
      android:layout_centerHorizontal="true"                //水平居中
android:src="@drawable/a1" />                               //图片为 a1.jpg
  <ImageButton                                              //定义一个 ImageButton 组件
      android:id="@+id/myimg"                               //命名为 myimg
      android:layout_width="wrap_content"
      android:layout_height="wrap_content"
      android:layout_centerHorizontal="true"                //水平居中
      android:layout_marginTop="100dp"/>
 </RelativeLayout>
```

2）修改 MainActivity.java 文件：

```java
package org.hnist.imagebuttondemo;
import android.os.Bundle;
import android.app.Activity;
import android.widget.ImageButton;              //导入 widget.ImageButton 类
public class MainActivity extends Activity {
   private ImageButton img= null;               //定义一个 ImageButton 对象 img
   @Override
   protected void onCreate(Bundle savedInstanceState) {
      super.onCreate(savedInstanceState);
      setContentView(R.layout.activity_main);
      this.img=(ImageButton) super.findViewById(R.id.myimg);  //获得 ImageButton 组件
      this.img.setImageResource(R.drawable.aa);  } }          //设置按钮上的图片为
                                                              aa.jpg
```

保存文件，运行结果如图 3.13 所示，发现显示了 2 个图片按钮，ImageButton 组件与 ImageView 显示在屏幕上的主要区别是单击上面的图片时显示的效果不同。

3.7 单选按钮组件 RadioGroup

单选按钮组件在开发中提供了一种多选一的操作模式，例如在选择学历时，用户需要从众多选择中选定一个选项，实现这个需求可在 Android 程序中使用 RadioGroup 组件来定义单选按钮并提供多个选项存放的容器，RadioGroup 组件的层次关系如下：

图 3.13　exam3_7 运行结果

```
java.lang.Object
    android.view.View
        android.view.ViewGroup
            android.widget.LinearLayout
                android.widget.RadioGroup
```

RadioGroup 组件提供的只是一个单选按钮的容器，在 Android 平台中提供了 RadioButton 类，可通过该类配置多个选项，设置单选按钮的内容，RadioButton 组件的层次关系如下：

```
java.lang.Object
    android.view.View
        android.widget.TextView
```

```
        android.widget.Button
            android.widget.CompoundButton
                android.widget.RadioButton
```

要在 Android 程序中使用 RadioGroup 组件，必须在程序中使用下面的语句。：

```
import android.widget.RadioGroup;            //导入 widget.RadioGroup 类
```

3.7.1 RadioGroup 组件常见的属性

RadioGroup 组件是 TextView 的子类，所以 TextView 组件的属性和方法它都继承了，其属性见表 3-3，方法见表 3-4，RadioGroup 组件还有如表 3-10 列出的常见方法。

表 3-10 RadioGroup 组件常见方法列表

方法	描述
public void check (int id)	设置要选中的单选按钮编号
public void clearCheck ()	清空选中状态
public int getCheckedRadioButtonId()	取得选中按钮的 RadioButton 的 ID
public void setOnCheckedChangeListener(RadioGroup.OnCheckedChangeListener listener)	设置单选按钮选中的操作事件
public void setOnHierarchyChangeListener (ViewGroup.OnHierarchyChangeListener listener)	注册一个当该单选按钮组中的勾选状态发生改变时所要调用的回调函数

3.7.2 RadioGroup 组件使用实例

实例 3-8：RadioGroup 组件使用举例

新建一个项目，项目的名称为：exam3_8，包名称为：org.hnist.demo，该程序的目标是显示一个单项选择题的界面。

1）修改 activity_main.xml 文件，定义一个 RadioGroup 组件，里面包含四个 RadioButton 组件，其中 2 个选项有文字提示信息，另外 2 个没有，具体代码如下：

```xml
<RelativeLayout xmlns:android="http://schemas.android.com/apk/res/android"
    xmlns:tools="http://schemas.android.com/tools"
    android:layout_width="match_parent"
    android:layout_height="match_parent"
    tools:context=".MainActivity" >
    <TextView
        android:layout_width="wrap_content"
        android:layout_height="wrap_content"
        android:textSize="18dp"
        android:text="下面的计算哪个是正确的：" />
    <RadioGroup                              //定义一个 RadioGroup 组件
        android:id="@+id/Radio1"             //给组件命名为 Radio1
        android:layout_width="wrap_content"
        android:layout_height="wrap_content"
        android:orientation="vertical">      //里面 RadioButton 对齐方式为垂直对齐
    <RadioButton                             //定义一个 RadioButton 组件
        android:id="@+id/RBut1"              //命名为 RBut1
        android:layout_width="wrap_content"
        android:layout_height="wrap_content"
        android:layout_marginTop="30dp"
        android:text="   1+2=1"/>            //显示内容为：1+2=1
    <RadioButton                             //定义一个 RadioButton 组件
```

```
                android:id="@+id/RBut2"            //命名为RBut2
                android:layout_width="wrap_content"
                android:layout_height="wrap_content"
                android:text="    1+2=2"/>         //显示内容为：1+2=1
        <RadioButton                               //定义一个RadioButton组件
                android:id="@+id/RBut3"            //命名为RBut3
                android:layout_width="wrap_content"
                android:layout_height="wrap_content"/>
        <RadioButton                               //定义一个RadioButton组件
                android:id="@+id/RBut4"            //命名为RBut4
                android:layout_width="wrap_content"
                android:layout_height="wrap_content"/>
    </RadioGroup>
</RelativeLayout>
```

2）修改 MainActivity.java 文件，在 RBut3 和 RBut4 选项处也显示提示信息，代码如下：

```
package org.hnist.demo;
import android.os.Bundle;
import android.app.Activity;
import android.widget.RadioButton;              //导入widget.RadioButton类
import android.widget.RadioGroup;               //导入widget.RadioGroup类
public class MainActivity extends Activity {
    private RadioButton rbut1=null;             //定义一个RadioButton
    private RadioButton rbut2=null;             //定义一个RadioButton
    @Override
    protected void onCreate(Bundle savedInstanceState) {
        super.onCreate(savedInstanceState);
        setContentView(R.layout.activity_main);
        this.rbut1=(RadioButton) super.findViewById(R.id.RBut3); //获得RadioButton
        this.rbut2=(RadioButton) super.findViewById(R.id.RBut4); //获得RadioButton
        this.rbut1.setText("    1+2=3");        //设置RadioButton的内容为"1+2=3"
        this.rbut2.setText("    1+2=4");}}      //设置RadioButton的内容为"1+2=4"
```

保存文件，程序运行结果如图 3.14 所示，当然各选项的内容也可以在 string.xml 文件中定义好，然后在 Main_Activity.java 文件和 activity_main.xml 文件中进行调用，请读者自行完成。

图 3.14　exam3_8 运行结果

3.8　复选框组件 CheckBox

CheckBox 组件的主要功能是完成多项选择的操作，例如，在选择兴趣爱好的时候可能会存在多个选择的情况，Android 平台提供的 CheckBox 组件可以实现这样的功能，该组件层次关系如下：

```
java.lang.Object
    android.view.View
        android.widget.TextView
            android.widget.Button
                android.widget.CompoundButton
                    android.widget.CheckBox
```

要在 Android 程序中使用 CheckBox 组件，必须在程序中使用下面的语句：

```
    import android.widget.CheckBox;                       //导入widget.CheckBox类
```

3.8.1 CheckBox 组件常见的属性和方法

CheckBox 组件是 TextView 的子类，其属性见表 3-3 所示，另外 CheckBox 组件还有以下常用的方法，见表 3-11 所示。

表 3-11 CheckBox 常见方法列表

方法	描述
public CheckBox(Context context)	实例化 CheckBox 组件
public void setChecked (boolean checked)	设置默认选中
public int getCheckedRadioButtonId()	取得选中按钮的 RadioButton 的 ID
public Boolean isChecked()	判断组件状态是否勾选
public void onRestoreInstanceState()	设置视图恢复以前的状态
public Boolean iperformClick()	执行 click 动作触发事件监听器
public void setButtonDrawable()	根据 Drawable 对象设置组件的背景
public void setChecked()	设置组件的状态
public void setOnCheckedChangeListener()	设置事件监听器
public void toggle()	改变按钮当前的状态
public CharSequence onCreateDrawableState()	获取文本框为空时文本框里面的内容
public int onCreateDrawableState()	为当前视图生成新的 Drawable 状态

3.8.2 CheckBox 组件使用实例

实例 3-9：CheckBox 的使用举例

新建一个项目，项目的名称为：exam3_9，包名称为：org.hnist.demo。该项目显示的界面是用户进行个人爱好的多项选择。修改布局管理器文件 activity_main.xml，进行一个选项的定义，另外两个选项的数据来源于 string.xml 文件。

1）修改 activity_main.xml 文件：

```xml
<RelativeLayout xmlns:android="http://schemas.android.com/apk/res/android"
    xmlns:tools="http://schemas.android.com/tools"
    android:layout_width="match_parent"
    android:layout_height="match_parent"
    tools:context=".MainActivity" >
<TextView
    android:id="@+id/txt1"
    android:layout_width="wrap_content"
    android:layout_height="wrap_content"
    android:text="您的喜爱的体育运动有："
    android:textSize="18dp" />              //设置文字的大小为 18dp
<CheckBox                                    //定义一个 CheckBox
    android:id="@+id/mybox1"                 //命名为 mybox1
    android:layout_width="wrap_content"
    android:layout_height="wrap_content"
    android:layout_below="@+id/txt1"
    android:layout_marginTop="10dp"
    android:text="@string/hobby1" />         //显示的内容来源于 string.xml 定义的 hobby2
<CheckBox                                    //定义一个 CheckBox
    android:id="@+id/mybox2"                 //命名为 mybox2
    android:layout_width="wrap_content"
```

```
            android:layout_height="wrap_content"
            android:layout_alignBottom="@+id/mybox1"
            android:layout_toRightOf="@+id/mybox1"
            android:text="@string/hobby2" />//显示的内容来源于string .xml定义的hobby2
    <CheckBox                                    //定义一个CheckBox
            android:id="@+id/mybox3"             //命名为mybox3
            android:layout_width="wrap_content"
            android:layout_height="wrap_content"
            android:layout_alignBottom="@+id/mybox2"
            android:layout_toRightOf="@+id/mybox2"
            android:checked="true"               //处于被选中状态
            android:text="羽毛球" />              //显示的内容为"羽毛球"
</RelativeLayout>
```

2）修改 string.xml 文件：

```
<?xml version="1.0" encoding="utf-8"?>
<resources>
    <string name="app_name">exam3_9</string>
    <string name="action_settings">Settings</string>
    <string name="hello_world">Hello world!</string>
    <string name="hobby1">篮球</string>          //定义hobby1的值为篮球
    <string name="hobby2">排球</string>          //定义hobby2的值为排球
</resources>
```

保存文件，程序运行结果如图 3.15 所示。

当然，我们也可以在 Activity 的 MainActivity.java 文件中利用表 3-11 提供的方法来定义 CheckBox，设置各个选项的内容，请用户自行完成。

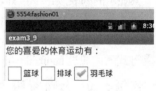

图 3.15　exam3_9 运行结果

3.9　下拉列表框组件 Spinner

Spinner 组件的功能类似 RadioGroup 组件，它可以为用户提供列表的选择方式，一个 Spinner 对象包含多个子项，子项来源于与之相关联的适配器中，每个子项只有两种状态，选中或未被选中。在 Android 平台中可以使用 android.widget.Spinner 类实现该组件。它的层次关系如下：

```
java.lang.Object
    android.view.View
        android.view.ViewGroup
            android.widget.AdapterView<T extends android.widget.Adapter>
                android.widget.AbsSpinner
                    android.widget.Spinner
```

要在 Android 程序中使用 Spinner 组件，必须在程序中使用下面的语句：

```
import android.widget.Spinner;                   //导入widget.Spinner类
```

3.9.1　Spinner 组件常见的属性和方法

Spinner 组件是 View 的子类，所以 Spinner 组件的属性可参见表 3-3。此外 Spinner 组件还有一个常用的属性 android:prompt，该属性在下拉列表对话框显示时显示，也就是显示对话框的标题。Spinner 常见方法列表见表 3-12。

表 3-12 Spinner 常见方法列表

方法	描述
public CharSequence getPrompt ()	取得提示文字
public void setPrompt (CharSequence prompt)	设置组件的提示文字
public void setAdapter (SpinnerAdapter adapter)	设置下拉列表项
public CharSequence getPrompt()	得到提示信息
public void setOnItemClickListener(AdapterView.OnItemClickListener l)	设置选项单击事件
public void onClick(DialogInterface dialog, int which)	当单击弹出框中的项时这个方法将被调用
public Boolean performClick()	如果它被定义就调用此视图的 OnClickListener
public void setPromptId(CharSequence prompt)	设置弹出视图上的标题字
public ArrayAdapter (Context context, int textViewResourceId, List<T> objects)	定义 ArrayAdapter 对象，传入一个 Activity 实例、列表项的显示风格、List 集合数据
public ArrayAdapter (Context context, int textViewResourceId, T[] objects)	定义 ArrayAdapter 对象，传入一个 Activity 实例、列表项的显示风格、数组数据
public static ArrayAdapter<CharSequence> createFromResource (Context context, int textArrayResId, int textViewResId)	通过静态方法取得 ArrayAdapter 对象，传入 Activity 实例、资源文件的 id、列表项的显示风格
public void setDropDownViewResource (int resource)	设置下拉列表项的显示风格

3.9.2 Spinner 组件使用实例

在 Android 平台中，可以在 activity_main.xml 文件中定义 Spinner 组件，但是不能在该文件中直接设置其显示的列表项，下拉列表框中的列表项有以下几种方式进行配置。

方式一：直接通过资源文件配置；

方式二：通过 android.widget.ArrayAdapter 类读取资源文件的数据；

方式三：通过 android.widget.ArrayAdapter 类设置具体的数据。

实例 3-10：Spinner 组件使用举例

新建一个项目，项目的名称为：exam3_10，包名称为：org.hnist.demo，该项目显示的界面是让用户选择学历、专业和班级，我们分别用上面的三种方式来设置具体的列表项数据。

1）修改 activity_main.xml 文件，具体代码如下：

```xml
<RelativeLayout xmlns:android="http://schemas.android.com/apk/res/android"
    xmlns:tools="http://schemas.android.com/tools"
    android:layout_width="match_parent"
    android:layout_height="match_parent"
    tools:context=".MainActivity" >
<TextView
    android:id="@+id/txt1"
    android:layout_width="wrap_content"
    android:layout_height="wrap_content"
    android:textSize="20dp"
    android:text="您的学历是" />
<Spinner                                    //定义一个 Spinner 组件
    android:id="@+id/spin1"                 //命名为 spin1
    android:layout_width="wrap_content"
    android:layout_height="wrap_content"
    android:layout_below="@+id/txt1"
    android:prompt="@string/eduprompt"  //设置提示信息,在 string.xml 中定义
    android:entries="@array/edu"/>       //选项的列表项来源于 edu.xml 文件
<TextView
    android:id="@+id/txt2"
    android:layout_width="wrap_content"
```

```
        android:layout_height="wrap_content"
        android:layout_below="@+id/spin1"
        android:textSize="20dp"
        android:text="您的专业是"/>
    <Spinner                                    //定义一个Spinner组件
        android:id="@+id/spin2"                 //命名为spin2
        android:layout_width="wrap_content"
        android:layout_height="wrap_content"
        android:layout_below="@+id/txt2" />
    <TextView
        android:id="@+id/txt3"
        android:layout_width="wrap_content"
        android:layout_height="wrap_content"
        android:layout_below="@+id/spin2"
        android:textSize="20dp"
        android:text="您的班级是"/>
    <Spinner                                    //定义一个Spinner组件
        android:id="@+id/spin3"                 //命名为spin3
        android:layout_width="wrap_content"
        android:layout_height="wrap_content"
        android:layout_below="@+id/txt3"/>
</RelativeLayout>
```

2）打开 string.xml 文件，在</resources>标签前面添加一行语句："<string name="eduprompt">请输选择您的学历：</string>"。具体代码如下：

```
<?xml version="1.0" encoding="utf-8"?>
<resources>
    <string name="app_name">exam3_10</string>
    <string name="action_settings">Settings</string>
    <string name="hello_world">Hello world!</string>
    <string name="eduprompt">请输选择您的学历：</string>
</resources>
```

3）代码中的"android:entries="@array/edu""语句表示这个选项的列表数据项来源于 edu.xml 文件，edu.xml 文件要事先建立完成，以免程序报错。

① 找到 res/values 文件夹，右键单击该文件夹，选择"New→Android XML File"选项，如图 3.16 所示。

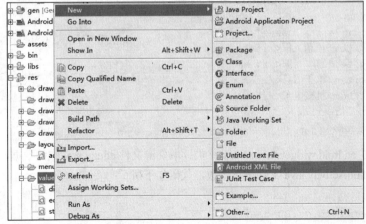

图 3.16　新建一个 XML 文件

② 在出现的对话框中的 File 输出框中输入 "edu",然后单击 "Finish" 按钮,如图 3.17 所示。

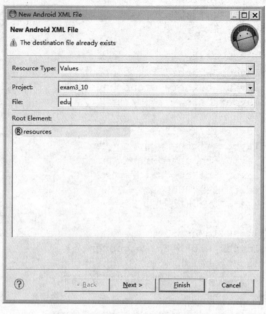

图 3.17 为新建的 XML 文件命名

③ 进入到 edu.xml 文件的编辑状态,输入如下的内容。

```xml
<?xml version="1.0" encoding="utf-8"?>
<resources>
    <string-array name="edu">
        <item>高中</item>
        <item>大专</item>
        <item>本科</item>
        <item>硕士</item>
        <item>博士</item>
    </string-array>
</resources>
```

类似地,新建 eduzy.xml 文件,文件内容如下。

```xml
<?xml version="1.0" encoding="utf-8"?>
<resources>
    <string-array name="eduzy">
        <item>信息工程</item>
        <item>通    信</item>
        <item>电子信息</item>
        <item>机械电子</item>
    </string-array>
</resources>
```

4) 接下来通过 android.widget.ArrayAdapter 类读取资源文件 eduzy.xml 中的数据,并通过该类设置具体的数据,将 MainActivity.java 文件进行修改,代码如下所示:

```java
package org.hnist.demo;
import android.os.Bundle;
import android.app.Activity;
```

第3章 Android 开发常用组件

```java
import android.widget.Spinner;            //导入 widget.Spinner 类
import android.widget.ArrayAdapter;       //导入 widget.ArrayAdapter 类
import java.util.ArrayList;               //导入 java.util.ArrayList 类
import java.util.List;                    //导入 java.util.List 类
public class MainActivity extends Activity {
    private Spinner eduzy=null;           //定义表示专业的 Spinner
    private Spinner classid=null;         //定义表示班级的 Spinner
    private ArrayAdapter<CharSequence> adaptereduzy=null;  //定义下拉列表内容适配器
    private ArrayAdapter<CharSequence> adapterclassid=null; //定义下拉列表内
                                                           容适配器
    private List<CharSequence> dataclassid=null;  //定义 List 集合来保存下拉
                                                   列表选项
    @Override
    protected void onCreate(Bundle savedInstanceState) {
        super.onCreate(savedInstanceState);
        setContentView(R.layout.activity_main);
        this.eduzy=(Spinner) super.findViewById(R.id.spin2);   //获得 Spinner 组件
        this.classid=(Spinner) super.findViewById(R.id.spin3); //获得 Spinner 组件
        this.eduzy.setPrompt("请选择您的专业:");   //设置 eduzy 下拉列表框的提示信息
        this.classid.setPrompt("请选择您的班级: "); //设置 classid 下拉列表框的提
                                                    示信息
//从资源文件 eduzy 中读取选项内容
this.adaptereduzy=ArrayAdapter.createFromResource(this,R.array.eduzy,andr
oid.R.layout.simple_spinner_item);
//设置下拉列表的风格
this.adaptereduzy.setDropDownViewResource(android.R.layout.simple_spinner
_dropdown_item);
this.eduzy.setAdapter(this.adaptereduzy);          //设置下拉列表的选项内容
this.dataclassid=new ArrayList<CharSequence>();    //实例化 List 集合
    this.dataclassid.add("信工 12-2BF");            //往 List 集合添加选项内容
    this.dataclassid.add("通信 12-2BF");            //往 List 集合添加选项内容
    this.dataclassid.add("电信 12-2BF");            //往 List 集合添加选项内容
    this.dataclassid.add("机电 12-2BF");            //往 List 集合添加选项内容
        //获得一个下拉列表适配器
this.adapterclassid=new ArrayAdapter<CharSequence>(this,android.R.layout.
simple_spinner_item,this.dataclassid);
//设置下拉列表的风格
this.adapterclassid.setDropDownViewResource(android.R.layout.simple_spinn
er_dropdown_item);
        this.classid.setAdapter(this.adapterclassid); }}   //设置下拉列表选项内容
```

上述代码中我们设置了 3 个下拉列表框，其中表示学历的下拉列表框在布局文件中定义，数据来源于 edu.xml；表示专业的下拉列表框在布局文件中定义，在 Activity 程序中引用，数据来源于 eduzy.xml；表示班级的下拉列表框在布局文件中定义，在 Activity 程序中引用，数据也在 Activity 程序中添加。

保存所有文件，程序运行后，可以进行学历、专业和班级的选择，如图 3.18 所示。

图 3.18　exam3_10 运行结果

3.10 信息提示框组件 Toast

Android 平台中提供了提示界面效果，这种提示不会打断用户的正常操作，这就是信息提示框组件 Toast，它以浮于应用程序之上的形式呈现给用户。因为它并不获得焦点，即使用户正在进行内容输入也不会受到影响。它的目标是尽可能以不显眼的方式，使用户看到你提供的信息。该组件直接继承 java.lang.Object，其类层次结构如下：

```
java.lang.Object
    android.widget.Toast
```

要在 Android 程序中使用 Toast 组件必须在程序中使用下面的语句：

```
import android.widget.Toast;                    //导入 widget.Toast 类
```

3.10.1 Toast 组件常见的属性和方法

信息提示框 Toast 组件的属性不多，最常用的两个常量是 Toast 组件常用的方法和属性如表 3-13 所示。

表 3-13 Toast 组件常见的属性和方法列表

方法	描述
public static final int LENGTH_LONG	常量，持续显示视图或文本提示较长时间
public static final int LENGTH_SHORT	常量，显示时间较短，为默认值
public Toast(Context context)	创建 Toast 对象
public static Toast makeText(Context context, int resId, int duration)	创建一个 Toast 对象，并指定显示文本资源的 ID 和信息的显示时间
public static Toast makeText(Context context, CharSequence text, int duration)	创建一个 Toast 对象，并指定显示文本资源和信息的显示时间
public void show()	显示信息
public void setDuration(int duration)	设置显示的时间
public void setView(View view)	设置显示的 View 组件
public void setText(int resId)	设置显示的文字资源 ID
public void setText(CharSequence s)	直接设置要显示的文字
public void setGravity(int gravity, int xOffset, int yOffset)	设置组件的对齐方式
public View getView()	取得内部包含的 View 组件
public int getXOffset()	返回组件的 X 坐标位置
public int getYOffset()	返回组件的 Y 坐标位置
public void cancel()	取消显示

3.10.2 Toast 组件使用实例

实例 3-11：Toast 的使用举例

新建一个项目，项目的命名为：exam3_11，包名称为：org.hnist.demo，该程序的目标是显示一个长时间提示的演示。

1）修改 activity_main.xml 文件，代码如下：

```
<RelativeLayout xmlns:android="http://schemas.android.com/apk/res/android"
    xmlns:tools="http://schemas.android.com/tools"
    android:layout_width="match_parent"
```

```
        android:layout_height="match_parent"
        tools:context=".MainActivity" >
        <TextView
            android:id="@+id/txt1"
            android:layout_width="wrap_content"
            android:layout_height="wrap_content"
            android:layout_centerHorizontal="true"
            android:textSize="20dp"
            android:text="Toast 演示实例" />
    </RelativeLayout>
```

2）修改 MainActivity.java 文件，代码如下：

```
package org.hnist.demo;
import android.os.Bundle;
import android.app.Activity;
import android.widget.Toast;                        //导入 widget.Toast 类
public class MainActivity extends Activity {
    @Override
    protected void onCreate(Bundle savedInstanceState) {
        super.onCreate(savedInstanceState);
        setContentView(R.layout.activity_main);
        //创建 Toast，设置提示信息和显示时间长短
Toast myToast = Toast.makeText(this, "长时间显示的 Toast 信息提示框", Toast.LENGTH_LONG) ;
        myToast.setGravity(BIND_AUTO_CREATE, 5, 10);   //设置组件的对齐方式
        myToast.show();   }}   //显示这个提示信息
```

保存文件，默认的提示信息显示在屏幕的底部，可以通过 setGravity()方法修改提示信息的位置，程序运行结果如图 3.19 所示，程序界面显示几秒钟后，下面的提示信息会自动消失。

图 3.19　exam3_11 运行结果

在 Android 平台中有四种常用的布局管理器组件。

LinearLayout 组件：线性布局管理器，分为水平和垂直两种，只能进行单行布局；

FrameLayout 组件：所有的组件放在左上角，一个覆盖一个；

TableLayout 组件：任意行和列的表格布局管理器，其中 TableRow 代表一行，可以向行中增加组件；

RelativeLayout 组件：相对布局管理器，根据最近一个视图组件，或是顶层父组件来确定下一个组件的位置。它是 Android 平台建议采用的布局管理器。

除了上面介绍的布局管理器外，还有一种绝对布局管理器（AbsoluteLayout），它可以设置任意控件在屏幕中（x,y）坐标点，然后绘制的控件会覆盖住之前绘制的控件，不建议使用绝对布局，因为 Android 的手机分辨率各式各样，使用绝对布局管理器设计完成后，在其他分辨率的手机上就可能无法正常显示，在这里不做介绍。

3.11　相对布局管理器组件 RelativeLayout

相对布局管理器是指参考其他控件进行组件的放置，可以将组件摆放在一个指定参考组件的上、下、左、右等位置，也可以直接通过各个组件提供的属性完成。组件的层次关系如下：

```
java.lang.Object
    android.view.View
        android.view.ViewGroup
            android.widget.RelativeLayout
```

Android 手机屏幕的分辨率五花八门，为了考虑屏幕自适应的情况在开发中建议使用相对布局，它的坐标取值范围都是相对的，所以使用它来做自适应屏幕是正确的，在 Android 项目设计中默认布局管理器就是相对布局管理器。

要在 Android 程序中使用 RelativeLayout 组件必须在程序中使用下面的语句：

```
import android.widget.RelativeLayout;         //导入 widget.RelativeLayout 类
```

3.11.1 RelativeLayout 组件常用的属性和方法

RelativeLayout 组件常用的属性和方法见表 3-14。

表 3-14 RelativeLayout 组件常用的属性

属性名称	对应的常量	描述
android:layout_below	RelativeLayout.BELOW	放置在指定组件的下边
android:layout_toLeftOf	RelativeLayout.LEFT_OF	放置在指定组件的左边
android:layout_toRightOf	RelativeLayout.RIGHT_OF	放置在指定组件的右边
android:layout_alignTop	RelativeLayout.ALIGN_TOP	以指定组件为参考进行上对齐
android:layout_alignBottom	RelativeLayout.ALIGN_BOTTOM	以指定组件为参考进行下对齐
android:layout_alignLeft	RelativeLayout.ALIGN_LEFT	以指定组件为参考进行左对齐
android:layout_alignRight	RelativeLayout.ALIGN_RIGHT	以指定组件为参考进行右对齐

如果要在程序中控制 RelativeLayout 组件的操作，需要对一些布局参数进行配置，布局参数保存在 RelativeLayout.LayoutParams 类中。

```
RelativeLayout.LayoutParams 层次关系如下:
java.lang.Object
    android.view.ViewGroup.LayoutParams
        android.view.ViewGroup.MarginLayoutParams
            android.widget. RelativeLayout.LayoutParams
```

RelativeLayout.LayoutParams 类提供了以下操作方法，如表 3-15 所示。

表 3-15 RelativeLayout.LayoutParams 类常用的方法

方法	描述
public RelativeLayout.LayoutParams (int w, int h)	指定 RelativeLayout 组件布局的宽度和高度
public void addRule (int verb, int anchor)	增加指定的参数规则
public int[] getRules ()	取得一个组件的全部参数规则

3.11.2 RelativeLayout 组件使用实例

实例 3-12：RelativeLayout 的使用举例 1

新建一个项目，项目的命名为：exam3_12，包名称为：org.hnist.demo，利用相对布局管理器对组件进行放置。

1）修改 activity_main.xml 文件，代码如下：

```xml
<?xml version="1.0" encoding="utf-8"?>
<RelativeLayout                                          //采用相对布局管理器进行布局
    xmlns:android="http://schemas.android.com/apk/res/android"
    android:id="@+id/Layout1"                            //给这个相对布局管理器命名为 Layout1
    android:layout_width="fill_parent"
    android:layout_height="fill_parent">
    <ImageView                                           //定义一个 ImageView 组件
        android:id="@+id/img1"                           //命名为 img1
        android:src="@drawable/a1"                       //图片来源于 a1.jpg
        android:layout_width="wrap_content"              //组件宽度为图片宽度
        android:layout_height="wrap_content" />          //组件高度为图片高度
    <ImageView                                           //定义一个 ImageView 组件
        android:id="@+id/img2"                           //命名为 img2
        android:src="@drawable/a2"                       //图片来源于 a2.jpg
        android:layout_width="wrap_content"              //组件宽度为图片宽度
        android:layout_height="wrap_content"             //组件高度为图片高度
        android:layout_toRightOf="@id/img1"/>            //位于 img1 的右边摆放
    <TextView                                            //定义一个 TextView 组件
        android:text="湖南理工学院"                        //显示的内容
        android:id="@+id/mytxt"                          //命名为 mytxt
        android:layout_width="wrap_content"              //组件宽度为文字宽度
        android:layout_height="wrap_content"             //组件高度为文字高度
        android:layout_toRightOf="@id/img1"              //位于 img1 右边摆放
        android:layout_below="@id/img2" />               //位于 img2 下方摆放
    <Button                                              //定义一个 Button 组件
        android:text="http://www.hnist.cn"               //显示的内容
        android:id="@+id/mybut"                          //命名为 mytxt
        android:layout_width="wrap_content"              //组件宽度为文字宽度
        android:layout_height="wrap_content"             //组件高度为文字高度
        android:layout_below="@id/mytxt" />              //位于 mytxt 下方摆放
</RelativeLayout>
```

保存文件，程序运行结果如图 3.20 所示。

实例 3-13：RelativeLayout 的使用举例 2

新建一个项目，项目的命为：exam3_13，包名称为：org.hnist.demo，在 Activity 中动态的生成 RelativeLayout 组件，这里在上例的基础上来操作。

1）新建布局文件 activity_main.xml 文件，代码与实例 3-12 中 activity_main.xml 一样。

2）修改 MainActivity.java，代码如下：

图 3.20　exam3_12 运行结果

```java
package org.hnist.demo;
import android.app.Activity;
import android.os.Bundle;
import android.view.ViewGroup;
import android.widget.EditText;
```

```
import android.widget.RelativeLayout;        //导入widget.RelativeLayout类
public class MainActivity extends Activity {
    @Override
    public void onCreate(Bundle savedInstanceState) {
        super.onCreate(savedInstanceState);
        super.setContentView(R.layout.activity_main);
        //定义一个RelativeLayout组件rl,并从布局管理器中获得具体的值Layout01
        RelativeLayout rl = (RelativeLayout) super.findViewById(R.id.Layout01);
        //定义一个LayoutParams对象param,并设置参数
        RelativeLayout.LayoutParams param = new RelativeLayout.LayoutParams(
            ViewGroup.LayoutParams.FILL_PARENT,     //布局管理器宽度为屏幕宽度
            ViewGroup.LayoutParams.FILL_PARENT);    //布局管理器高度为屏幕高度
        param.addRule(RelativeLayout.BELOW, R.id.mybut);  //放在mybut组件之下
        param.addRule(RelativeLayout.RIGHT_OF, R.id.img1);//放在img1组件右边
        EditText text = new EditText(this) ;              //在其中定义文本输入框
        rl.addView(text,param) ; }}                       //将设置的参数加入组件rl
```

保存文件,程序运行结果如图3.21所示。

图3.21 exam3_13运行结果

3.12 线性布局管理器组件 LinearLayout

线性布局的形式可以分为两种:第一种是横向线性布局,第二种是纵向线性布局。以线性的形式一个个排列出来的,纯线性布局的缺点是不方便修改控件的显示位置,在开发中经常会以线性布局与相对布局嵌套的形式设置布局。

LinearLayout层次关系如下:

```
java.lang.Object
    android.view.View
        android.view.ViewGroup
            android.widget.LinearLayout
```

要在 Android 程序中使用 LinearLayout 组件必须在程序中使用下面的语句。

```
import android.widget.LinearLayout;          //导入 widget.LinearLayout 类
```

3.12.1 LinearLayout 组件常用的属性和方法

LinearLayout 组件常用的属性和方法见表 3-16。

表 3-16 LinearLayout 组件常用的属性和方法

方法及常量	描述
public static final int HORIZONTAL	常量，设置水平对齐
public static final int VERTICAL	常量，设置垂直对齐
public LinearLayout(Context context)	创建 LinearLayout 类的对象
public void addView(View child, ViewGroup.LayoutParams params)	增加组件并且指定布局参数
public void addView(View child)	增加组件
protected void onDraw(Canvas canvas)	用于图形绘制的方法
public void setOrientation(int orientation)	设置对齐方式

如果要在程序中控制 LinearLayout 的操作，需要对一些布局参数进行配置，布局参数保存在 LinearLayout.LayoutParams 类中。其层次关系如下：

```
java.lang.Object
    android.view.ViewGroup.LayoutParams
        android.view.ViewGroup.MarginLayoutParams
            android.widget.LinearLayout.LayoutParams
```

LinearLayout.LayoutParams 类提供了以下一个构造方法：

public LinearLayout.LayoutParams (int width, int height)，创建布局参数时需要传递布局参数的宽度和高度，这两个参数可以通过下面的两个常量来提供。

```
public static final int FILL_PARENT,填充全部父控件
public static final int WRAP_CONTENT,包裹自身控件
```

3.12.2 LinearLayout 组件使用实例

实例 3-14：LinearLayout 的使用举例 1

新建一个项目，项目的命名为：exam3_14，包名称为：org.hnist.demo，利用线性布局管理器对组件进行放置。

1）建立一个布局文件 activity_main.xml 文件，代码如下：

```xml
<?xml version="1.0" encoding="utf-8"?>
<LinearLayout                                         //采用线性布局管理器
    xmlns:android="http://schemas.android.com/apk/res/android"
    android:orientation="vertical"                    //组件垂直放置
    android:layout_width="fill_parent"                //此布局管理器填充整个屏幕宽度
    android:layout_height="fill_parent">              //此布局管理器填充整个屏幕高度
    <TextView                                         //定义一个文本显示组件
        android:id="@+id/mytxt"                       //命名为 mytxt
        android:layout_width="fill_parent"            //文本显示组件的宽度为屏幕宽度
        android:layout_height="wrap_content"          //文本显示的高度为文字的高度
        android:text="更多精彩尽在按钮中......"        //文本显示的内容
        android:textSize="25dp" />                    //文字显示的大小为 25dp
```

```
        <Button                                     //定义一个按钮组件
            android:id="@+id/mybut"                 //命名为 mybut
            android:layout_width="fill_parent"      //按钮组件的宽度为屏幕宽度
            android:layout_height="wrap_content"    //按钮组件的高度为文字的高度
            android:text="去看看"                    //设置按钮上的文字
            android:textSize="20dp"/>               //按钮上文字大小为 20dp
</LinearLayout>                                     //结束采用线性布局管理器
```

保存文件，程序运行结果如图 3.21 所示。

实例 3-15：LinearLayout 的使用举例 2

新建一个项目，项目的命名为：exam3_15，包名称为：org.hnist.demo，在 Activity 程序中动态的生成 LinearLayout 组件，要求程序运行结果与实例 3-14 一样。

图 3.21　exam3_14 运行结果

1）建立一个 MainActivity.java，代码如下：

```
package org.hnist.demo;
import android.app.Activity;
import android.os.Bundle;
import android.view.ViewGroup;
import android.widget.LinearLayout;                 //导入 widget.LinearLayout 类
import android.widget.TextView;
import android.widget.Button;
public class MainActivity extends Activity {
@Override
public void onCreate(Bundle savedInstanceState) {
    super.onCreate(savedInstanceState);
    LinearLayout layout = new LinearLayout(this);   //创建线性布局对象 layout
    //定义一个 LayoutParams 对象 param，并设置参数
LinearLayout.LayoutParams param = new LinearLayout.LayoutParams(
        ViewGroup.LayoutParams.FILL_PARENT,         //布局管理器宽度为屏幕宽度
        ViewGroup.LayoutParams.FILL_PARENT);        //布局管理器高度为屏幕高度
    layout.setOrientation(LinearLayout.VERTICAL);   //垂直摆放组件
        LinearLayout.LayoutParams txtParam = new LinearLayout.LayoutParams(
            ViewGroup.LayoutParams.FILL_PARENT,     //组件宽度为屏幕宽度
            ViewGroup.LayoutParams.WRAP_CONTENT);   //组件高度为文字高度
    TextView txt = new TextView(this);              //定义文本显示组件 txt
    txt.setLayoutParams(txtParam);                  //设置文本显示布局参数
    txt.setText("更多精彩尽在按钮中......");          //设置显示内容
    txt.setTextSize(25);                            //设置文字大小
    layout.addView(txt, txtParam);                  //增加文本显示组件
    Button but = new Button(this);                  //定义按钮组件
    but.setLayoutParams(txtParam);                  //设置按钮布局参数
    but.setText("去看看");                           //设置按钮提示内容
    but.setTextSize(20);                            //设置按钮文字大小
    layout.addView(but, txtParam);                  //增加按钮组件
    super.addContentView(layout, param) ; }}        //显示布局管理器
```

保存文件，程序运行结果如图 3.21 所示，与实例 3-14 运行结果一样，从这个实例可以看出，采用布局管理器的方式来设置组件比在 Activity 程序中设置组件要轻松多了。一般的 Android 项目设计也建议采用布局管理器来添加组件，在 Activity 程序中尽量不要添加及放置组件。

3.13 框架布局管理器组件 FrameLayout

FrameLayout 布局是在屏幕上开辟一个区域放置组件，它会将所有的组件都放在屏幕的左上角，任何一个组件都会被后绘制的组件覆盖，所有的组件以层叠的方式显示出来，如果组件一样大，在同一时刻只能看到最上面的组件。FrameLayout 组件的层次关系如下：

```
java.lang.Object
    android.view.View
        android.view.ViewGroup
            android.widget.FrameLayout
```

要在 Android 程序中使用 FrameLayout 组件必须在程序中使用下面的语句：

```
import android.widget.FrameLayout;        //导入 widget.FrameLayout 类
```

3.13.1 FrameLayout 组件常用的属性和方法

FrameLayout 组件是 View 类的子类，因此 View 类的属性和方法，它都具备，另外还有如表 3-17 所示的方法和属性。

表 3-17 FrameLayout 组件常用的属性和方法

属性名称	对应方法	描述
android:foreground	setForeground(Drawable)	设置绘制在所有子控件之上的内容
android:foregroundGravity	setForegroundGravity(int)	设置绘制在所有子控件之上内容的 gravity 属性
public FrameLayout (Context context)		创建 FrameLayout 类的对象
public void addView(View child, ViewGroup.LayoutParams params)		增加组件并且指定布局参数
public void addView(View child)		增加组件

如果要在程序中控制 FrameLayout 的操作，需要对一些布局参数进行配置，布局参数保存在 FrameLayout.LayoutParams 中。其层次关系如下：

```
java.lang.Object
    android.view.ViewGroup.LayoutParams
        android.view.ViewGroup.MarginLayoutParams
            android.widget.FrameLayout.LayoutParams
```

3.13.2 FrameLayout 的使用举例

实例 3-16：FrameLayout 的使用举例 1

新建一个项目，项目的命为：exam3_16，包名称为：org.hnist.demo，利用 FrameLayout 对组件进行放置。

1）建立一个布局文件 activity_main.xml 文件，代码如下：

```
<?xml version="1.0" encoding="utf-8"?>
<FrameLayout                              //采用框架布局管理器
xmlns:android="http://schemas.android.com/apk/res/android"
android:id="@+id/Layout01"               //命名为 Layout01
android:layout_width="wrap_content"
android:layout_height="wrap_content">
<ImageView
```

```
            android:id="@+id/myimg"
            android:src="@drawable/a2"//注意要先将 a2.jpg 文件复制到 res\drawable 目录
            android:layout_width="wrap_content"
            android:layout_height="wrap_content"/>
    <TextView
            android:id="@+id/myinput"
            android:text="大家好!我是 FrameLayout"
            android:textSize="20dp"
            android:layout_width="wrap_content"
            android:layout_height="wrap_content"/>
    <Button
            android:id="@+id/mybut"
            android:text="确定"
            android:layout_width="wrap_content"
            android:layout_height="wrap_content"/>
</FrameLayout>                                        //结束采用框架布局管理器
```

保存文件,程序运行结果如图 3.23 所示。

可以发现在 FrameLayout 布局管理器中,增加了 3 个组件,一张图片、一个文本显示框、一个按钮,它们都放置在屏幕的左上角,按钮在最上方,图片在最下方,如果设置图片在最上方,那么按钮会被遮住看不见,用户可以自己去修改实验。当然也可以在 Activity 程序中动态地生成 FrameLayout 组件,详细代码见 exam3_17。

3.14 表格布局管理器组件 TableLayout

TableLayout 是采用表格的形式对组件的布局进行管理的,在表格布局中使用 TableRow 进行表格行的控制,通过设置 TableRow 组件,可以设置表格中每一行显示的内容及位置等,所有的组件的增加都要在 TableRow 组件中进行。TableLayout 组件层次关系如下:

图 3.23　exam3_16 运行结果

```
java.lang.Object
    android.view.View
        android.view.ViewGroup
            android.widget.LinearLayout
                android.widget.TableLayout
```

要在 Android 程序中使用 TableLayout 组件必须在程序中使用下面的语句:

```
import android.widget.TableLayout;            //导入 widget.TableLayout 类
```

3.14.1　TableLayout 组件常用的属性和方法

TableLayout 组件是 LinearLayout 的子类,因此 LinearLayout 组件的属性和方法它都具备,另外还有如表 3-18 所示的属性和表 3-19 所示的方法。

表 3-18　TableLayout 组件的常用属性

属性名称	对应方法	描述
android:collapseColumns	setColumnCollapsed(int,boolean)	隐藏从 0 开始的索引列。列之间必须用逗号隔开,如:1,2,5
android:shrinkColumns	setColumnCollapsed(int,boolean)	收缩从 0 开始的索引列。列之间必须用逗号隔开,如:1,2,5
android:stretchColumns	setColumnCollapsed(int,boolean)	拉伸从 0 开始的索引列。列之间必须用逗号隔开,如:1,2,5

表 3-19 TableLayout 组件常用的方法

方法	描述
public TableLayout (Context context)	创建表格布局
public void addView (View child)	添加子视图
public void addView (View child, int index)	添加子视图
public void addView (View child, int index, ViewGroup.LayoutParams params)	根据指定参数，添加子视图
public TableLayout.LayoutParams generateLayoutParams (AttributeSet attrs)	返回一组基于提供的属性集合的布局参数集合
public boolean isColumnCollapsed (int columnIndex)	返回指定列的折叠状态
public boolean isColumnShrinkable (int columnIndex)	返回指定的列是否可收缩
public boolean isColumnStretchable (int columnIndex)	返回指定的列是否可拉伸
public boolean isShrinkAllColumns ()	指示是否所有的列都是可收缩的
public boolean isStretchAllColumns ()	指示是否所有的列都是可拉伸的

在表格布局中使用 TableRow 设置表格中每一行显示的内容及位置等，一般组件的增加都要在 TableRow 中进行。

例如：表格中，要在一行显示一个文本，后面接着显示一个文本编辑框，代码如下：

```
<TableRow>                                              //增加一行
    <TextView                                           //在这行添加一个文本组件
        android:id="@+id/myinput"                       //给文本组件命名
        android:layout_width="wrap_content"             //组件宽度为文字宽度
        android:layout_height="wrap_content"            //组件高度为文字高度
        android:text="请输入您的姓名："/>               //设置显示内容
    <EditText                                           //在这行添加一个文本编辑框组件
        android:id="@+id/myinput"                       //给文本编辑组件命名
        android:layout_width="wrap_content"             //组件宽度为文字宽度
        android:layout_height="wrap_content"            //组件高度为文字高度
</TableRow>                                             //一行结束
```

有时文本内容太长，会有一些内容看不到，可以将长的列定义为可收缩列，它会根据文字信息调整显示格式，可用如下代码定义可收缩列：

```
<TableLayout                                            //采用表格布局管理器
    xmlns:android="http://schemas.android.com/apk/res/android"
    android:id="@+id/Layout01"                          //定义名字为 Layout01，程序中使用
    android:layout_width="fill_parent"                  //布局管理器宽度为屏幕宽度
    android:layout_height="fill_parent"                 //布局管理器高度为屏幕高度
    android:shrinkColumns="2">                          //设置第二列为可收缩列
```

在表格布局管理器内还可以设置某些列不显示，要注意的是列数的计数是从第 0 列开始的，可用如下代码设置第 1 列和第 3 列不显示：

```
<TableLayout                                            //采用表格布局管理器
    xmlns:android="http://schemas.android.com/apk/res/android"
    android:id="@+id/Layout01"                          //定义名字为 Layout01，程序中使用
    android:layout_width="fill_parent"                  //布局管理器宽度为屏幕宽度
    android:layout_height="fill_parent"                 //布局管理器高度为屏幕高度
    android:collapseColumns="0,2">                      //设置第 1 列和第 3 列不显示
```

3.14.2 TableLayout 的使用举例

实例 3-18：TableLayout 的使用举例

新建一个项目，项目的命名为：exam3_18，包名称为：org.hnist.demo，利用 TableLayout 对组件进行放置。

建立一个布局文件 activity_main.xml 文件，代码如下：

```xml
<?xml version="1.0" encoding="utf-8"?>
<TableLayout                                              //采用表格布局管理器
xmlns:android="http://schemas.android.com/apk/res/android"
android:id="@+id/Layout01"                                //定义名字为 Layout01
android:layout_width="fill_parent"                        //布局管理器宽度为屏幕宽度
android:layout_height="fill_parent"                       //布局管理器高度为屏幕高度
<TextView
        android:layout_width="fill_parent"
        android:layout_height="wrap_content"
        android:textSize="20dp"
        android:text="个人信息登记表"
        android:gravity="center"                          //文本居中显示
        android:paddingBottom="12dp"   />                 //距离下边缘为 12dp
<TableRow>                                                //增加一行
    <TextView                                             //在这行添加一个文本组件
        android:id="@+id/myinput"                         //给文本组件命名
        android:layout_width="wrap_content"
        android:layout_height="wrap_content"
        android:text="请输入您的姓名："/>
    <EditText                                             //在这行添加一个文本编辑框组件
        android:id="@+id/myinput"
        android:layout_width="wrap_content"
        android:layout_height="wrap_content"/>
</TableRow>                                               //一行结束
<View
        android:layout_height="1dip"                      //画一根高度为 1dip 的线
        android:background="#909090" />                   //设置线的颜色
<TableRow>                                                //增加一行
    <TextView                                             //在这行添加一个文本组件
        android:id="@+id/info"
        android:text="请选择您的性别："
        android:layout_width="wrap_content"
        android:layout_height="wrap_content"/>
    <RadioGroup                                           //在这行添加一个单选按钮组
        android:id="@+id/sex"
        android:layout_width="wrap_content"
        android:layout_height="wrap_content"
        android:orientation="horizontal"                  //组件水平摆放
        android:checkedButton="@+id/man">                 //设置被选中的选项
        <RadioButton                                      //设置单选按钮选项
            android:id="@+id/man"
            android:text=" 男"/>
        <RadioButton                                      //设置单选按钮选项
            android:id="@+id/woman"
```

```
            android:text="女"/>
    </RadioGroup>                              //单选按钮组结束
</TableRow>                                    //一行结束
<View
    android:layout_height="1dip"
    android:background="#909090" />
    <Button                                    //添加一个按钮
        android:id="@+id/ok"
        android:layout_width="wrap_content"
        android:layout_height="wrap_content"
        android:text="确 定"/>
</TableLayout>                                 //表格布局结束
```

保存文件，程序运行结果如图 3.24 所示。

也可以在 Activity 程序中动态的生成 TableLayout 组件，（详细代码见 exam3_19）但是在设计时通常不建议这样做，一般的做法是：先在布局管理器中定义各个要显示的组件，如果需要较多数据，要定义一个 XML 文件，从 XML 文件中获取数据，然后在 Activity 程序中设计对这些组件的操作。

图 3.24　exam3_18 运行结果

3.15　布局管理器的嵌套

在使用布局管理器进行组件布局时，也可以将各个布局管理器嵌套在一起使用，以达到更好的效果。把布局管理器当做一个类似于 TextView 组件一样放到另一个布局管理中，就可以实现布局管理器的嵌套。

实例 3-20：布局管理器的嵌套使用举例

新建一个项目，项目的命名为：exam3_20，包名称为：org.hnist.demo，利用布局管理器的嵌套对组件进行放置。

建立一个布局文件 activity_main.xml 文件，代码如下：

```
<?xml version="1.0" encoding="utf-8"?>
<LinearLayout                                  //采用线性布局管理器
xmlns:android="http://schemas.android.com/apk/res/android"
android:orientation="vertical" android:layout_width="fill_parent"
android:layout_height="fill_parent">
<TextView                                      //增加一个文本显示组件
    android:layout_width="fill_parent"
    android:layout_height="wrap_content"
    android:text="检索你感兴趣的内容......"
    android:textSize="22dp"/>
<TableLayout                                   //增加一个表格布局组件
    android:layout_width="wrap_content"
    android:layout_height="wrap_content">
    <TableRow>                                 //在表格中增加一行
        <TextView                              //第一行的第一列内容
            android:layout_width="wrap_content"
            android:layout_height="wrap_content"
            android:text="历史"
            android:textSize="20dp"
            android:padding="10px"/>           //与四周的间隔距离为 10px
```

```xml
    <TextView                                    //第一行的第二列内容
        android:layout_width="wrap_content"
        android:layout_height="wrap_content"
        android:text="地理"
        android:textSize="20dp"
        android:padding="10px"/>
    <TextView                                    //第一行的第三列内容
        android:layout_width="wrap_content"
        android:layout_height="wrap_content"
        android:text="军事"
        android:textSize="20dp"
        android:padding="10px"/>
    <TextView                                    //第一行的第四列内容
        android:layout_width="wrap_content"
        android:layout_height="wrap_content"
        android:text="影视"
        android:textSize="20dp"
        android:padding="10px"/>
    <TextView                                    //第一行的第五列内容
        android:layout_width="wrap_content"
        android:layout_height="wrap_content"
        android:text="旅游"
        android:textSize="20dp"
        android:padding="10px"/>
</TableRow>                                      //第一行结束
<View                                            //行之间的分割线
    android:layout_height="1px"                  //高度为1px
    android:background="#909090"/>               //设置线的颜色
<TableRow>                                       //表格的第二行
    <TextView                                    //第二行的第一列内容
        android:layout_width="wrap_content"
        android:layout_height="wrap_content"
        android:text="音乐"
        android:textSize="20dp"
        android:padding="10px"/>
    <TextView                                    //第二行的第二列内容
        android:layout_width="wrap_content"
        android:layout_height="wrap_content"
        android:text="体育"
        android:textSize="20dp"
        android:padding="10px"/>
    <TextView                                    //第二行的第三列内容
        android:layout_width="wrap_content"
        android:layout_height="wrap_content"
        android:text="卫生"
        android:textSize="20dp"
        android:padding="10px"/>
    <TextView                                    //第二行的第四列内容
        android:layout_width="wrap_content"
        android:layout_height="wrap_content"
        android:text="美食"
        android:textSize="20dp"
        android:padding="10px"/>
    <TextView                                    //第二行的第五列内容
        android:layout_width="wrap_content"
        android:layout_height="wrap_content"
```

第3章 Android 开发常用组件

```
                android:text="......"
                android:textSize="20dp"
                android:padding="10px"/>
        </TableRow>                                    //第二行结束
        <View                                          //行之间的分割线
            android:layout_height="1px"                //高度为 1px
            android:background="#909090"/>             //设置线的颜色
    </TableLayout>                                     //表格结束
    <TableLayout                                       //新增加一个表格
        android:layout_width="wrap_content"            //设置表格的参数
        android:layout_height="wrap_content">          //设置表格的参数
        <TableRow>                                     //增加一行
            <EditText                                  //第一行第一列内容
                android:id="@+id/myinput"
                android:src="@drawable/a2"
                android:layout_width="wrap_content"
                android:layout_height="wrap_content"
                android:text="请输入检索关键字..." />
            <Button                                    //第一行第二列内容
                android:id="@+id/search"
                android:text="检索"
                android:layout_width="wrap_content"
                android:layout_height="wrap_content" />
        </TableRow>                                    //第一行结束
    </TableLayout>                                     //表格管理器结束
</LinearLayout>                                        //线性布局管理结束
```

保存文件，程序运行结果如图 3.25 所示。

最外围的布局管理器是线性布局管理器，在里面定义两个表格布局管理器，如果单一地用某一个布局管理器要达到图 3.25 的效果，还是比较困难的，多个布局管理的结合使用能让屏幕布局实现起来更简单、快捷、美观。

图 3.25 exam3_20 运行结果

本章小结

本章着重介绍了 Android 中常用的基本组件 View、TextView、EditText、Button、ImageButton、ImageView、RadioGroup、CheckBox、Spinner、Toast 及其基本操作，并详细介绍了 Android 中 4 种布局管理器及其基本使用，建议先在布局管理器中定义各个要显示的组件，一般不要在 Activity 程序中生成布局管理器组件。

习题

1. 查找 Android API 了解 View 组件的属性和方法，列举 10 种常用的属性和对应的方法。
2. TextView 组件的什么属性能够实现跑马灯的效果，写出相关代码。
3. 修改 EditText 组件的什么属性值，能让 EditText 组件里的值不能编辑？
4. 是否可以在 Button 组件上同时显示一个图片和文字？如果可以该设置什么属性？写出相关代码。
5. 怎样在 Activity 程序中自动生成 ImageView 组件，并显示一张居中的图片？写出相关代码。
6. 编写程序，利用 RadioGroup 组件实现性别的选择。
7. CheckBox 组件是通过什么属性知道选中了这个选项？
8. 编写程序，利用 Spinner 组件实现职称的选择，职称有：高级、中级、初级三种情况。
9. 新建一个项目，在布局管理中，利用相对布局管理器实现图 3.25 所示的效果。

第 4 章　Android 中的事件处理

学习目标：
- 了解 Android 中的事件处理的原理；
- 掌握 Android 中的主要事件：单击事件、长按事件、焦点改变事件、键盘事件、触摸事件、菜单事件的基本操作。

在前面的章节中介绍了组件的布局，事实上在图形界面（UI）的开发中，有两个非常重要的内容：一个是组件的布局，另一个是事件处理。本章主要对 Android 中几种不同事件的处理进行分析。

4.1　Android 中的事件处理基础

Android 在事件处理过程中主要涉及 3 个概念。

1）事件（Event）：表示在图形界面操作的描述，通常是封装成各种类，比如，单击事件、触摸事件、键盘事件等。

2）事件源（Event Source）：事件源是指事件发生的场所，通常是指各个组件，例如，Button、EditText 等控件。

3）事件监听器（Event Listener）：事件监听器是指接收事件对象并对其进行处理的对象，事件处理一般是一个实现某些特定接口类创建的对象。

例如，单击按钮后，屏幕显示"你好!"，在这个事件处理中，事件是"单击事件"，事件源是"按钮"，事件监听器是定义的"OnClickListener"对象，由它来实现具体的操作。

4.1.1　事件处理的过程

1）为事件源对象添加监听，当某个事件被触发时，系统才会知道该通知谁来处理这个事件，如图 4.1(a)所示。

2）当事件发生时，系统会将事件封装成相应类型的事件对象，并发送给注册到事件源的事件监听器，如图 4.1(b)所示。

3）当监听器对象接收到事件对象之后，系统会调用监听器中相应的事件处理方法来处理事件并给出响应，如图 4.1(c)所示。

事件处理有以下几个关键的操作：注册监听程序、根据指定的事件编写处理程序、在事件处理类中完成事件的处理操作。

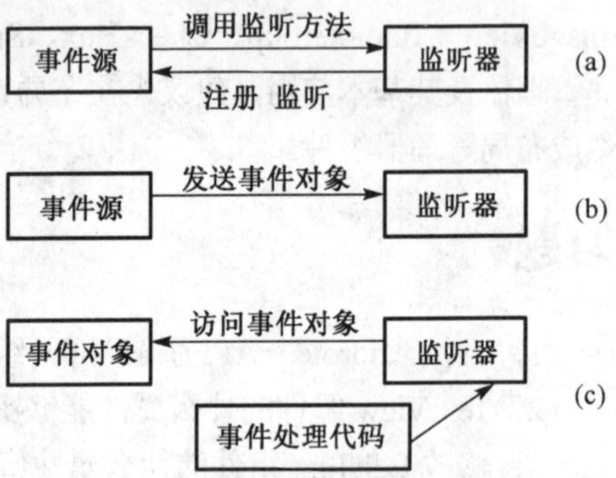

图 4.1　事件处理的过程示意图

4.1.2　事件处理模型

Android 中的事件处理模型常用的有：基于监听接口的事件处理，基于回调的事件处理。

1. 基于监听接口的事件处理

基于监听的事件处理模型的编程步骤如下：

1）获取普通界面组件；
2）实现事件监听类，该监听类是一个特殊的 java 类，必须实现一个 XXXListener 接口；
3）调用事件源的 setXXXListener 方法注册事件监听器。

Android 中提供了以下几种基于监听接口的事件处理模型。

1）OnClickListener 接口：单击事件。
2）OnLongClickListener 接口：长按事件。
3）OnFocusChangeListener 接口：焦点改变事件。
4）OnKeyListener 接口：键盘事件。
5）OnTouchListener 接口：触摸事件。
6）OnCreateContextMenuListener 接口：上下文菜单事件。

2．基于回调机制的事件处理

在 Android 平台中，每个 View 类都有自己的处理事件的回调方法，可以通过重写 View 类中的回调方法来实现需要的响应事件，Android 类提供了以下回调方法供用户使用。

1）onKeyDown：用来捕捉手机键盘被按下的事件。
2）onKeyUp：用来捕捉手机键盘按键抬起的事件。
3）onTouchEvent：用来处理手机屏幕的触摸事件。
4）onTrackBallEvent：用来处理轨迹球事件。
5）onFocusChanged：用来处理焦点改变的事件。

4.2 单击事件 OnClickListener

4.2.1 单击事件基础

单击事件需要注册相应的监听器（setOnClickListener）监听事件的来源，利用 OnClickListener 接口中的 onClick 方法，当事件发生时作出相应的处理。

单击事件使用 View.OnClickListener 接口进行事件的处理，此接口定义如下：

```
    public static interface View.OnClickListener{
        public void onClick(View v) ;}
```

当单击事件触发后自动使用该接口中的 public void onClick(View v)方法进行事件处理。
说明：需要实现 onClick 方法，参数 v 为事件发生的事件源。
要在 Android 程序中使用单击事件，必须在程序中使用下面的语句：

```
    import android.view.View.OnClickListener;        //导入 View.OnClickListener 类
```

单击事件的实现步骤如下：
1）通过组件 ID 获取组件实例；
例如：this.mybut=(Button)super.findViewById(R.id.mybut);   //获得按钮
2）为该组件注册 OnClickListener 监听；
例如：mybut.setOnClickListener(new ShowListener()) ; //注册监听
3）实现 onClick 方法。
例如：

```
    private class ShowListener implements OnClickListener {//定义监听处理程序
```

```
public void onClick(View v) {                    //执行具体操作
    ……}}
```

4.2.2 单击事件实例

实例 4-1：单击事件实例

新建一个项目，项目的命名为：exam4_1，包名称为：org.hnist.demo，编程实现：手机界面显示 3 个组件 EditText、Button 和 TextView，单击"Button"按钮时会在手机上显示输入的信息。

1）修改 activity_main.xml 文件，代码如下：

```xml
<RelativeLayout xmlns:android="http://schemas.android.com/apk/res/android"
    xmlns:tools="http://schemas.android.com/tools"
    android:layout_width="match_parent"
    android:layout_height="match_parent"
    tools:context=".MainActivity" >
<EditText
    android:id="@+id/myedit"
    android:layout_width="wrap_content"
    android:layout_height="wrap_content"
    android:textSize="20px"
    android:phoneNumber="true"
    android:selectAllOnFocus="true"
    android:text="请输入您的手机号码："/>
<Button
    android:id="@+id/mybut"
    android:layout_width="wrap_content"
    android:layout_height="wrap_content"
    android:layout_below="@+id/myedit"
    android:textSize="20px"
    android:layout_marginTop="20dp"
    android:text="显示您输入的号码" />
<TextView
    android:id="@+id/mytxt"
    android:layout_width="wrap_content"
    android:layout_height="wrap_content"
    android:layout_below="@+id/mybut"
    android:layout_marginTop="20dp"
    android:textSize="20px"/>
</RelativeLayout>
```

2）修改 MainActivity.java，代码如下：

```java
package org.hnist.demo;
import android.os.Bundle;
import android.app.Activity;
import android.view.View;
import android.view.View.OnClickListener;    //导入 View.OnClickListener 类
import android.widget.EditText;
import android.widget.Button;
import android.widget.TextView;
public class MainActivity extends Activity {
```

```
        private TextView mytxt=null;
        private Button mybut=null;
        private EditText myedit=null;
    @Override
    protected void onCreate(Bundle savedInstanceState) {
        super.onCreate(savedInstanceState);
        setContentView(R.layout.activity_main);
        this.mytxt=(TextView)super.findViewById(R.id.mytxt);
        this.mybut=(Button)super.findViewById(R.id.mybut);
        this.myedit=(EditText)super.findViewById(R.id.myedit);
        mybut.setOnClickListener(new ShowListener() ;}        //注册监听
        private class ShowListener implements OnClickListener {   //定义监听处理程序
        public void onClick(View v) {                             //回调方法，执行具体的操作
            String info = myedit.getText().toString() ;
            mytxt.setText("您的手机号码是: " + info) ; }}}
```

保存文件，程序运行会显示一个文本编辑框，一个按钮。在文本编辑框中输入内容，然后单击按钮，会在下面显示相应的信息，如图 4.2 所示。

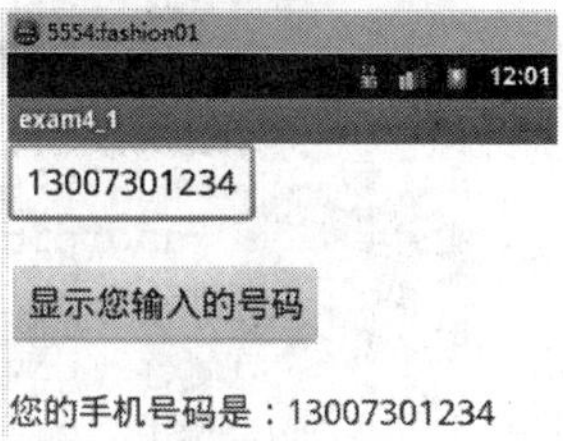

4.3 长按事件 OnLongClickListener

4.3.1 长按事件基础

图 4.2　exam4_1 运行结果

在 Android 中提供了长按事件的处理操作，长按事件只有在触发 2 秒之后才会有反应，长按事件使用 View.OnLongClickListener 接口进行事件的处理操作。

此接口定义如下：

```
public static interface View.OnLongClickListener{
    public boolean onLongClick(View v) ; }
```

当长按事件触发后自动使用该接口中的 public boolean onLongClick(View v)方法进行事件处理。

说明：需要实现 onLongClick 方法，该方法的返回值为一个 boolean 类型的变量，当返回 true 时，表示已经完整地处理了这个事件，并不希望其他的回调方法再次进行处理；当返回 false 时，表示并没有完全处理完该事件希望其他方法继续对其进行处理。

要在 Android 程序中使用长按事件必须在程序中使用下面的语句：

```
    import android.view.View.OnLongClickListener; //导入View.OnLongClickListener类
```

长按事件的实现步骤：
1）通过组件 ID 获取组件实例；
例如，bgimg=(ImageView)findViewById(R.id.bgimg);
2）为该组件注册 OnLongClickListener 监听；
例如，bgimg.setOnLongClickListener(new OnLongClickListener(){……}
3）实现 onLongClick 方法。
例如，public boolean onLongClick(View v) {
        ……}

4.3.2 长按事件实例

实例 4-2：长按事件实例

新建一个项目，项目的命名为：exam4_2，包名称为：org.hnist.demo，编程实现：长按一张图片，将这张图片设置为背景。

将准备好的图片复制到 res\drawablw-hdpi 文件夹下。

1）修改 activity_main.xml 文件，代码如下：

```xml
<LinearLayout xmlns:android="http://schemas.android.com/apk/res/android"
    xmlns:tools="http://schemas.android.com/tools"
    android:layout_width="match_parent"
    android:layout_height="match_parent"
    android:orientation="vertical"
    tools:context=".MainActivity" >
    <TextView
        android:id="@+id/mytxt"
        android:layout_width="fill_parent"
        android:layout_height="wrap_content"
        android:text="长按图片将其设为屏幕背景" />
    <ImageView
        android:id="@+id/bgimg"
        android:layout_width="wrap_content"
        android:layout_height="wrap_content"
        android:layout_gravity="center"
        android:src="@drawable/love" />
</LinearLayout>
```

2）修改 MainActivity.java，代码如下：

```java
package org.hnist.demo;
import java.io.IOException;
import android.os.Bundle;
import android.app.Activity;
import android.view.View;
import android.view.View.OnLongClickListener; //导入View.OnLongClickListener类
import android.widget.ImageView;
import android.widget.TextView;
public class MainActivity extends Activity {
private ImageView bgimg=null;
private TextView showInfo=null;
@Override
protected void onCreate(Bundle savedInstanceState) {
    super.onCreate(savedInstanceState);
    setContentView(R.layout.activity_main);
    showInfo=(TextView)findViewById(R.id.mytxt);
    bgimg=(ImageView)findViewById(R.id.bgimg);
    bgimg.setOnLongClickListener(new OnLongClickListener(){
        @Override
        public boolean onLongClick(View v) {
            try {
```

```
                    clearWallpaper();
setWallpaper(bgimg.getResources().openRawResource(R.drawable.love));//设为桌面背景
                    showInfo.setText("图片已经设置为手机桌面");
                } catch (IOException e) {
                    //TODO Auto-generated catch block
                    e.printStackTrace();
                    showInfo.setText("手机桌面背景设置没有成功！");}
                return true;      }});}}        //表示已经完整地处理了这个事件
```

3）设置手机桌面背景的操作属于手机的支持服务，首先必须得到相关的授权后才可以执行，双击打开 AndroidManifest.xml 文件，在</manifest>之前增加以下语句进行授权：

```
<uses-permission android:name="android.permission.SET_WALLPAPER">
</uses-permission>
```

保存所有文件并运行程序，结果如图 4.3 所示；长按图片结果如图 4.4 所示。单击右侧的"返回"按钮，显示已经将这张图片设置成了手机桌面背景。

图 4.3　exam4_2 运行结果 1

图 4.4　exam4_2 运行结果 2

4.4　焦点改变事件 OnFocusChangeListener

4.4.1　焦点改变事件基础

焦点改变事件是指对一个组件状态的监听，是在组件获得或失去焦点时进行处理操作，所有的组件都存在有监听焦点变化的方法，利用 OnFocusChangeListener 接口来监听焦点改变事件。单击事件或长按时间就是指单击或长按某个组件时触发的事情，而焦点事件则是只要该组件获得或失去焦点，不用单击就会触发相应的事件。

此接口定义如下：

```
public void setOnFocusChangeListener(View.OnFocusChangeListener l)
```

当焦点改变事件触发之后自动使用该接口中的 public void onFocusChange(View v, Boolean hasFocus)方法进行事件处理。

说明：需要实现 onFocusChange 方法；参数 v：为触发该事件的事件源；参数 hasFocus：表示 v 的新状态，即 v 是否获得焦点。

要在 Android 程序中使用焦点改变事件必须在程序中使用下面的语句：

```
import android.view.View.OnFocusChangeListener;//导入View.OnFocusChangeListener类
```

焦点改变事件的实现步骤如下：

1）通过组件 ID 获取组件实例；

例如：this.edit = (EditText) super.findViewById(R.id.edit1);

2）为该组件注册 OnFocusChangeListener 监听；

例如：this.edit.setOnFocusChangeListener(new OnFocusChangeListenerImpl());

3）实现 onFocusChange 方法。

例如：public void onFocusChange(View v, boolean hasFocus) {……

列出一些与焦点有关的常用方法。

setFocusable 方法：设置 View 类是否可以拥有焦点。

isFocusable 方法：监测此 View 类是否可以拥有焦点。

setNextFocusDownId 方法：设置 View 类的焦点向下移动后获得焦点 View 的 ID。

hasFocus 方法：返回了 View 类的父控件是否获得了焦点。

requestFocus 方法：尝试让此 View 类获得焦点。

isFocusableTouchMode 方法：设置 View 类是否可以在触摸模式下获得焦点，在默认情况下是不可以获得的。

4.4.2 焦点改变事件实例

实例 4-3：焦点改变事件实例

一般输入信息时会要求对输入的信息的合法性做出判断，例如，输入的邮箱地址要符合相应的规范，类似这样的操作就可以用焦点改变事件来实现。

新建一个项目，项目的命名为：exam4_3，包名称为：org.hnist.demo。

1）修改 activity_main.xml 文件，代码如下：

```xml
<?xml version="1.0" encoding="utf-8"?>
<LinearLayout xmlns:android="http://schemas.android.com/apk/res/android"
android:orientation="vertical" android:layout_width="fill_parent"
android:layout_height="fill_parent">
<TextView
    android:id="@+id/txt1"
    android:layout_width="fill_parent"
    android:layout_height="wrap_content"
    android:textSize="20dp"
    android:text="请输入您的邮箱" />
<EditText
    android:id="@+id/edit1"
    android:layout_width="fill_parent"
    android:layout_height="wrap_content"
    android:focusable="true"
      android:selectAllOnFocus="true"
      android:text="第一个文本输入框组件"/>
<EditText
    android:id="@+id/edit2"
    android:layout_width="fill_parent"
    android:layout_height="wrap_content"
    android:text="第二个文本输入框组件" />
```

```xml
<TextView
    android:id="@+id/txt2"
    android:layout_width="fill_parent"
    android:layout_height="wrap_content"
    android:textSize="15dp"/>
</LinearLayout>
```

2）修改 MainActivity.java，代码如下：

```java
package org.hnist.demo;
import android.app.Activity;
import android.os.Bundle;
import android.view.View;
import android.view.View.OnFocusChangeListener;
import android.widget.EditText;
import android.widget.TextView;
public class MainActivity extends Activity {
private EditText edit = null;
private TextView txt = null;
@Override
public void onCreate(Bundle savedInstanceState) {
    super.onCreate(savedInstanceState);
    super.setContentView(R.layout.activity_main);
    this.edit = (EditText) super.findViewById(R.id.edit1);    //取得组件
    this.txt = (TextView) super.findViewById(R.id.txt2);       //取得组件
    //注册监听焦点事件
    this.edit.setOnFocusChangeListener(new OnFocusChangeListenerImpl());}
    //设置焦点事件
    private class OnFocusChangeListenerImpl implements OnFocusChangeListener {
        @Override
        public void onFocusChange(View v, boolean hasFocus) {    //具体要实现的部分
          String inputmsg=MainActivity.this.edit.getText().toString();
            if (v.getId() == MainActivity.this.edit.getId()) {
                if (hasFocus) {        //判断是否获得焦点
                    MainActivity.this.txt.setText("第一个文本输入框组件获得了焦点。");
                } else {
                    if(inputmsg.matches("\\w+@\\w+\\.\\w+")) {//正则表达式判断
                MainActivity.this.txt.setText("您输入的邮箱是:"+inputmsg+"符合要求!");
                    } else {
                MainActivity.this.txt.setText("您输入的邮箱是:"+inputmsg+"不符合要求!");
}}}}}}
```

保存文件并运行程序，结果如图 4.5 和图 4.6 所示，可以对输入的信息进行判断并输出。

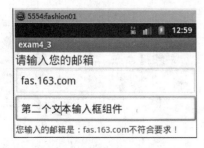

图 4.5　exam4_3 运行结果 1

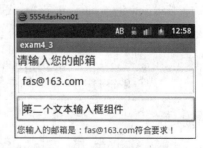

图 4.6　exam4_3 运行结果 2

4.5 键盘事件 OnKeyListener

4.5.1 键盘事件基础

键盘事件是用户在利用键盘输入数据时所触发的操作,主要功能是用于键盘的监听处理操作,键盘事件使用 OnKeyListener 接口进行事件的处理。OnKeyListener 接口定义如下:

```
public static interface View.OnKeyListener{
public boolean onKey(View v, int keyCode, KeyEvent event) ;}
```

当键盘事件触发之后自动使用该接口中的回调方法 public boolean onKey(View v, int keyCode, KeyEvent event)进行事件处理。

说明:需要实现 onKey 方法;参数 v:为事件的事件源控件;参数 keyCode:为手机键盘的键码;参数 event 便为键盘事件封装类的对象。其中包含了事件的详细信息,例如发生的事件、事件的类型等。

要在 Android 程序中使用键盘事件必须在程序中使用下面的语句:

```
import android.view.View.OnKeyListener;    //导入 View.OnKeyListener 类
```

还有一个与键盘事件联系很密切的类 view.KeyEvent:

```
import android.view.KeyEvent;              //导入 view.KeyEvent 类
```

键盘事件的实现步骤如下:
1)通过组件 ID 获取组件实例;
例如:this.edit = (EditText) super.findViewById(R.id.edit1);
2)为该组件注册 OnKeyListener 监听;
例如:this.edit.setOnKeyListener(new OnKeyListenerImpl());
3)实现 onKey 方法。
例如:public boolean onKey(View v, int keyCode, KeyEvent event) {……

4.5.2 键盘事件实例

实例 4-4:键盘事件实例 1

前面介绍了使用焦点改变事件对输入的信息的合法性做出判断,键盘事件也可以对输入信息做合法性判断,下面介绍对输入的年龄做判断,小于 200 的认为合法,否则认为是不合法的年龄。

新建一个项目,项目的命名为:exam4_4,包名称为:org.hnist.demo。
1)修改 activity_main.xml 文件,代码如下:

```xml
<?xml version="1.0" encoding="utf-8"?>
<LinearLayout
xmlns:android="http://schemas.android.com/apk/res/android"
android:orientation="vertical"
android:layout_width="fill_parent"
android:layout_height="fill_parent">
<EditText
    android:id="@+id/edit1"
    android:layout_width="wrap_content"
```

```xml
        android:layout_height="wrap_content"
        android:selectAllOnFocus="true"
        android:numeric="integer"
        android:text="请输入您的年龄: "/>
<TextView
        android:id="@+id/txt1"
        android:layout_width="wrap_content"
        android:layout_height="wrap_content" />
</LinearLayout>
```

2) 修改 MainActivity.java 文件，代码如下：

```java
package org.hnist.demo;
import android.app.Activity;
import android.os.Bundle;
import android.view.KeyEvent;                     //导入view.KeyEvent类
import android.view.View;
import android.view.View.OnKeyListener;           //导入View.OnKeyListener类
import android.widget.EditText;
import android.widget.TextView;
public class MainActivity extends Activity {
private EditText edit = null;
    private TextView txt=null;
@Override
public void onCreate(Bundle savedInstanceState) {
    super.onCreate(savedInstanceState);
    super.setContentView(R.layout.activity_main);
    this.edit = (EditText) super.findViewById(R.id.edit1);
    this.txt = (TextView) super.findViewById(R.id.txt1);
    this.edit.setOnKeyListener(new OnKeyListenerImpl());   } //设置键盘事件监听
    private class OnKeyListenerImpl implements OnKeyListener {
    @Override
    public boolean onKey(View v, int keyCode, KeyEvent event) {  //实现onKey方法
        switch(event.getAction()) {
        case KeyEvent.ACTION_UP:                              //键盘松开触发
          String inputmsg = MainActivity.this.edit.getText().toString();
          //取出已输入内容
            int a=Integer.parseInt(inputmsg); //将取出的信息转换为数字型数据
            if (a>0 && a<200) {               //判断年龄是否是大于0且小于200
               MainActivity.this.txt.setText("您输入的年龄是:"+inputmsg+", 符合
                                             要求! ");}
            else {
               MainActivity.this.txt.setText("您输入的年龄是:"+inputmsg+", 不符
                                             合要求! ");}
        case KeyEvent.ACTION_DOWN:                            //键盘按下触发
        default:
            break ; }
        return false;    }}}                                  //继续事件应有的流程
```

保存文件，程序运行结果如图 4.7 和图 4.8 所示，可以对输入的信息进行判断并输出。

图4.7 exam4_4 运行结果1　　　　　　　　　图4.8 exam4_4 运行结果2

实例4-5：键盘事件实例2

游戏时经常会用到方向键来进行控制，如何知道按下了这些键呢？下面通过编写程序来实现提取用户键盘输入的信息。

注意，在模拟器设置时要设置"Keyboard"和"DPad"可用，如图4.9所示，否则你无论怎么按键都不会有反应的。

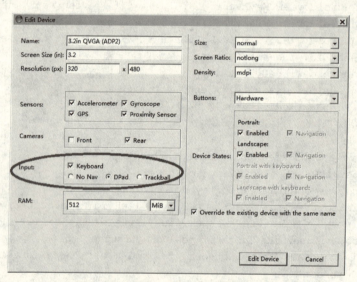

图4.9 设置模拟器的 Keyboard 和 DPad 可以用

新建一个项目，项目的命名为：exam4_5，包名称为：org.hnist.demo。
1）修改 activity_main.xml 文件，代码如下：

```xml
<RelativeLayout
    xmlns:android="http://schemas.android.com/apk/res/android"
    xmlns:tools="http://schemas.android.com/tools"
    android:layout_width="match_parent"
    android:layout_height="match_parent"
    tools:context=".MainActivity" >
<Button
    android:id="@+id/mybut1"
    android:layout_width="wrap_content"
    android:layout_height="wrap_content"
    android:layout_centerHorizontal="true"
    android:layout_marginTop="10dp"
    android:text="确    定" />
</RelativeLayout>
```

2) 修改 MainActivity.java，代码如下：

```java
package org.hnist.demo;
import android.app.Activity;
import android.os.Bundle;
import android.view.KeyEvent;              //导入 view.KeyEvent 类
import android.view.MotionEvent;
import android.view.View;
import android.widget.Button;
import android.widget.Toast;
public class MainActivity extends Activity {
    private Button mybut;
    @Override
    public void onCreate(Bundle savedInstanceState) {
        super.onCreate(savedInstanceState);
        setContentView(R.layout.activity_main);
        mybut = (Button)findViewById(R.id.mybut1);
        mybut.setOnClickListener(new Button.OnClickListener()
        {   @Override
            public void onClick(View arg0) {
                //TODO Auto-generated method stub
                ShowMessage("单击了Button按钮");
            } });  }
    //重写 onKeyDown 方法
    public boolean onKeyDown(int keyCode, KeyEvent event) {
        switch(keyCode)            {
        case KeyEvent.KEYCODE_DPAD_CENTER:
            ShowMessage("按下:中键");break;
        case KeyEvent.KEYCODE_DPAD_UP:
            ShowMessage("按下:上方向键");break;
        case KeyEvent.KEYCODE_DPAD_DOWN:
            ShowMessage("按下:下方向键");break;
        case KeyEvent.KEYCODE_DPAD_LEFT:
            ShowMessage("按下:左方向键");break;
        case KeyEvent.KEYCODE_DPAD_RIGHT:
            ShowMessage("按下:右方向键");break;    }
        return super.onKeyDown(keyCode, event);        }
    //重写 onKeyUp 方法
    public boolean onKeyUp(int keyCode, KeyEvent event) {
        switch(keyCode) {
        case KeyEvent.KEYCODE_DPAD_CENTER:
            ShowMessage("弹起:中键");
            break;
        case KeyEvent.KEYCODE_DPAD_DOWN:
            ShowMessage("弹起:下方向键");
            break;
        case KeyEvent.KEYCODE_DPAD_UP:
            ShowMessage("弹起:上方向键");
            break;
        case KeyEvent.KEYCODE_DPAD_LEFT:
            ShowMessage("弹起：左方向键");
```

```
            break;
        case KeyEvent.KEYCODE_DPAD_RIGHT:
            ShowMessage("弹起:右方向键");
            break;        }
        return super.onKeyUp(keyCode, event);         }
/**按键重复点击事件*/
public boolean onKeyMultiple(int keyCode, int repeatCount, KeyEvent event)
{     return onKeyMultiple(keyCode, repeatCount, event);          }
   /**显示触发事件的信息*/
public void ShowMessage(String str)
{    Toast toast = Toast.makeText(this, str, Toast.LENGTH_SHORT);
     toast.show();         }  }
```

保存有文件并运行程序，可以对手机上方向键的按下或弹起状态进行判断并输出。当按下"确定"按钮时，结果如图 4.10 所示。

当按下或松开左方向键时，显示如图 4.11 和图 4.12 所示界面。

图 4.10 单击"确定"按钮

图 4.11 按下左方向键

图 4.12 松开左方向键

4.6 触摸事件 onTouchEvent

4.6.1 触摸事件基础

触摸事件指的是当用户接触到屏幕之后所产生的一种事件形式，当用户在屏幕上划过时，可以使用触摸事件取得用户当前的坐标，OnTouchListener 接口定义如下：

```
public interface View.OnTouchListener {
    public abstract boolean onTouch (View v, MotionEvent event) ;}
```

当触摸事件触发之后自动使用该接口中的 public boolean onTouch(View v, MotionEvent event)方法进行事件处理。

说明：需要实现 onTouch 方法；参数 v：为事件源对象；参数 event：为事件封装类的对象。其中封装了了触发事件的详细信息，同样包括事件的类型、触发时间等信息。

返回值：该方法的返回值机理与键盘响应事件相同，同样是当已经完整地处理了该事件且不希望其他回调方法再次处理时返回 true，否则返回 false。

该方法并不像之前介绍过的方法只处理一种事件，一般情况下以下 3 种情况的事件全部由 onTouchEvent 方法处理，只是 3 种情况中的动作值不同。

第 4 章 Android 中的事件处理

屏幕被按下：当屏幕被按下时，会自动调用该方法来处理事件，此时 MotionEvent.getAction()的值为 MotionEvent.ACTION_DOWN，如果在应用程序中需要处理屏幕被按下的事件，只需重新该回调方法，然后在方法中进行动作的判断即可。

屏幕被抬起：当触控笔离开屏幕时触发的事件，该事件同样需要 onTouchEvent 方法来捕捉，然后在方法中进行动作判断。当 MotionEvent.getAction()的值为 MotionEvent.ACTION_UP 时，表示屏幕被抬起的事件。

在屏幕中拖动：还负责处理触控笔在屏幕上滑动事件，调用 MotionEvent.getAction()方法来判断动作值是否为 MotionEvent.ACTION_MOVE 再进行处理。

要在 Android 程序中使用触摸事件必须在程序中使用下面的语句：

```
import android.view.View.OnTouchListener;    //导入 View.OnTouchListener 类
```

还有一个与键盘事件联系很密切的类 view.MotionEvent：

```
import android.view.MotionEvent;             //导入 view.MotionEvent 类
```

触摸事件的实现步骤如下：

1）通过组件 ID 获取组件实例；

例如：this.edit = (EditText) super.findViewById(R.id.edit1);

2）为该组件注册 OnTouchEvent Listener 监听；

例如：this.locate.setOnTouchListener(new OnTouchListenerImpl());

3）实现 onTouchEvent 方法。

例如：public boolean onTouch (View v, MotionEvent event) {……

4.6.2 触摸事件实例

实例 4-6：触摸事件实例

新建一个项目，项目的命名为：exam4_6，包名称为：org.hnist.demo。

1）修改 activity_main.xml 文件，代码如下：

```xml
<?xml version="1.0" encoding="utf-8"?>
<LinearLayout
    xmlns:android="http://schemas.android.com/apk/res/android"
    android:orientation="vertical"
    android:layout_width="fill_parent"
    android:layout_height="fill_parent">
<TextView
    android:id="@+id/msg"
    android:layout_width="fill_parent"
    android:layout_height="fill_parent" />
</LinearLayout>
```

2）修改 MainActivity.java，代码如下：

```java
package org.hnist.ontouchdemo;
import android.app.Activity;
import android.os.Bundle;
import android.view.MotionEvent;                    //导入 view.MotionEvent 类
import android.view.View;
import android.view.View.OnTouchListener;           //导入 View.OnTouchListener 类
```

```
import android.widget.TextView;
public class MainActivity extends Activity {
private TextView locate = null;
@Override
public void onCreate(Bundle savedInstanceState) {
    super.onCreate(savedInstanceState);
    super.setContentView(R.layout.activity_main);
    this.locate = (TextView) super.findViewById(R.id.txt1);    //取得组件
    this.locate.setOnTouchListener(new OnTouchListenerImpl());}    //设置事件监听
private class OnTouchListenerImpl implements OnTouchListener {
    @Override
    public boolean onTouch(View v, MotionEvent event) {    //实现onTouchEvent 方法
        MainActivity.this.locate.setText("您当前的位置是: "+"X = " + event.getX()
                    + ", Y = "+ event.getY());    //设置文本
        return false; }}}
```

保存文件并运行程序,在屏幕任意空白处单击、触摸,会显示出触摸位置的坐标,结果如图 4.13 所示。

还有一个上下文菜单事件(onCreateContextMenu),在某个 View 类中显示上下文菜单时被调用,可以处理上下文菜单显示时的一些操作,在后续章节介绍菜单组件时再进行介绍。

图 4.13 触摸屏幕后显示的坐标

4.7 选择改变事件 OnCheckedChange

4.7.1 选择改变事件基础

在 RadioGroup、RadioButton(单选按钮)、CheckBox 等组件上也可以进行事件的处理操作,当用户选中某选项后也将触发相应的监听器进行相应的处理操作。在 Android 平台的组件中提供了选择改变事件的处理操作方法,使用 View.OnCheckedChangeListener 接口可选择并改变事件的处理操作方法。View 类指的 RadioGroup 组件或 CheckBox 组件。此接口定义如下:

```
View.setOnCheckedChangeListener(new view.OnCheckedChangeListener() {
    public void onCheckedChanged(View view, int checkedId) {……}}
```

当选择改变事件触发之后,自动使用该接口中的 public void onCheckedChanged(View view, int checkedId)方法进行事件处理。

要在 Android 程序中使用选择改变事件必须在程序中使用下面的语句:

```
import android.widget.RadioGroup.OnCheckedChangeListener;
import android.widget.CompoundButton.OnCheckedChangeListener;
```

选择改变事件的实现步骤如下:

1)通过组件 ID 获取组件实例;

例如:group = (RadioGroup)findViewById(R.id.radiogroup1);

2)为该组件注册 OnCheckedChangeListener 监听;

如:group.setOnCheckedChangeListener(new RadioGroup.OnCheckedChangeListener();

3)实现 onCheckedChanged 方法。

例如:public void onCheckedChanged(RadioGroup group, int checkedId) { ……

4.7.2　RadioGroup 选择改变事件实例

实例 4-7：选择改变事件实例 1

前面介绍了一个选择题的实例，但是没有对选择的情况进行正确或错误的判断，下面介绍对选择的结果进行正确或错误的判断。

新建一个项目，项目的命名为：exam4_7，包名称为：org.hnist.demo。

1）修改 activity_main.xml 文件，代码如下：

```xml
<?xml version="1.0" encoding="utf-8"?>
<LinearLayout xmlns:android="http://schemas.android.com/apk/res/android"
    android:orientation="vertical"
    android:layout_width="fill_parent"
    android:layout_height="fill_parent" >
    <TextView
        android:id="@+id/mytextview"
        android:layout_width="fill_parent"
        android:layout_height="wrap_content"
        android:text="下面的哪个选项是正确的？ "/>
    <RadioGroup
        android:id="@+id/radiogroup1"
        android:layout_width="wrap_content"
        android:layout_height="wrap_content"
        android:orientation="vertical" >
        <RadioButton
            android:id="@+id/but1"
            android:layout_width="wrap_content"
            android:layout_height="wrap_content"
            android:text="1+2=2"  />
        <RadioButton
            android:id="@+id/but2"
            android:layout_width="wrap_content"
            android:layout_height="wrap_content"
            android:text="1+2=3"  />
        <RadioButton
            android:id="@+id/but3"
            android:layout_width="wrap_content"
            android:layout_height="wrap_content"
            android:text="1+2=4"  />
        <RadioButton
            android:id="@+id/but4"
            android:layout_width="wrap_content"
            android:layout_height="wrap_content"
            android:text="1+2=1"  />
    </RadioGroup>
</LinearLayout>
```

2）修改 MainActivity.java，代码如下：

```java
package org.hnist.demo;
import android.app.Activity;
```

```java
import android.os.Bundle;
import android.view.Gravity;
import android.widget.RadioButton;
import android.widget.RadioGroup;
import android.widget.Toast;
public class MainActivity extends Activity {
    private RadioGroup group;
    private RadioButton radio1,radio2,radio3,radio4;
    @Override
    public void onCreate(Bundle savedInstanceState) {
        super.onCreate(savedInstanceState);
        setContentView(R.layout.activity_main);
        group = (RadioGroup)findViewById(R.id.radiogroup1);  //获得RadioGroup组件
        radio1 = (RadioButton)findViewById(R.id.but1);
        radio2 = (RadioButton)findViewById(R.id.but2);
        radio3 = (RadioButton)findViewById(R.id.but3);
        radio4 = (RadioButton)findViewById(R.id.but4);
        group.setOnCheckedChangeListener(new RadioGroup. OnCheckedChangeListener()
        {    //定义监听
        //实现onCheckedChanged方法
        public void onCheckedChanged(RadioGroup group, int checkedId) {
           if (checkedId == radio2.getId())  {
               showMessage("正确答案: " + radio2.getText()+",恭喜你,答对了"); }
           else {
               showMessage("对不起,您的选择错误,再仔细思考下。"); } } });}
    public void showMessage(String str)
    {   Toast toast = Toast.makeText(this, str, Toast.LENGTH_SHORT);
        toast.setGravity(Gravity.TOP, 0, 220);
        toast.show();      }  }
```

保存文件并运行程序,分别选择正确和错误的选项,结果如图4.14和图4.15所示。

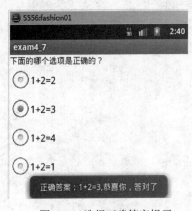

图 4.14 选择正确答案提示

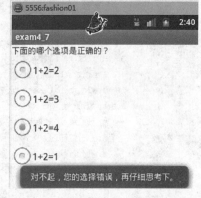

图 4.15 选择错误答案提示

4.7.3 CheckBox 选择改变事件实例

实例 4-8:选择改变事件实例 2

前面章节介绍了一个复选框的实例,但是没有对选择的结果进行判断或操作,下面就来介绍对这些选择的结果进行一些操作。

新建一个项目,项目的命名为:exam4_8,包名称为:org.hnist.demo。

1) 修改 activity_main.xml 文件，代码如下：

```xml
<RelativeLayout
    xmlns:android="http://schemas.android.com/apk/res/android"
    xmlns:tools="http://schemas.android.com/tools"
    android:layout_width="match_parent"
    android:layout_height="match_parent"
    tools:context=".MainActivity" >
 <TextView
        android:id="@+id/txt1"
        android:layout_width="wrap_content"
        android:layout_height="wrap_content"
        android:textSize="16dp"
        android:text="你喜欢体育运动有哪些：" />
 <CheckBox
        android:id="@+id/check1"
        android:layout_width="wrap_content"
        android:layout_height="wrap_content"
        android:layout_below="@+id/txt1"
        android:layout_marginTop="20dp"
        android:text="游泳" />
 <CheckBox
        android:id="@+id/check2"
        android:layout_width="wrap_content"
        android:layout_height="wrap_content"
        android:layout_toRightOf="@+id/check1"
        android:layout_alignBaseline="@+id/check1"
        android:text="登山" />
 <CheckBox
        android:id="@+id/check3"
        android:layout_width="wrap_content"
        android:layout_height="wrap_content"
        android:layout_toRightOf="@+id/check2"
        android:layout_alignBaseline="@+id/check1"
        android:text="跑步" />
 <CheckBox
        android:id="@+id/check4"
        android:layout_width="wrap_content"
        android:layout_height="wrap_content"
        android:layout_toRightOf="@+id/check3"
        android:layout_alignBaseline="@+id/check1"
        android:text="篮球" />
 <CheckBox
        android:id="@+id/check5"
        android:layout_width="wrap_content"
        android:layout_height="wrap_content"
        android:layout_below="@+id/check1"
        android:layout_marginTop="20dp"
        android:text="乒乓球" />
  <CheckBox
        android:id="@+id/check6"
```

```xml
        android:layout_width="wrap_content"
        android:layout_height="wrap_content"
        android:layout_toRightOf="@+id/check5"
        android:layout_alignBaseline="@+id/check5"
        android:text="羽毛球" />
<Button
        android:id="@+id/mybut"
        android:layout_width="wrap_content"
        android:layout_height="wrap_content"
        android:layout_alignRight="@+id/check3"
        android:layout_below="@+id/check5"
        android:layout_marginTop="23dp"
        android:text=" 提    交 " />
</RelativeLayout>
```

2）修改 MainActivity.java，代码如下：

```java
package org.hnist.demo;
import android.app.Activity;
import android.os.Bundle;
import android.view.Gravity;
import android.view.View;
import android.widget.Button;
import android.widget.CheckBox;
import android.widget.CompoundButton;
import android.widget.CompoundButton.OnCheckedChangeListener;
import android.widget.Toast;
public class MainActivity extends Activity {
    private CheckBox chk1,chk2,chk3,chk4,chk5,chk6;
    private Button mybutton;
    @Override
    public void onCreate(Bundle savedInstanceState) {
        super.onCreate(savedInstanceState);
        setContentView(R.layout.activity_main);
        mybutton = (Button)findViewById(R.id.mybut);
        chk1 = (CheckBox)findViewById(R.id.check1);
        chk2 = (CheckBox)findViewById(R.id.check2);
        chk3 = (CheckBox)findViewById(R.id.check3);
        chk4 = (CheckBox)findViewById(R.id.check4);
        chk5 = (CheckBox)findViewById(R.id.check5);
        chk6 = (CheckBox)findViewById(R.id.check6);
        chk1.setOnCheckedChangeListener(new CheckBox.OnCheckedChangeListener() {
            public void onCheckedChanged(CompoundButton arg0, boolean arg1) {
                if(chk1.isChecked())
                { showMessage("你刚才选择了"+chk1.getText());
                } } });
        chk2.setOnCheckedChangeListener(new CheckBox.OnCheckedChangeListener() {
            public void onCheckedChanged(CompoundButton arg0, boolean arg1) {
                if(chk3.isChecked())
                { showMessage("你刚才选择了"+chk2.getText()); } } });
        chk3.setOnCheckedChangeListener(new CheckBox.OnCheckedChangeListener() {
            public void onCheckedChanged(CompoundButton arg0, boolean arg1) {
                if(chk3.isChecked())
                { showMessage("你刚才选择了"+chk3.getText()); } } });
        chk4.setOnCheckedChangeListener(new CheckBox.OnCheckedChangeListener() {
```

```
            public void onCheckedChanged(CompoundButton arg0, boolean arg1) {
                if(chk4.isChecked())
                { showMessage("你刚才选择了"+chk4.getText()); } } });
        chk5.setOnCheckedChangeListener(new CheckBox.OnCheckedChangeListener() {
            public void onCheckedChanged(CompoundButton arg0, boolean arg1) {
                if(chk5.isChecked())
{showMessage("你刚才选择了"+chk5.getText());}}});
        chk6.setOnCheckedChangeListener(new CheckBox.OnCheckedChangeListener() {
            public void onCheckedChanged(CompoundButton arg0, boolean arg1) {
                if(chk6.isChecked())
                { showMessage("你刚才选择了"+chk6.getText());}}});
                mybutton.setOnClickListener(new Button.OnClickListener(){
            public void onClick(View arg0) {
                int num = 0;
                String str="";
                if(chk1.isChecked())
                { str=chk1.getText()+","+str;
                   num++; }
                if(chk2.isChecked())
                { str=chk2.getText()+","+str;
                   num++; }
                if(chk3.isChecked())
                { str=chk3.getText()+","+str;
                   num++; }
                if(chk4.isChecked())
                { str=chk4.getText()+","+str;
                   num++; }
                if(chk5.isChecked())
                { str=chk5.getText()+","+str;
                   num++; }
                if(chk6.isChecked())
                { str=chk6.getText()+","+str;
                   num++; }
                showMessage("您选择了"+str+"共"+num+"项");     }}); }
      void showMessage(String str)
      { Toast toast = Toast.makeText(this, str, Toast.LENGTH_SHORT);
        toast.setGravity(Gravity.TOP, 0, 220);
        toast.show();    } }
```

保存文件并运行程序，选择相应的选项会有提示，最后单击"提交"按钮，会弹出所有选择的情况提示，如图 4.16 和图 4.17 所示。

图 4.16　选择某个选项后提示

图 4.17　选择提交后提示

4.8 选项选中事件 OnItemSelected

4.8.1 选项选中事件基础

Spinner 组件的主要功能是用于进行下拉列表显示的功能，当用户选中下拉列表中的某个选项后，可以使用 Spinner 类中提供的下面 3 个接口进行相应的处理操作：

1）当列表项被选中或者被单击时触发的事件：

```
setOnItemClickListener(AdapterView.OnItemClickListener listener);
```

2）当列表项改变时所触发的事件：

```
setOnItemSelectedListener(AdapterView.OnItemSelectedListener listener)
```

3）当列表项被长时间按住时所触发的事件：

```
setOnItemLongClickListener(AdapterView.OnItemLongClickListener listener)
```

当事件触发后，自动使用该接口中的 public void onItemSelected()方法进行事件处理。

选项选中事件的实现步骤如下：

1）通过组件 ID 获取组件实例；例如：

```
spin=(Spinner)findViewById(R.id.spin);
```

2）为该组件注册监听；例如：

```
spin.setOnItemSelectedListener(new Spinner.OnItemSelectedListener(){…
```

3）实现 onItemSelected 方法。例如：

```
public void onItemSelected(AdapterView<?> arg0, View arg1, int arg2, long arg3) {…
```

4.8.2 OnItemSelected 选项选中事件实例

实例 4-9：选项选中事件实例 1

前面介绍 Spinner 组件的基本属性，没有对选择的结果进行判断或操作，下面介绍对这些选择的结果进行显示。

新建一个项目，项目的命名为：exam4_9，包名称为：org.hnist.demo。

1）修改 activity_main.xml 文件，代码如下：

```xml
<?xml version="1.0" encoding="utf-8"?>
<LinearLayout
    xmlns:android="http://schemas.android.com/apk/res/android"
    android:orientation="vertical"
    android:layout_width="fill_parent"
    android:layout_height="fill_parent"  >
<TextView
    android:id="@+id/txt"
    android:layout_width="wrap_content"
    android:layout_height="wrap_content"
    android:text="您心中的理想专业是：" />
<Spinner
    android:id="@+id/spin"
    android:layout_width="wrap_content"
```

```
       android:layout_height="wrap_content"
       android:layout_centerHorizontal="true" />
</LinearLayout>
```

2）修改 MainActivity.java，代码如下：

```java
package org.hnist.demo;
import android.app.Activity;
import android.os.Bundle;
import android.view.View;
import android.widget.AdapterView;
import android.widget.ArrayAdapter;
import android.widget.Spinner;
import android.widget.TextView;
public class MainActivity extends Activity {
    private static final String[] subjects={"计算机","数学","文学","英语","哲学"};
    private TextView txt;
    private Spinner spin;
    private ArrayAdapter<String> adapter;
    @Override
    public void onCreate(Bundle savedInstanceState) {
        super.onCreate(savedInstanceState);
        setContentView(R.layout.activity_main);
        txt=(TextView)findViewById(R.id.txt);
        spin=(Spinner)findViewById(R.id.spin);
        this.spin.setPrompt("请选择您的理想专业：");
        //将可选内容与ArrayAdapter连接
        adapter=new ArrayAdapter<String>(this,android.R.layout.simple_spinner_item,subjects);
        //设置下拉列表风格
        adapter.setDropDownViewResource(android.R.layout.simple_spinner_dropdown_item);
        spin.setAdapter(adapter);              //将adapter添加到spinner中
        //添加Spinner事件监听
        spin.setOnItemSelectedListener(new Spinner.OnItemSelectedListener()
        {public void onItemSelected(AdapterView<?> arg0, View arg1, int arg2, long arg3)
            {   txt.setText("您心中的理想专业是："+subjects[arg2]);
                arg0.setVisibility(View.VISIBLE); }   //设置显示当前选择的项
         public void onNothingSelected(AdapterView<?> arg0) { }     }); } }
```

保存文件并运行程序，弹出如图 4.18 所示的界面。单击下拉列表，弹出如图 4.19 所示的界面，选择相应的选项后会在屏幕上显示出来。

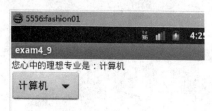

图 4.18　选择某个选项后的提示

图 4.19　选择下拉列表后的提示

实例 4-10：选项选中事件实例 2

Spinner 组件里面的选项一般是在程序中通过数组设置，或者是在布局管理器中设置完成后再在程序中调用，下面介绍一种动态添加或删除 Spinner 组件里面的选项的方法。

新建一个项目，项目的命名为：exam4_10，包名称为：org.hnist.demo。

1）修改 activity_main.xml 文件，代码如下：

```xml
<?xml version="1.0" encoding="utf-8"?>
<LinearLayout xmlns:android="http://schemas.android.com/apk/res/android"
  android:orientation="vertical"
  android:layout_width="fill_parent"
  android:layout_height="fill_parent" >
  <TextView
  android:id="@+id/myTextView"
  android:layout_width="fill_parent"
  android:layout_height="wrap_content" />
  <EditText
  android:id="@+id/myEditText"
  android:layout_width="fill_parent"
  android:layout_height="wrap_content" />
  <Button
  android:id="@+id/myButton_add"
  android:layout_width="fill_parent"
  android:layout_height="wrap_content"
  android:text="增加" />
  <Button
  android:id="@+id/myButton_remove"
  android:layout_width="fill_parent"
  android:layout_height="wrap_content"
  android:text="删除" />
  <Spinner
  android:id="@+id/mySpinner"
  android:layout_width="fill_parent"
  android:layout_height="wrap_content" />
</LinearLayout>
```

2）修改 MainActivity.java，代码如下：

```java
package org.hnist.demo;
import android.app.Activity;
import android.os.Bundle;
import android.view.View;
import android.widget.AdapterView;
import android.widget.ArrayAdapter;
import android.widget.Button;
import android.widget.EditText;
import android.widget.Spinner;
import android.widget.TextView;
import java.util.ArrayList;
import java.util.List;
public class MainActivity extends Activity {
```

```java
    private static final String[] countriesStr = { "北京市", "上海市"};
    private TextView myTextView;
    private EditText myEditText;
    private Button myButton_add;
    private Button myButton_remove;
    private Spinner mySpinner;
    private ArrayAdapter adapter;
    private List allCountries;
    public void onCreate(Bundle savedInstanceState) {
      super.onCreate(savedInstanceState);
      setContentView(R.layout.activity_main);
      allCountries = new ArrayList();
      for (int i = 0; i < countriesStr.length; i++){
        allCountries.add(countriesStr[i]); }
      adapter = new ArrayAdapter(this, android.R.layout.simple_spinner_item,
              allCountries);
adapter .setDropDownViewResource(android.R.layout.simple_spinner_dropdown_item);
      myTextView = (TextView) findViewById(R.id.myTextView);
      myEditText = (EditText) findViewById(R.id.myEditText);
      myButton_add = (Button) findViewById(R.id.myButton_add);
      myButton_remove = (Button) findViewById(R.id.myButton_remove);
      mySpinner = (Spinner) findViewById(R.id.mySpinner);
      mySpinner.setAdapter(adapter);       //将ArrayAdapter加入Spinner对象中
      //将myButton_add加入OnClickListener
      myButton_add.setOnClickListener(new Button.OnClickListener(){
        public void onClick(View arg0) {
          String newCountry = myEditText.getText().toString();
          //先比对新增的值是否已存在，不存在才可新增
          for (int i = 0; i < adapter.getCount(); i++) {
            if (newCountry.equals(adapter.getItem(i))) {
              return; } }
          if (!newCountry.equals("")){
            adapter.add(newCountry);              //将值新增至adapter
            int position = adapter.getPosition(newCountry);  //取得新增的值的位置
            mySpinner.setSelection(position);    //将Spinner选取在新增的值的位置
            myEditText.setText("");} } });       //将myEditText清空
      //将myButton_remove加入OnClickListener
      myButton_remove.setOnClickListener(new Button.OnClickListener()
      { public void onClick(View arg0) {
         if (mySpinner.getSelectedItem() != null) {
           //移除mySpinner的值
           adapter.remove(mySpinner.getSelectedItem().toString());
           myEditText.setText("");             //将myEditText清空
           if (adapter.getCount() == 0) {
             myTextView.setText("");} } }});    //将myTextView清空
      //将mySpinner加入OnItemSelectedListener
      mySpinner.setOnItemSelectedListener(new Spinner.OnItemSelectedListener()
```

```
{ public void onItemSelected(AdapterView arg0, View arg1, int arg2, long arg3)
  { //将所选mySpinner的值带入myTextView中
    myTextView.setText("您选择的城市是："+arg0.getSelectedItem().toString());}
    public void onNothingSelected(AdapterView arg0) {   } }); } }
```

保存文件并运行程序，弹出如图 4.20 所示界面。在编辑框内输入文字"长沙"，然后单击"增加"按钮，弹出如图 4.21 所示界面，发现已经将"长沙"添加到 Spinner 组件，类似地也可以删除，请用户自己测试。

图 4.20　运行开始输入界面

图 4.21　单击"增加"后的界面

4.9　日期和时间监听事件

4.9.1　日期和时间选择器组件

日期选择器组件 DatePicker 是一个选择年月日的日历布局视图，使用它可以对年、月、日进行设置，其层次关系如下：

```
java.lang.Object
    android.view.View
        android.view.ViewGroup
            android.widget.FrameLayout
                android.widget.DatePicker
```

要在 Android 程序中使用 DatePicker 组件必须要在程序中使用下面的语句：

```
import android.widget.DatePicker;            //导入widget.DatePicker类
```

时间选择器组件 TimePicker 是用于选择一天中时间的视图，使用它可以进行时间的调整，其层次关系如下：

```
java.lang.Object
    android.view.View
        android.view.ViewGroup
            android.widget.FrameLayout
                android.widget.TimePicker
```

要在 Android 程序中使用 TimePicker 组件，必须要在程序中使用下面的语句：

```
import android.widget.TimePicker;            //导入widget.TimePicker类
```

和前面介绍的组件一样，DatePicker 和 TimePicker 组件也有其属性和方法，常用的方法如表 4-1 所示。

第 4 章 Android 中的事件处理

表 4-1 DatePicker 和 TimePicker 组件常用的方法

方法	描述
public Integer getCurrentHour()	返回当前设置的小时
public Integer getCurrentMinute()	返回当前设置的分钟
public boolean is24HourView()	判断是否是 24 小时制
public void setCurrentHour(Integer currentHour)	设置当前的小时数
public void setCurrentMinute(Integer currentMinute)	设置当前的分钟
public void setEnabled(boolean enabled)	设置是否可用
public void setIs24HourView(Boolean is24HourView)	设置时间为 24 小时制
public int getYear()	取得设置的年
public void setOnTimeChangedListener (TimePicker.OnTimeChangedListener onTimeChangedListener)	设置时间调整事件的回调函数
public int getYear ()	取得设置的年份
public int getMonth()	取得设置的月
public int getDayOfMonth()	取得设置的日
public void setEnabled(boolean enabled)	设置组件是否可用
public void updateDate(int year, int monthOfYear, int dayOfMonth)	设置一个指定的日期
public void init (int year, int monthOfYear, int dayOfMonth,DatePicker. OnDateChangedListener onDateChangedListener)	初始化状态（初始化年月日）

4.9.2 DatePicker 和 TimePicker 组件使用实例

实例 4-11：DatePicker 和 TimePicker 组件使用实例

新建一个项目，项目的命名为：exam4_11，包名称为：org.hnist.demo，利用 DatePicker 和 TimePicker 组件显示系统的日期和时间。

1）修改 activity_main.xml 文件，代码如下：

```
<RelativeLayout xmlns:android="http://schemas.android.com/apk/res/android"
    xmlns:tools="http://schemas.android.com/tools"
    android:layout_width="match_parent"
    android:layout_height="match_parent"
    tools:context=".MainActivity" >
    <DatePicker                                    //增加 DatePicker 组件
        android:id="@+id/datepick"
        android:layout_width="wrap_content"
        android:layout_height="wrap_content"
        android:layout_below="@+id/txt1" />
    <TimePicker                                    //增加 TimePicker 组件
        android:id="@+id/timepick"
        android:layout_width="wrap_content"
        android:layout_height="wrap_content"
        android:layout_alignParentLeft="true"
        android:layout_below="@+id/datepick" />
    <TextView
        android:id="@+id/time"
        android:layout_width="wrap_content"
        android:layout_height="wrap_content"
        android:layout_alignParentLeft="true"
        android:layout_below="@+id/timepick"
```

```
            android:layout_marginTop="25dp"
            android:text="系统当前时间为："
            android:textSize="20dp" />
</RelativeLayout>
```

2) 修改 MainActivity.java，代码如下：

```
package org.hnist.demo;
import android.os.Bundle;
import android.app.Activity;
import android.view.Menu;
import android.widget.DatePicker;
import android.widget.EditText;
import android.widget.TextView;
import android.widget.TimePicker;
public class MainActivity extends Activity {
private TextView time = null;
private DatePicker datepick = null;
private TimePicker timepick = null;
    @Override
    protected void onCreate(Bundle savedInstanceState) {
        super.onCreate(savedInstanceState);
        setContentView(R.layout.activity_main);
        this.time = (TextView) super.findViewById(R.id.time);   //取得组件
        this.datepick = (DatePicker) super.findViewById(R.id.datepick);//取得组件
    this.timepick = (TimePicker) super.findViewById(R.id.timepick);
    this.setDateTime();}                                           //设置日期
    public void setDateTime()  {
//设置文本内容
    this.time.setText(this.datepick.getYear() + "-"
            +(this.datepick.getMonth()+1)+"-"+this.datepick.getDayOfMonth()
            + " " + this.timepick.getCurrentHour() + ":"
            + this.timepick.getCurrentMinute());}
    @Override
    public boolean onCreateOptionsMenu(Menu menu) {
        //Inflate the menu; this adds items to the action bar if it is present.
        getMenuInflater().inflate(R.menu.main, menu);
        return true;     }      }
```

保存文件并运行程序，弹出如图 4.22 所示界面。

图 4.22　exam4_11 运行结果

4.9.3 日期和时间的设置

DatePicker 和 TimePicker 可以显示当前的系统日期和时间，要进行日期和时间的设置，Android 平台提供了日期对话框 DatePickerDialog 及时间对话框 TimePickerDialog，其层次关系如下：

```
java.lang.Object
    android.app.Dialog
        android.app.AlertDialog
            android.app.DatePickerDialog、android.app.TimePickerDialog
```

类似地要在 Android 程序中使用 DatePickerDialo 和 TimePickerDialog 组件必须要在程序中使用下面的语句：

```
import android.app.DatePickerDialog; //导入widget.DatePickerDialog类
import android.app.TimePickerDialog; //导入widget.TimePickerDialog类
```

在 DatePickerDialog 组件内有一个 OnDateSetListener，可以指定监听操作，在 TimePickerDialog 组件内有一个 OnTimeSetListener，可以指定监听操作，具体方法见表 4-2。

表 4-2 DatePickerDialog、TimePickerDialog 的常用方法

方法	描述
public DatePickerDialog (Context context, DatePickerDialog.OnDateSetListener callBack, int year, int monthOfYear, int dayOfMonth)	创建 DatePickerDialog 对象，同时指定监听操作、要设置的年、月、日等信息
public void updateDate (int year, int monthOfYear, int dayOfMonth)	更新显示组件上的年、月、日信息
public TimePickerDialog (Context context, TimePickerDialog.OnTimeSetListener callBack, int hourOfDay, int minute, boolean is24HourView)	创建时间对话框，同时设置时间改变的事件操作、小时、分以及是否是 24 小时制
public void updateTime (int hourOfDay, int minutOfHour)	更新时、分

实例 4-12：DatePickerDialog 和 TimePickerDialog 组件使用实例

新建一个项目，项目的命名为：exam4_12，包名称为：org.hnist.demo，利用 DatePickerDialog 和 TimePickerDialog 组件设置系统的日期和时间。

1）修改 activity_main.xml 文件，代码如下：

```xml
<?xml version="1.0" encoding="utf-8"?>
<LinearLayout
xmlns:android="http://schemas.android.com/apk/res/android"
android:id="@+id/MyLayout"
android:orientation="vertical"
android:layout_width="fill_parent"
android:layout_height="fill_parent">
<TextView
    android:id="@+id/txt"
    android:layout_width="wrap_content"
    android:layout_height="wrap_content"/>
<Button
    android:id="@+id/datebut"
    android:text="设置日期"
    android:layout_width="wrap_content"
    android:layout_height="wrap_content"/>
<Button
```

```xml
        android:id="@+id/timebut"
        android:text="设置时间"
        android:layout_width="wrap_content"
        android:layout_height="wrap_content"/>
</LinearLayout>
```

2）修改 MainActivity.java，代码如下：

```java
package org.hnist.demo;
import android.app.Activity;
import android.app.Dialog;
import android.app.DatePickerDialog;
import android.app.TimePickerDialog;
import android.os.Bundle;
import android.view.View;
import android.view.View.OnClickListener;
import android.widget.Button;
import android.widget.DatePicker;
import android.widget.TimePicker;
import android.widget.TextView;
public class MainActivity extends Activity {
    private Button datebut = null ;            //定义按钮组件
    private Button timebut = null ;            //定义按钮组件
    @Override
    public void onCreate(Bundle savedInstanceState) {
        super.onCreate(savedInstanceState);
        super.setContentView(R.layout.activity_main);
        this.datebut = (Button) super.findViewById(R.id.datebut); //取得按钮组件
        this.timebut = (Button) super.findViewById(R.id.timebut) ;   //取得按钮组件
        this.datebut.setOnClickListener(new OnClickListenerDate()) ;//设置日期单击事件
        this.timebut.setOnClickListener(new OnClickListenerTime()) ; }//设置日
                                                                      期单击事件
    private class OnClickListenerDate implements OnClickListener {   //注册单击事件监听
        @Override
        public void onClick(View v) {              //实现onClick()方法
            Dialog dialog = new DatePickerDialog(MainActivity.this,new
            DatePickerDialog.OnDateSetListener() {      //日期事件监听
                public void onDateSet(DatePicker view, int year, int monthOfYear,
                            int dayOfMonth) {       //日期改变时触发
                    TextView text = (TextView)findViewById(R.id.txt);
                    text.setText("更新的日期为: " + (year+1) + "-" + monthOfYear + "-"+
                            dayOfMonth);            //设置文本内容
                }}, 2013, 8, 15);    //默认为2013年8月15日
            dialog.show() ; }}            //显示对话框
    private class OnClickListenerTime implements OnClickListener {
        @Override
        public void onClick(View v) {
            Dialog dialog1= new TimePickerDialog(MainActivity.this,new
                    TimePickerDialog.OnTimeSetListener() {//时间事件监听
    public void onTimeSet(TimePicker view, int hourOfDay, int minute) {
    //时间改变时触发
            TextView text = (TextView)findViewById(R.id.txt);
```

```
            text.setText("更新的时间为: " +hourOfDay + ": " + minute);//设置
                                                                    文本内容
         }}, 9, 30,true);//默认为 9:30
         dialog1.show() ; }}}//显示对话框
```

保存文件并运行程序,弹出如图 4.23 所示界面,单击"设置日期"按钮,弹出如图 4.24 所示界面,设置完毕,单击"Set"按钮弹出如图 4.25 所示界面。

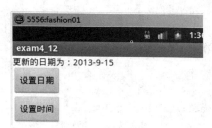

图 4.23 设置前的界面　　　图 4.25 设置完毕的界面　　　图 4.24 设置日期的界面

类似地可以设置时间,请用户自行实验。

4.9.4 日期和时间监听事件

日期和时间选择器可以对日期和事件进行调整,当日期和时间发生变化时也可以进行事件的触发,Android 平台提供日期、时间监听器接口来实现日期和时间监听事件。

日期监听器接口定义如下:

```
    View.setOnDateChangedListener(new OnDateChangedListenerImpl());
```

事件触发之后,自动使用该接口中的 public void onDateChanged()方法进行事件处理。

时间监听器接口定义如下:

```
    View.setOnTimeChangedListener(new OnTimeChangedListenerImpl());
```

事件触发后,自动使用该接口中的 public void onTimeChanged()方法进行事件处理。

要在 Android 的 java 程序中使用日期和时间监听事件必须在程序中使用下面的语句。

```
    import android.widget.DatePicker.OnDateChangedListener;   //导入日期监听所需的类
    import android.widget.TimePicker.OnTimeChangedListener;   //导入时间监听所需的类
```

日期和时间监听事件的实现步骤,以时间监听事件为例进行介绍:

1) 通过组件 ID 获取组件实例;

例如:time = (DatePicker) super.findViewById(R.id.time);

2) 为该组件注册监听;

例如:time.setOnTimeChangedListener(new OnTimeChangedListenerImpl());...

3) 实现 onTimeChanged 方法。

例如:public void onTimeChanged(TimePicker view, int hourOfDay, int minute) {...

实例 4-13：日期和时间监听事件实例

新建一个项目，项目的命名为：exam4_13，包名称为：org.hnist.demo，利用 DatePickerDialog 和 TimePickerDialog 组件设置系统的日期和时间。

1）修改 activity_main.xml 文件，代码如下：

```xml
<LinearLayout
    xmlns:android="http://schemas.android.com/apk/res/android"
    android:orientation="vertical"
    android:layout_width="fill_parent"
    android:layout_height="fill_parent">
<EditText
    android:id="@+id/input"
    android:layout_width="fill_parent"
    android:layout_height="wrap_content"/>
<LinearLayout
android:layout_below="@+id/input"
android:orientation=" vertical "
    android:layout_width="fill_parent"
    android:layout_height="fill_parent">
    <DatePicker                                    //增加一个日期选择器组件
        android:id="@+id/date"                     //定义组件 ID，程序中使用
        android:layout_width="wrap_content"
        android:layout_height="wrap_content" />
    <TimePicker                                    //增加一个时间选择器组件
        android:id="@+id/time"                     //定义组件 ID，程序中使用
        android:layout_width="wrap_content"
        android:layout_height="wrap_content" />
</LinearLayout>
</LinearLayout>
```

2）修改 MainActivity.java，代码如下：

```java
package org.hnist.demo;
import android.app.Activity;
import android.os.Bundle;
import android.widget.DatePicker;
import android.widget.DatePicker.OnDateChangedListener;
import android.widget.EditText;
import android.widget.TimePicker;
import android.widget.TimePicker.OnTimeChangedListener;
public class MainActivity extends Activity {
private EditText input = null;
private DatePicker date = null;
private TimePicker time = null;
@Override
public void onCreate(Bundle savedInstanceState) {
    super.onCreate(savedInstanceState);
    super.setContentView(R.layout.activity_main);
    this.input = (EditText) super.findViewById(R.id.input);
    this.date = (DatePicker) super.findViewById(R.id.date);
    this.time = (TimePicker) super.findViewById(R.id.time);
    this.time.setIs24HourView(true);                        //设置为 24 小时制
//设置时间改变监听
    this.time.setOnTimeChangedListener(new OnTimeChangedListenerImpl());
    this.date.init(this.date.getYear(), this.date.getMonth(),
```

```
            this.date.getDayOfMonth(),
            new OnDateChangedListenerImpl());           //设置日期改变监听
        this.setDateTime(); }                           //设置文本日期
    public void setDateTime() {                         //定义一个方法,设置文本内容
        this.input.setText(this.date.getYear() +"-"+ (this.date.getMonth() + 1)
            + "-" + this.date.getDayOfMonth()+ " " + this.time.getCurrentHour()
            + ":"+ this.time.getCurrentMinute());}
    private class OnDateChangedListenerImpl implements OnDateChangedListener {
        public void onDateChanged(DatePicker view, int year, int monthOfYear,int
            dayOfMonth) {
            MainActivity.this.setDateTime(); }}        //日期改变时调用方法修改文本
    private class OnTimeChangedListenerImpl implements OnTimeChangedListener {
        public void onTimeChanged(TimePicker view, int hourOfDay, int minute) {
            MainActivity.this.setDateTime(); }}}       //时间改变时调用方法修改文本
```

保存文件并运行程序,弹出如图 4.26 所示界面,修改日期和时间上方的提示会随着改变。

4.10 菜单事件

菜单是应用程序中非常重要的组成部分,能够在不占用界面空间的前提下,为应用程序提供统一的功能和设置界面,并为程序开发人员提供了易于使用的编程接口。在 Android 手机上有一个 "Menu" 键,当选择后会在屏幕的底部显示系统的菜单。

图 4.26 日期和时间监听

4.10.1 菜单事件基础

Android 系统支持三种菜单:选项菜单(OptionsMenu)、上下文菜单(ContextMenu)和子菜单(SubMenu)。

尽管每一种类型菜单的创建方法不同,但是 Android 系统提供了丰富的菜单操作方法,如表 4-3～表 4-5 所示,因此创建菜单在 Android 系统中并不复杂。

表 4-3 常用的菜单操作方法

方法	描述
public void closeContextMenu()	关闭上下文菜单
public void closeOptionsMenu()	关闭选项菜单
public boolean onContextItemSelected(MenuItem item)	设置上下文菜单项
public void onContextMenuClosed(Menu menu)	上下文菜单关闭时触发
public void onCreateContextMenu(ContextMenu menu, View v, ContextMenu.ContextMenuInfo menuInfo)	创建上下文菜单
public boolean onCreateOptionsMenu(Menu menu)	当用户选择 "Menu" 按钮时调用此操作,可以生成一个选项菜单
public boolean onMenuItemSelected(int featureId, MenuItem item)	设置选项菜单项
public boolean onOptionsItemSelected(MenuItem item)	当一个选项菜单中的某个菜单项被选中时触发此操作
public void onOptionsMenuClosed(Menu menu)	当选项菜单关闭时触发此操作
public boolean onPrepareOptionsMenu(Menu menu)	当选项菜单显示之前操作触发此操作
public void openOptionsMenu()	打开选项菜单
public MenuInflater getMenuInflater()	取得 MenuInflater 类的对象
public void registerForContextMenu(View view)	注册上下文菜单

表 4-4 Menu 接口的常用方法及常量

方法及常量	描述
public static final int FIRST	常量，用于定义菜单项的编号
public static final int NONE	常量，表示菜单不分组
public abstract MenuItem add(int groupId, int itemId, int order, CharSequence title)	此方法用于向菜单之中添加菜单项
public abstract MenuItem add(int groupId, int itemId, int order, int titleRes)	增加菜单项
public abstract SubMenu addSubMenu(int groupId, int itemId, int order, int titleRes)	增加子菜单
public abstract SubMenu addSubMenu(int groupId, int itemId, int order, CharSequence title)	增加子菜单
public abstract void removeGroup(int groupId)	删除一个菜单组
public abstract void removeItem(int id)	删除一个菜单项
public abstract void clear()	清空菜单
public abstract void close()	关闭菜单
public abstract MenuItem getItem(int index)	返回指定的菜单项
public abstract int size()	返回菜单项的个数

表 4-5 MenuItem 接口的常用方法

方法	描述
public abstract int getGroupId()	得到菜单组编号
public abstract Drawable getIcon()	得到菜单项上的图标
public abstract int getItemId()	得到菜单项上的 ID
public abstract int getOrder()	得到菜单项上的编号
public abstract SubMenu getSubMenu()	取得子菜单
public abstract CharSequence getTitle()	得到菜单项上的标题
public abstract boolean isCheckable()	判断菜单项是否可用
public abstract boolean isChecked()	判断此菜单项是否被选中
public abstract boolean isEnabled()	判断此菜单项是否可用
public abstract boolean isVisible()	判断此菜单项是否可见
public abstract MenuItem setCheckable(boolean checkable)	设置此菜单项是否可用
public abstract MenuItem setChecked(boolean checked)	设置此菜单项是否默认选中
public abstract MenuItem setEnabled(boolean enabled)	设置此菜单项是否可用
public abstract MenuItem setIcon(Drawable icon)	设置此菜单项的图标
public abstract MenuItem setIcon(int iconRes)	设置此菜单项的图标
public abstract MenuItem setOnMenuItemClickListener(MenuItem.OnMenuItemClickListener menuItemClickListener)	设置此菜单项的监听操作
public abstract MenuItem setTitle(CharSequence title)	设置此菜单项的标题
public abstract MenuItem setVisible(boolean visible)	设置此菜单项是否可见
public abstract ContextMenu.ContextMenuInfo getMenuInfo()	得到菜单中的内容

4.10.2 选项菜单 OptionsMenu

选项菜单是一种经常被使用的 Android 系统菜单，也是用户在使用手机时最常见的一种形式。选项菜单通过 "菜单键"（MENU）打开。选项菜单分为：图标菜单（Icon Menu）和扩展菜单（Expanded Menu）。

图标菜单能够同时显示文字和图标的菜单，不支持单选框和复选框，在一个菜单之中最多只会显示 6 个菜单项（MenuItem），如果菜单项超出了 6 个，则超出部分会自动隐藏，而且会自动出现一个 "More" 菜单项提示用户，单击 "More" 按钮弹出第六项及以后的菜单项，这些菜单项被称作扩展菜单，扩展菜单是垂直的列表型菜单，不显示图标，支持单选按钮和复选框。

要在 Android 程序中使用选项菜单必须要在程序中使用下面的语句:

```
import android.view.Menu;                //导入 view.Menu 类
import android.view.MenuItem;            //导入 view.MenuItem 类
```

在 OptionsMenu 中常用到如下几个操作方法:

1) public boolean onCreateOptionsMenu(Menu menu): 设置多个菜单项 (MenuItem);

如果要实现选项菜单,需要重载 Activity 程序的 onCreateOptionMenu()函数,才能够实现,返回 true 则显示菜单,返回 false 则不显示菜单。初次使用选项菜单时,会调用 onCreateOptionMenu()函数,用来初始化菜单子项的相关内容,设置菜单子项自身的子项的 ID 和组 ID,菜单子项显示的文字和图片等,例如:

```
public boolean onCreateOptionsMenu(Menu menu) {             //显示菜单
    menu.add(Menu.NONE,                                     //菜单不分组
        Menu.FIRST + 1,                                     //菜单项 ID
        5,                                                  //菜单编号
        "邮件")                                              //显示显示标题
        .setIcon(android.R.drawable. sym_action_email);     //设置显示图标
    ……
        return true;}  //函数的返回值,值为 true 显示设置的菜单,否则不能够显示
```

其中 Menu.FIRST 是常量,值为 1。

sym_action_email 是Android 系统自带图标库 android.R.drawable 中的图标,在这个库中含有大量的常用图标,用户可以查找相关资料进行了解。

使用 setIcon()函数设置显示的图标,也可以使用 setShortcut()函数添加菜单子项的快捷键。例如:

```
menu.add(Menu.NONE, Menu.FIRST + 2, 3, "保存")            //添加选项"保存"
    .setShortcut('3','s');                                //"添加快捷键's'和'3'
```

setShortcut 中用两个参数来设定两个快捷键是为了应对不同的手机键盘。

第一个参数: 数字快捷键为 12 键键盘 (0~9,*,#,共 12 个按键),第二个参数: 全键盘。任何键不区分大小写。

2) public boolean onOptionsItemSelected(MenuItem item): 判断菜单项的操作;

该函数能够处理菜单选择事件,每次单击菜单子项时都会被调用,下面的代码说明了如何通过菜单子项的子项 ID 执行不同的操作。

```
public boolean onOptionsItemSelected(MenuItem item) {      //选中某个菜单项
    switch (item.getItemId()) {                            //判断菜单项 ID
    case Menu.FIRST + 1:                                   //ID 为 FIRST + 1 时
        MenuemailCounter=MenuemailCounter+1;               //选中该项就增加 1
        Toast.makeText(this, "您选择的是"邮件菜单"项。", Toast.LENGTH
                    LONG).show();                          //显示选中的选项名
        break;                                             //退出
    case Menu.FIRST + 2:
    …… }
    return false; }                                        //返回 false
```

onOptionsItemSelected ()的返回值表示是否对菜单的选择事件进行处理,如果已经处理过则返回 true,否则返回 false。

3) public boolean onPrepareOptionsMenu(Menu menu): 在菜单显示前触发此操作;

重载 onPrepareOptionsMenu()函数,能够动态的添加、删除菜单子项,或修改菜单的标题、图标和可见性等内容,函数返回值为 true 则继续调用 onCreateOptionsMenu()方法,反之则不再调用。

下面的代码是在用户每次打开选项菜单时,在菜单子项中显示用户打开该子项的次数:

```
static int MenuemailCounter = 0;                        //统计选项选中的计数器
public boolean onPrepareOptionsMenu(Menu menu) {        //菜单显示前调用
    MenuItem emailItem = menu.findItem(Menu.FIRST + 1); //获得Menu.FIRST + 1项
    emailItem.setTitle("邮件选项:" +String.valueOf(MenuemailCounter)); }
        //设置邮件标题为"邮件选项"与MenuemailCounter的组合
```

4)public void onOptionsMenuClosed(Menu menu):当菜单关闭时触发此操作;例如:

```
public void onOptionsMenuClosed(Menu menu) {            //菜单退出时调用
    Toast.makeText(this, "注意,选项现在菜单关闭!!", Toast.LENGTH_LONG).show();}
```

5)如果希望从配置文件之中取出数据,需要使用到下面两个常用方法:

```
public MenuInflater(Context context)              //创建MenuInflater类对象
public void inflate(int menuRes, Menu menu)       //将配置的资源填充到菜单之中
```

inflater 在 Android 系统中建立了从资源文件到对象的桥梁,MenuInflater 即把菜单 xml 资源转换为对象并添加到 menu 对象中,它可以通过 activity 的 getMenuInflater()得到。在 MainActivity 中重写 onCreateOptionsMenu(...)方法。

```
public boolean onCreateOptionsMenu(Menu menu) {
    MenuInflater inflater = getMenuInflater();
    inflater.inflate(R.menu.mainmenu, menu);
    return true;}
```

实例 4-14:选项菜单实例

新建一个项目,项目的命名为:exam4_14,包名称为:org.hnist.demo,新建一个选项菜单 OptionsMenu。

1)修改 activity_main.xml 文件,代码如下:

```xml
<?xml version="1.0" encoding="utf-8"?>
<LinearLayout
xmlns:android="http://schemas.android.com/apk/res/android"
android:id="@+id/MyLayout"
android:orientation="vertical"
android:layout_width="fill_parent"
android:layout_height="fill_parent">
<TextView
    android:id="@+id/mytxt"
    android:layout_width="wrap_content"
    android:layout_height="wrap_content"
    android:text="按下手机上的Menu键"/>
</LinearLayout>
```

2)修改 MainActivity.java,代码如下:

```java
package org.hnist.demo;
import android.app.Activity;
import android.os.Bundle;
import android.view.Menu;
import android.view.MenuItem;
import android.widget.TextView;
import android.widget.Toast;
```

```java
public class MainActivity extends Activity {
public void onCreate(Bundle savedInstanceState) {
    super.onCreate(savedInstanceState);
    super.setContentView(R.layout.activity_main);      }
public boolean onCreateOptionsMenu(Menu menu) {              //显示菜单
    menu.add(Menu.NONE,                                      //菜单不分组
            Menu.FIRST + 1,                                  //菜单项ID
            5,                                               //菜单编号
            "邮件")                                          //显示标题
         .setIcon(android.R.drawable.sym_action_email);   //设置图标
        //.setShortcut('3','e');                           //设置快捷键
    menu.add(Menu.NONE, Menu.FIRST + 2, 3, "保存").setIcon(
            android.R.drawable.ic_menu_save);             //设置菜单项
    menu.add(Menu.NONE, Menu.FIRST + 3, 6, "删除").setIcon(
            android.R.drawable.ic_menu_delete);           //设置菜单项
    menu.add(Menu.NONE, Menu.FIRST + 4, 1, "返回").setIcon(
            android.R.drawable.ic_menu_revert);           //设置菜单项
    menu.add(Menu.NONE, Menu.FIRST + 5, 4, "查找").setIcon(
            android.R.drawable.ic_menu_search);           //设置菜单项
    menu.add(Menu.NONE, Menu.FIRST + 6, 7, "管理").setIcon(
            android.R.drawable.ic_menu_manage);           //设置菜单项
    menu.add(Menu.NONE, Menu.FIRST + 7, 2, "编辑").setIcon(
            android.R.drawable.ic_menu_edit);             //设置菜单项
    MenuItem item1 =menu.add(Menu.NONE, Menu.FIRST + 8, 8, "帮助");
    item1.setIcon(android.R.drawable.ic_menu_help);
    return true;     }                                       //菜单显示
public boolean onOptionsItemSelected(MenuItem item) {        //选中某个菜单项
    switch (item.getItemId()) {                              //判断菜单项ID
    case Menu.FIRST + 1:
        MenuemailCounter=MenuemailCounter+1;
        Toast.makeText(this,"您选择的是"邮件菜单"项。",Toast.LENGTH_LONG).show();
        break;
    case Menu.FIRST + 2:
        Toast.makeText(this,"您选择的是"保存菜单"项。",Toast.LENGTH_LONG).show();
        break;
    case Menu.FIRST + 3:
        Toast.makeText(this,"您选择的是"删除菜单"项。",Toast.LENGTH_LONG).show();
        break;
    case Menu.FIRST + 4:
        Toast.makeText(this,"您选择的是"返回菜单"项。",Toast.LENGTH_LONG).show();
        break;
    case Menu.FIRST + 5:
        Toast.makeText(this,"您选择的是"查找菜单"项。",Toast.LENGTH_LONG).show();
        break;
    case Menu.FIRST + 6:
        Toast.makeText(this,"您选择的是"管理菜单"项。",Toast.LENGTH_LONG).show();
        break;
    case Menu.FIRST + 7:
        Toast.makeText(this,"您选择的是"编辑菜单"项。",Toast.LENGTH_LONG).show();
        break;            }
    return false;    }                                       //返回false
```

```
public void onOptionsMenuClosed(Menu menu) {            //菜单退出时调用
    Toast.makeText(this, "注意,选项现在菜单关闭!!", Toast.LENGTH_LONG).show();}
static int MenuemailCounter = 0;                        //统计选项选中的计数器
public boolean onPrepareOptionsMenu(Menu menu) {        //菜单显示前调用
    MenuItem emailItem = menu.findItem(Menu.FIRST + 1); //获得Menu.FIRST+1项
    emailItem.setTitle("邮件选项:" +String.valueOf(MenuemailCounter)); //设置标题
    Toast.makeText(this,                                //调用Toast显示提示信息
    "在菜单显示之前会先执行此操作,您可以在这里进行一些预处理操作.",Toast.LENGTH_
        LONG).show();
    return true;      }}                                //调用onCreateOptionsMenu()
```

保存文件,按手机上的"Menu"键,程序运行结果如图4.27所示。

单击"邮件选项:0"选项,弹出如图4.28所示界面,再次单击手机上的"Menu"键,弹出如图4.29所示界面,程序能够自动统计按了多少次该选项。

图4.27 exam4_14运行界面   图4.28 选择"邮件菜单"项

这里邮件选项的标题发生了变化,按下"More"选项,可以进入扩展菜单,如图4.30所示(注意:扩展菜单的选项上没有图标)。

图4.29 统计次数   图4.30 选择"More"后的界面

4.10.3 上下文菜单 ContextMenu

上下文菜单 ContextMenu 也称为快捷菜单,类似于 Windows 系统中的右键菜单,不过 Android 系统是通过长按(按住不动约两秒钟)某个组件来弹出上下文菜单的,上下文菜单项不支持图标或快捷键。

要进行上下文菜单的操作有几个常用的方法:

1)将快捷菜单注册到界面控件上,注册后长按就会弹出上下文菜单:

```
public void registerForContextMenu(View v,);
```

2）设置需要显示的所有菜单项：

```
public void onCreateContextMenu(ContextMenu menu,View v,ContextMenu.
                ContextMenuInfo menuInfo);
```

其中，参数 menu 是需要显示的快捷菜单；参数 v 是用户选择的界面元素；参数 menuInfo 是所选择界面元素的额外信息。

3）当某一个菜单项被选中时触发此操作：

```
public boolean onContextItemSelected(MenuItem item);
```

4）当菜单项关闭时触发此操作：

```
public void onContextMenuClosed(Menu menu);
```

5）menu.add

要在 Android 程序中使用上下文菜单必须在程序中使用下面的语句：

```
import android.view.ContextMenu;              //导入 view.ContextMenu 类
```

还有一个与 ContextMenu 联系密切的类：

```
import android.view.ContextMenu.ContextMenuInfo;
```

上下文菜单实现步骤如下。

1）通过组件 ID 获取组件实例，注册。

例如：TextView txt = (TextView) this.findViewById(R.id.txt1);

```
this.registerForContextMenu(txt) ;                  //将快捷菜单注册到 txt 组件上
```

2）为该组件注册 onCreateContextMenuListener 监听。

```
public void onCreateContextMenu(ContextMenu menu, View txt, ContextMenu.
        ContextMenuInfo menuInfo) {  //显示菜单
    super.onCreateContextMenu(menu, txt, menuInfo) ;
    menu.setHeaderTitle("信息操作") ;                           //设置显示信息头
    menu.add(Menu.NONE, Menu.FIRST + 1, 1, "添加信息");        //设置菜单项
    menu.add(Menu.NONE, Menu.FIRST + 2, 2, "查看信息");        //设置菜单项
    menu.add(Menu.NONE, Menu.FIRST + 3, 3, "删除信息");        //设置菜单项
    menu.add(Menu.NONE, Menu.FIRST + 4, 4, "信息另存");        //设置菜单项
    menu.add(Menu.NONE, Menu.FIRST + 5, 5, "编辑信息");    }   //设置菜单项
```

3）有列表操作，就该有相应的事件。上面的事件一般会与下面方法结合使用实现 onCreateContextMenu 方法。

```
public boolean onContextItemSelected(MenuItem item) {       //选中某个菜单项
    switch (item.getItemId()) {                              //判断菜单项 ID
    case Menu.FIRST + 1:
        Toast.makeText(this, "您选择的是"添加信息"。", Toast.LENGTH_LONG).show();
        break;
    case Menu.FIRST + 2:
        Toast.makeText(this, "您选择的是"查看信息"。", Toast.LENGTH_LONG).show();
        break;
    case Menu.FIRST + 3:
        Toast.makeText(this, "您选择的是"删除信息"。", Toast.LENGTH_LONG).show();
        break;
```

```
        case Menu.FIRST + 4:
            Toast.makeText(this,"您选择的是"另存信息"。",Toast.LENGTH_LONG).show();
            break;
        case Menu.FIRST + 5:
            Toast.makeText(this,"您选择的是"编辑信息"。",Toast.LENGTH_LONG).show();
            break;}
```

实例 4-15：上下文菜单实例

新建一个项目，项目的命名为：exam4_15，包名称为：org.hnist.demo，新建一个上下文菜单。

1）修改 activity_main.xml 文件，代码如下：

```xml
<?xml version="1.0" encoding="utf-8"?>
<LinearLayout xmlns:android="http://schemas.android.com/apk/res/android"
    android:orientation="vertical"
    android:layout_width="fill_parent"
    android:layout_height="fill_parent"   >
<TextView
    android:id="@+id/txt1"
    android:layout_width="fill_parent"
    android:layout_height="wrap_content"
    android:text="请长按触发快捷菜单"
    android:textSize="20dp" />
</LinearLayout>
```

2）修改 MainActivity.java，代码如下：

```java
package org.hnist.demo;
import android.app.Activity;
import android.os.Bundle;
import android.view.ContextMenu;
import android.view.Menu;
import android.view.MenuItem;
import android.view.View;
import android.widget.TextView;
import android.widget.Toast;
public class MainActivity extends Activity {
private TextView txt=null;
public void onCreate(Bundle savedInstanceState) {
    super.onCreate(savedInstanceState);
    setContentView(R.layout.activity_main);
    TextView txt = (TextView) this.findViewById(R.id.txt1);
    this.registerForContextMenu(txt) ;   }         //将快捷菜单注册到txt组件上
public void onCreateContextMenu(ContextMenu menu, View txt, ContextMenu.
        ContextMenuInfo menuInfo) {                             //显示菜单
    super.onCreateContextMenu(menu, txt, menuInfo) ;
    menu.setHeaderTitle("信息操作") ;                      //设置显示信息头
    menu.add(Menu.NONE, Menu.FIRST + 1, 1, "添加信息");    //设置菜单项
    menu.add(Menu.NONE, Menu.FIRST + 2, 2, "查看信息");    //设置菜单项
    menu.add(Menu.NONE, Menu.FIRST + 3, 3, "删除信息");    //设置菜单项
    menu.add(Menu.NONE, Menu.FIRST + 4, 4, "信息另存");    //设置菜单项
    menu.add(Menu.NONE, Menu.FIRST + 5, 5, "编辑信息");}   //设置菜单项
public boolean onContextItemSelected(MenuItem item) {      //选中某个菜单项
```

```java
        switch (item.getItemId()) {                              //判断菜单项 ID
        case Menu.FIRST + 1:
            Toast.makeText(this,"您选择的是"添加信息"。",Toast.LENGTH_LONG).show();
            break;
        case Menu.FIRST + 2:
            Toast.makeText(this,"您选择的是"查看信息"。",Toast.LENGTH_LONG).show();
            break;
        case Menu.FIRST + 3:
            Toast.makeText(this,"您选择的是"删除信息"。",Toast.LENGTH_LONG).show();
            break;
        case Menu.FIRST + 4:
            Toast.makeText(this,"您选择的是"另存信息"。",Toast.LENGTH_LONG).show();
            break;
        case Menu.FIRST + 5:
            Toast.makeText(this, "您选择的是"编辑信息"。", Toast.LENGTH_LONG).show();
            break;          }
        return false;       }
    public void onContextMenuClosed(Menu menu) {                 //菜单退出时调用
        Toast.makeText(this, "上下文菜单关闭了", Toast.LENGTH_LONG).show(); }}
```

保存文件并运行程序，长按屏幕上的文字，结果如图 4.31 所示。

在 Android 系统中，菜单也可以在 XML 文件中进行定义，使用 XML 文件定义界面菜单，下面来实现利用 XML 中的数据建立上下文菜单中的菜单选项。详细代码见 exam4_16。

如果希望从配置文件中取出数据，需要使用到下面两个常用方法：

```
public MenuInflater(Context context)   //创建 MenuInflater
        类对象
public void inflate(int menuRes, Menu menu)    //将配置
        的资源填充到菜单之中
```

图 4.31　显示上下文菜单

4.10.4　子菜单 SubMenu

Android 系统的子菜单使用非常灵活，可以在选项菜单或上下文菜单中使用子菜单。菜单子项使用浮动窗体的显示形式，能够更好地适应小屏幕的显示方式，子菜单不支持嵌套、不支持显示图标。

子菜单一般是在选项菜单或上下文菜单建立好的基础上，然后通过 addSubMenu(int groupId, int itemId, int order, int titleRes)方法非常方便地创建和响应子菜单。

例如：

```java
public boolean onCreateOptionsMenu(Menu menu) {          //建立选项菜单
    int base = Menu.FIRST;                               //设置 base 初值
    //在 onCreateOptionsMenu()函数传递的 menu 对象上调用 addSubMenu()函数，在选项菜单
      中添加两个菜单子项，用户单击后可以打开子菜单
    SubMenu subMenu1= menu.addSubMenu(base, base+1, Menu.NONE, "系统设置");
    SubMenu subMenu2 = menu.addSubMenu(base, base+2, Menu.NONE, "文件操作");
      //子菜单可以包括多个菜单项
      MenuItem menuitem1 = subMenu1.add(base, base+1, base+1, "显示设置");
      subMenu1.add(base, base+2, base+2, "网络设置");
      subMenu1.add(base, base+3, base+3, "高级设置");
      subMenu1.add(base, base+4, base+4, "安全设置");
      ……
```

表 4-6 SubMenu 接口的常用方法

方法	描述
public abstract MenuItem getItem()	得到一个子菜单所属的父菜单对象
public abstract SubMenu setHeaderIcon(int iconRes)	设置菜单的显示图标
public abstract SubMenu setHeaderTitle(int titleRes)	设置子菜单的显示标题
public abstract SubMenu setHeaderTitle (CharSequence title)	设置子菜单的显示标题
public abstract SubMenu setIcon(int iconRes)	设置每个子菜单项的图标

实例 4-16：子菜单实例

新建一个项目，项目的命名为：exam4_16，包名称为：org.hnist.demo，新建一个带有子菜单的选项菜单。

1）修改 activity_main.xml 文件，代码如下：

```xml
<?xml version="1.0" encoding="utf-8"?>
<LinearLayout xmlns:android="http://schemas.android.com/apk/res/android"
    android:orientation="vertical"
    android:layout_width="fill_parent"
    android:layout_height="fill_parent"   >
<Button
    android:id="@+id/but1"
    android:layout_width="fill_parent"
    android:layout_height="wrap_content"
    android:text="请长按触发快捷菜单"    />
</LinearLayout>
```

2）修改 MainActivity.java，代码如下：

```java
package org.hnist.demo;
import android.os.Bundle;
import android.app.Activity;
import android.view.ContextMenu;
import android.view.ContextMenu.ContextMenuInfo;
import android.view.Menu;
import android.view.MenuItem;
import android.view.SubMenu;
import android.view.View;
import android.widget.Button;
public class MainActivity extends Activity {
    public void onCreate(Bundle savedInstanceState) {
        super.onCreate(savedInstanceState);
        setContentView(R.layout.activity_main);
        Button but1 = (Button) this.findViewById(R.id.but1);
        this.registerForContextMenu(but1);    }
    // 重写  onCreateContextMenu  用以创建上下文菜单
    public void onCreateContextMenu(ContextMenu menu, View v,ContextMenuInfo menuInfo){
        super.onCreateContextMenu(menu, v, menuInfo);  //创建 R.id.txt1  的上下文菜单
        int base = Menu.FIRST;                                      //设置base 初值
    //建立上下文菜单的菜单项
    SubMenu subMenu1 = menu.addSubMenu(base, base+1, Menu.NONE, "系统设置");
    SubMenu subMenu2 = menu.addSubMenu(base, base+2, Menu.NONE, "文件操作");
    //子菜单 1 可以包括多个菜单项
    MenuItem menuitem1 = subMenu1.add(base, base+1, base+1, "显示设置");
        subMenu1.add(base, base+2, base+2, "网络设置");         //添加菜单项
        subMenu1.add(base, base+3, base+3, "高级设置");
```

```
            subMenu1.add(base, base+4, base+4, "安全设置");
        //子菜单 2 可以包括多个菜单项
        MenuItem menuitem2 = subMenu2.add(base, base+1, base+1, "打开文件");
            subMenu2.add(base, base+2, base+2, "保存文件");
            subMenu2.add(base, base+3, base+3, "关闭文件");           } }
```

保存文件并运行程序,长按屏幕上的按钮,结果如图 4.32 所示;选择菜单项,会弹出子菜单,如图 4.33 所示。

图 4.32　显示上下文菜单

图 4.33　显示子菜单

也可以利用 XML 中的数据建立上下文菜单中的菜单选项。

本章小结

本章着重分析了 6 种常见事件：单击事件、长按事件、焦点改变事件、键盘事件、触摸事件、菜单事件的基本操作,通过实例介绍加深对各个事件处理操作步骤的理解,对各事件的回调方法也详细做了分析。对选择改变事件、下拉列表选择事件、日期和时间改变触发的事件也做了简要介绍。

习题

1. 将一个按钮如何实现双击事件,给出具体步骤和关键代码。
2. 一个按钮和一个文本显示组件要实现单击事件,主要区别在哪里？
3. 焦点改变事件的操作步骤有哪些？
4. 在键盘事件中如何判断按下了方向键的右键？写出关键代码。
5. 单选按钮组中如何知道某个选项被选中？
6. 结合实例 4-9,哪些代码是判断下拉列表选择事件中某个选项被选中了的代码？
7. 编写代码将当前系统时间修改为 2013 年 10 月 20 日,12:30。
8. 当按下模拟器上的"Menu"按钮时,弹出菜单是什么菜单？长按某组件后弹出的菜单是什么菜单？
9. 子菜单是通过什么方法添加的？
10. 编程实现长按某个文本框,弹出一个菜单,该菜单有 3 个选项,其中第一个含有 4 个选项的子菜单。

第 5 章 Android 常用高级组件

学习目标：
- 了解 Android 平台中的 ScrollView 组件的使用；
- 掌握 Android 平台中的 ListView 组件的使用；
- 了解 Android 平台中的 ExpandableListView 组件的使用；
- 掌握 Android 平台中的 ProgressBar、SeekBar、RatingBar 组件的使用；
- 掌握 Android 平台中的 ImageSwitcher 、Gallery 组件的使用；
- 了解 Android 平台中的 AutoCompleteTextView 组件的使用；
- 掌握 Android 平台中的 Dialog 组件的使用；
- 掌握 Android 平台中的 TabHost 组件的使用。

在第 4 章介绍了 Android 平台的基本组件，事实上，Android 平台提供的组件还有很多，本章就介绍一些 Android 平台常用的高级组件，例如：LiseView、TabHost、Gallery 等，掌握这些常用组件的使用在 Android 程序设计中会起到事半功倍的效果。

5.1 滚动视图组件 ScrollView

手机屏幕的高度有限，当需要显示多组信息时，ScrollView 视图（滚动视图）可以合理安排这些组件，浏览时可以自动进行滚屏显示。ScrollView 是一个实现滚屏的组件，只要将需要滚动的组件添加到 ScrollView 中即可。ScrollView 只支持垂直滚动，如果需要水平滚动可采用 HorizontalScrollView 组件。

ScrollView 的层次关系如下所示：

```
java.lang.Object
    android.view.View
        android.view.ViewGroup
            android.widget.FrameLayout
                android.widget.ScrollView
```

要在 Android 程序中使用 ScrollView 组件必须要在程序中使用下面的语句：

```
import android.widget.ScrollView;                    //导入 widget.ScrollView 类
```

ScrollView 组件可以在代码中进行设置，也可以在 XML 布局文件中进行设置，其使用形式与布局管理器的操作形式类似，不同点在于，布局管理器中可以包含多个组件，而滚动视图里只能有一个组件，在这个组件里面可以容纳多于屏幕高度的组件。

5.1.1 ScrollView 组件常见的属性和方法

由于 ScrollView 组件是 View 类的间接子类，所以第 3 章介绍的 View 类属性它都具备，除此之外，还有表 5-1 所列出的常见属性，表 5-2 所列出的常见方法。

第 5 章 Android 常用高级组件

表 5-1 ScrollView 组件常用属性

属性	描述
android:scrollbars	设置滚动条显示
android:scrollbarFadeDuration	设置滚动条淡出效果
android:scrollbarSize	设置滚动条的宽度
android:scrollbarStyle	设置滚动条的风格和位置
android:scrollbarThumbHorizontal	设置水平滚动条的 drawable
android:scrollbarThumbVertical	设置垂直滚动条的 drawable
android:scrollbarTrackHorizontal	设置水平滚动条背景（轨迹）的色 drawable
android:soundEffectsEnabled	设置点击或触摸时是否有声音效果

表 5-2 ScrollView 组件常用方法

方法	描述
public ScrollView (Context context)	创建一个默认属性的 ScrollView 实例
public ScrollView (Context context, AttributeSet attrs)	创建一个带有 attrs 属性的 ScrollView 实例
public ScrollView (Context context, AttributeSet attrs, int defStyle)	创建一个带有 attrs 属性，并且指定其默认样式的 ScrollView 实例
public void addView(View child)	添加子视图
public void addView(View child, int index, ViewGroup.LayoutParams params)	根据指定的 layout 参数添加子视图
public boolean isFillViewport()	指示当前 ScrollView 的内容是否被拉伸以填充视图可视范围
public boolean onInterceptTouchEvent(MotionEvent ev)	拦截所有触摸屏幕时的运动事件
public boolean onTouchEvent(MotionEvent ev)	处理触摸屏幕的运动事件
public void setFillViewport(boolean fillViewport)	设置当前滚动视图是否将内容高度拉伸以填充视图可视范围
boolean requestChildRectangleOnScreen (View child, Rect rectangle,boolean immediate)	将某个子视图定位在屏幕的某个矩形范围
public void setSmoothScrollingEnabled(boolean smoothScrollingEnabled)	用来设置箭头滚动是否可以引发视图滚动

5.1.2 ScrollView 组件使用实例

实例 5-1：ScrollView 组件使用实例

新建一个项目，项目的命名为：exam5_1，包名称为：org.hnist.demo，利用 ScrollView 组件实现多个文本组件滚屏显示。

1）修改 activity_main.xml 文件：

```xml
<?xml version="1.0" encoding="utf-8"?>
<ScrollView                        //增加一个ScrollView组件
xmlns:android="http://schemas.android.com/apk/res/android"
android:id="@+id/myscroll"
android:layout_width="fill_parent"
android:layout_height="fill_parent">
<LinearLayout                      //增加一个线性布局管理器
    android:id="@+id/layout"       //该管理器的名字为layout
    android:orientation="vertical"
    android:layout_width="fill_parent"
    android:layout_height="fill_parent">
</LinearLayout>
</ScrollView>
```

2）修改 MainActivity.java 文件：

```java
package org.hnist.demo;
import android.app.Activity;
import android.os.Bundle;
import android.view.ViewGroup;
import android.widget.LinearLayout;
import android.widget.TextView;
public class MainActivity extends Activity {
private String scrolldata[] = { "信息学院", "机械学院","计算机学院","新闻学院",
        "化工学院", "美术学院","计算机学院","新闻学院", "化工学院", "美术学院",
        "体育学院", "音乐学院", "经济管理学院","南湖学院", "物理与电子学院",
        "机电工程学院", "法律学院","外语学院", "旅游学院","科技处",
        "图书馆", "教务处","网络中心", "学工处", "财务处"};      //定义显示的数据
public void onCreate(Bundle savedInstanceState) {
    super.onCreate(savedInstanceState);
    super.setContentView(R.layout.activity_main);
    LinearLayout layout = (LinearLayout) super.findViewById(R.id.layout) ;
    LinearLayout.LayoutParams param = new LinearLayout.LayoutParams(
            ViewGroup.LayoutParams.FILL_PARENT,
            ViewGroup.LayoutParams.WRAP_CONTENT);       //定义布局参数
    for (int x = 0; x < this.scrolldata.length; x++) {
        TextView msg = new TextView(this) ;             //创建文本组件
        msg.setTextSize(20);                            //设置字的大小
        msg.setText(this.scrolldata[x]);                //设置文本
        layout.addView(msg,param) ; }}}                 //增加组件
```

保存文件运行该项目，结果如图 5.1 所示，注意放到 LinearLayout 里的组件要足够多，否则看不到滚屏效果。

一般不在 ScrollView 内进行事件监听，如果需要选择某个选项有相应的事件发生，一般利用 ListView 组件来实现。

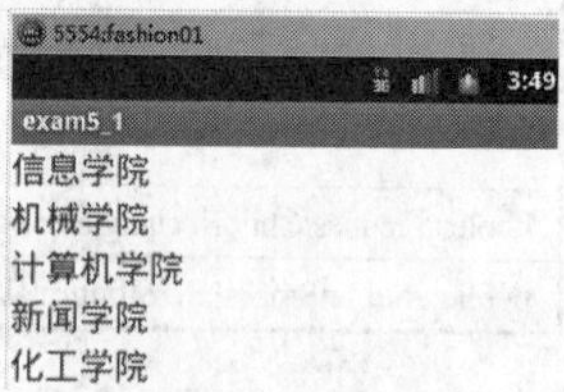

5.2 列表显示组件 ListView

图 5.1 实例 5_1 运行结果

与 ScrollView 类似的还有一种列表组件（ListView），它也可以将多个组件加入到 ListView 中以达到组件的滚动显示效果，ListView 组件本身也有对应的 ListView 类支持，可以通过操作 ListView 类以完成对此组件的操作，ListView 类的层次关系如下所示：

```
java.lang.Object
    android.view.View
        android.view.ViewGroup
            android.widget.AdapterView<T extends android.widget.Adapter>
                android.widget.AbsListView
                    android.widget.ListView
```

要在 Android 程序中使用 ListView 组件，必须要在程序中使用下面的语句：

```
import android.widget.ListView;                        //导入 widget.ListView 类
```

5.2.1 ListView 组件常见的属性和方法

ListView 组件是 View 组件的间接子类，所以第 3 章介绍的 View 类属性它都具备，除此之外，还有表 5-3 所列出的属性，表 5-4 所列出的常见方法。

表 5-3 ListView 组件属性

属性	描述
android:choiceMode	规定此 ListView 所使用的选择模式
android:divider	规定 List 项目之间用某个图形或颜色来分隔
android:dividerHeight	分隔符的高度
android:entries	引用一个将使用在此 ListView 里的数组
android:footerDividersEnabled	ListView 是否会在页脚视图前画分隔符,默认值为 true
android:headerDividersEnabled	ListView 是否会在页眉视图后画分隔符,默认值为 true

表 5-4 ListView 组件常用方法

方法	描述
public ListView(Context context)	创建 ListView 类的实例化对象
public void setAdapter(ListAdapter adapter)	设置显示的数据
public ListAdapter getAdapter()	返回 ListAdapter
public void clearChoices ()	取消之前设置的任何选择
public void addFooterView (View v)	加一个固定显示于 list 底部的视图
public void addHeaderView (View v)	加一个固定显示于 list 顶部的视图
public void addHeaderView (View v, Object data, boolean isSelectable)	加一个固定显示于 list 顶部的视图
public boolean onTouchEvent (MotionEvent ev)	用于处理触摸屏的动作事件
public void setOnItemSelectedListener(AdapterView.OnItemSelectedListener listener)	选项选中时触发
public void setOnItemClickListener(AdapterView.OnItemClickListener listener)	选项单击时触发
public void setOnItemLongClickListener(AdapterView.OnItemLongClickListener listener)	选项长按时触发

ListView 是一个经常用到的组件,ListView 里面的每个子项 Item 可以是一个字符串,也可以是一个组合组件。

要实现 ListView 组件,有如下步骤:

1)准备 ListView 要显示的数据;

2)构建适配器,简单来说,适配器就是 Item 数组,动态数组有多少元素就生成多少个 Item;

适配器类型常见的有 3 种:ArrayAdapter、SimpleAdapter 和 SimpleCursorAdapter。ArrayAdapter 最简单,只能展示一行字,SimpleAdapter 有最好的扩充性,可以自定义各种效果。SimpleCursorAdapter 是 SimpleAdapter 对数据库的简单结合,可以方便地把数据库记录以列表的形式展示出来。SimpleAdapter 继承自 AdapterView,可以通过一些方法给 ListView 添加监听器,当用户单击某一个列表项执行相应的操作。

3)把适配器添加 ListView,并显示出来。

实例 5-2:ListView 组件使用实例

新建一个项目,项目的命名为:exam5_2,包名称为:org.hnist.demo,利用 ListView 组件实现多个文本组件滚屏显示。

1)修改 activity_main.xml 文件:

```xml
<?xml version="1.0" encoding="utf-8"?>
<LinearLayout
    android:id="@+id/LinearLayout01"
    android:layout_width="fill_parent"
    android:layout_height="fill_parent"
```

```xml
xmlns:android="http://schemas.android.com/apk/res/android">
<ListView
    android:id="@+id/MyListView"
    android:layout_width="fill_parent"
    android:layout_height="wrap_content" />
</LinearLayout>
```

2）修改 MainActivity.java 文件：

```java
package org.hnist.demo;
package org.hnist.demo;
import android.app.Activity;
import android.os.Bundle;
import android.widget.ArrayAdapter;
import android.widget.ListView;
public class MainActivity extends Activity {
private String listdata[] = { "信息学院","机械学院","计算机学院","新闻学院",
    "化工学院","美术学院","计算机学院","新闻学院","化工学院","美术学院",
    "体育学院","音乐学院","经济管理学院","南湖学院","物理与电子学院",
    "机电工程学院","法律学院","外语学院","旅游学院","科技处",
    "图书馆","教务处","网络中心","学工处","财务处"};        //定义显示的数据
private ListView listview;                                //定义 ListView 组件
public void onCreate(Bundle savedInstanceState) {
    super.onCreate(savedInstanceState);
    super.setContentView(R.layout.activity_main);
    this.listview = (ListView)super.findViewById(R.id.MyListView); //获得
                                                                    ListView 组件
    this.listview.setAdapter(new ArrayAdapter<String>(this,    //将数据包装
        android.R.layout.simple_expandable_list_item_1,        //每行显示一条数据
        this.listdata));       }}                              //设置组件内容
```

保存文件并运行该项目，结果如图 5.2 所示，能实现滚屏效果。

5.2.2 SimpleAdapter 类

上面 ListView 实例的显示效果显得有些单一，如果希望在一行显示更多信息，就需要使用 SimpleAdapter，SimpleAdapter 类的主要功能是将 List 集合的数据转换为 ListView 可以支持的数据，定义出各种显示效果。其层次关系如下：

图 5.2 实例 5_2 运行结果

```
java.lang.Object
    android.widget.BaseAdapter
        android.widget.SimpleAdapter
```

要在 Android 程序中使用 SimpleAdapter 组件，必须在程序中使用下面的语句：

```java
import android.widget.SimpleAdapter;            //导入 widget.SimpleAdapter 类
```

SimpleAdapter 是一个简单的适配器，可以指定一个用于显示行的布局 XML 文件，通过关键字映射到指定的布局文件，其常用方法如表 5-5 所示。

一个比较重要的构造函数：public SimpleAdapter (Context context,List<? extends Map<String, ?>> data, int resource, String[] from, int[] to)

其中，参数 context 表示关联 SimpleAdapter 运行着的视图的上下文；参数 data 表示一个 Map 的列表。在列表中的每个条目对应列表中的一行，应该包含所有在 from 中指定的条目；参数 resource 表示一个定义列表项目的视图布局的资源唯一标识。布局文件将至少应包含哪些在 to 中定义了的名称；参数 from 表示一个将被添加到 Map 上关联每一个项目的列名称的列表；参数 to 表示应该在参数 from 显示列的视图。

表 5-5 SimpleAdapter 类的常用方法

方法	描述
public SimpleAdapter (Context context, List<? extends Map<String, ?>> data, int resource, String[] from, int[] to)	创建 SimpleAdapter 对象，需要传递 Context 对象，封装的 List 集合，要使用的布局文件 ID，需要显示的 key（对应 Map）、组件的 id
public int getCount()	得到保存集合的个数
public Object getItem(int position)	取得指定位置的对象
public long getItemId(int position)	取得指定位置对象的 ID
public void notifyDataSetChanged()	当列表项发生改变时，通知更新显示 ListView

一个 SimlpleAdapter 是这样工作的。假设将 SimpleAdapter 用于 ListView。那么 ListView 的每一个列表项就是 resource 参数值指定的布局。而 data 参数就是要加载到 ListView 中的数据。

例如：

```
//定义 ArrayList 对象命名为 list
ArrayList<HashMap<String,String>>list =new ArrayList<HashMap<String,String>>();
SimpleAdapter simpleAdapter=new SimpleAdapter(this,   //定义 SimpleAdapter
list,                                                 //要显示的数据
R.layout.user,                                        //按照 user.xml 布局摆放数据
new String[]{"userId","userName","userTel"},          //每一个记录的列名称的列表
new int[]{R.id.userId,R.id.userName,R.id.userTel});   //对应 user.xml 的 ID。
```

意思就是将 Map 对象中 key 为 userId 的 value 绑定到 R.id.userId 上，userName 和 userTel 也类似。SimpleAdapter 继承自 AdapterView，可以通过表 5-3 所示的方法给 ListView 添加监听器，当用户单击、长按或选中某一个列表项执行相应的操作。

以单击事件为例：

```
public abstract void onItemClick(AdapterView<?> parent,View view,int position,long id)
```

其中，AdapterView<?> parent：表示操作的 AdapterView 对象；View view：取得操作 AdapterView 的父组件，一般都是 ListView 显示时所使用的布局管理器；int position：取得 Adapter 的操作数据项的索引；long id：取得发生的 ListView 显示行的编号。

例如：

```
listView.setOnItemClickListener(new OnItemClickListener() {//设置事件监听
    public void onItemClick(AdapterView<?> arg0, View arg1, int arg2,long arg3) {
        ListView listView=(ListView)arg0;
        Toast.makeText(MainActivity.this, listView.getItemAtPosition(arg2).
            toString(), Toast.LENGTH_SHORT).show(); } }
```

实例 5-3：SimpleAdapter 类与 ListView 组件使用实例

新建一个项目，项目的命名为：exam5_3，包名称为：org.hnist.demo，利用 SimpleAdapter 类显示数据到 ListView 组件。

1）在 res 文件夹下选择 layout，右击 activity_main.xml，选择"copy"，再右击"layout"文件夹，选择"Paste"选项，在图 5-3 中输入 user.xml，单击"OK"按钮，新建一个 user.xml 布局文件用于显示每条记录的布局，作如下的修改：

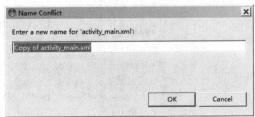

图 5.3 复制 activity_main.xml

```xml
<?xml version="1.0" encoding="utf-8"?>
<LinearLayout xmlns:android="http://schemas.android.com/apk/res/android"
    android:orientation="horizontal"
    android:layout_width="fill_parent"
    android:layout_height="fill_parent" >
    <TextView
        android:id="@+id/userId"
        android:layout_width="100dp"
        android:layout_height="wrap_content"
        android:textSize="16dp" />
    <TextView
        android:id="@+id/userName"
        android:layout_width="100dp"
        android:layout_height="wrap_content"
        android:textSize="16dp" />
    <TextView
        android:id="@+id/userTel"
        android:layout_width="100dp"
        android:layout_height="wrap_content"
        android:textSize="16dp" />
</LinearLayout>
```

2) 修改 activity_main.xml 文件，其代码如下：

```xml
<?xml version="1.0" encoding="utf-8"?>
<LinearLayout xmlns:android="http://schemas.android.com/apk/res/android"
    android:orientation="vertical"
    android:layout_width="fill_parent"
    android:layout_height="wrap_content" >
<TextView
    android:layout_width="fill_parent"
    android:layout_height="wrap_content"
    android:textSize="20dp"
    android:gravity="center"
    android:text="学生信息表" />
<TableLayout
    android:layout_width="fill_parent"
    android:layout_height="wrap_content"
    android:orientation="horizontal"    >
    <TableRow>
        <TextView
            android:text="学生编号"
            android:layout_width="100dp"
            android:layout_height="wrap_content"
            android:textSize="16dp" />
        <TextView
            android:text="学生姓名"
            android:layout_width="100dp"
            android:layout_height="wrap_content"
            android:textSize="16dp" />
        <TextView
            android:text="电话号码"
```

```xml
            android:layout_width="100dp"
            android:layout_height="wrap_content"
            android:textSize="16dp"  />
        </TableRow>
    </TableLayout>
    <ListView
        android:id="@+id/listView"
        android:layout_width="fill_parent"
        android:layout_height="wrap_content" />
</LinearLayout>
```

3）修改 MainActivity.java 文件：

```java
package org.hnist.demo;
import java.util.ArrayList;
import java.util.HashMap;
import android.app.Activity;
import android.os.Bundle;
import android.view.View;
import android.widget.AdapterView;
import android.widget.AdapterView.OnItemClickListener;
import android.widget.ListView;
import android.widget.SimpleAdapter;
import android.widget.Toast;
public class MainActivity extends Activity {
    public void onCreate(Bundle savedInstanceState) {
        super.onCreate(savedInstanceState);
        setContentView(R.layout.activity_main);
        ArrayList<HashMap<String,String>>list =new ArrayList<HashMap<String,
            String>>();
        HashMap<String,String> map1=new HashMap<String,String>();//定义map1 集合
        HashMap<String,String> map2=new HashMap<String,String>();
        HashMap<String,String> map3=new HashMap<String,String>();
        ListView listView=(ListView)findViewById(R.id.listView);
        map1.put("userId", "000001");          //设置 userId 组件显示的数据
        map1.put("userName", "陈飞翔");         //设置 userName 组件显示的数据
        map1.put("userTel", "8648871");        //设置 userTel 组件显示的数据
        list.add(map1);                        //增加数据到 list
        map2.put("userId", "000002");
        map2.put("userName", "张晓飞");
        map2.put("userTel", "8648872");
        list.add(map2);
        map3.put("userId", "000003");
        map3.put("userName", "王成功");
        map3.put("userTel", "8648873");
        list.add(map3);
        //定义一个 SimpleAdapter,每一个行有三个要显示的选项
        SimpleAdapter simpleAdapter=new SimpleAdapter(this,list,R.layout.
            user,new String[]{"userId","userName","userTel"},new int[]{R.id.
            userId,R.id.userName,R.id.userTel});
        //为 ListView 添加适配器
        listView.setAdapter(simpleAdapter);//设置 listView 数据为 simpleAdapter
```

```
listView.setOnItemClickListener(new OnItemClickListener() {//设置事件监听
        public void onItemClick(AdapterView<?> arg0, View arg1, int arg2, long
                                arg3) {
            ListView listView=(ListView)arg0;
            Toast.makeText(MainActivity.this, listView.getItemAtPosition
            (arg2).toString(), Toast.LENGTH_SHORT).show();    }    });    }    }
```

保存文件并运行该项目,单击某条记录会有提示信息,如图 5.4 所示。

在 ListView 中每行显示的不仅仅是文字,还可以是 ImageView、Button、CheckBox 等组件,按照实例 5-3 的方法读者可以自行编写程序实现。

图 5.4 实例 5_3 运行结果

5.3 可展开的列表组件 ExpandableListView

在 Android 系统中,ListView 组件可以为用户提供列表的显示功能,但在某些情况下,如果希望对列表项可以分组管理并实现收缩的列表,例如使用 QQ 的时候,有"我的好友"、"同学"、"家人"一样,单击某一项会扩展开,再单击又会收缩回去。那就要用到 Android 的 ExpandableListView 组件来实现,ExpandableListView 类的层次关系如下:

```
java.lang.Object
  android.view.View
    android.view.ViewGroup
      android.widget.AdapterView<T extends android.widget.Adapter>
        android.widget.AbsListView
          android.widget.ListView
            android.widget.ExpandableListView
```

要在 Android 程序中使用 ExpandableListView 组件,要在程序中使用下面的语句:

```
import android.widget.ExpandableListView;  //导入 widget.ExpandableListView 类
```

5.3.1 ExpandableListView 组件基础

1. ExpandableListView 组件常见的属性和方法

从上面的层次关系可以看出 ExpandableListView 组件是 ListView 组件的子类,所以 ListView 属性和方法它都具备,它还有表 5-6 所列出的属性,表 5-7 所列出的常见方法。

表 5-6 ExpandableListView 类的常用属性

属性	描述
android:childDivider	分离子列表项的图片或者颜色
android:childIndicator	在子列表项旁边显示的指示符
android:childIndicatorLeft	子列表项指示符的左边约束位置
android:childIndicatorRight	子列表项指示符的右边约束位置
android:groupIndicator	在组列表项旁边显示的指示符
android:indicatorLeft	组列表项指示器的左边约束位置
android:indicatorRight	组列表项指示器的右边约束位置

表 5-7 ExpandableListView 类的常用操作方法

方法	描述
public ExpandableListView (Context context)	实例化 ExpandableListView 类的对象
public boolean collapseGroup(int groupPos)	关闭指定的分组
public boolean expandGroup(int groupPos)	打开指定的分组
public ListAdapter getAdapter()	取得保存数据的 ListAdapter 对象
public ExpandableListAdapter getExpandableListAdapter()	取得保存数据的 ExpandableListAdapter 对象
public static int getPackedPositionType(long packedPosition)	取得操作的菜单项的类型（判断是菜单组，还是菜单项）
public static int getPackedPositionGroup(long packedPosition)	取得操作的所在的菜单组编号
public static int getPackedPositionChild(long packedPosition)	取得操作所在的菜单项编号
public long getSelectedId()	取得当前所操作的菜单 ID，如果没有则返回–1
public void setAdapter(ExpandableListAdapter adapter)	设置适配器数据对象
public void setAdapter(ListAdapter adapter)	设置适配器数据对象
public boolean setSelectedChild(int groupPosition, int childPosition, boolean shouldExpandGroup)	设置选中的菜单项
public void setSelectedGroup(int groupPosition)	设置选中的菜单组
public void setOnChildClickListener (ExpandableListView.OnChildClickListener onChildClickListener)	设置菜单组的单击事件处理
public void setOnGroupClickListener (ExpandableListView.OnGroupClickListener onGroupClickListener)	设置菜单组的单击事件处理
public void setOnGroupCollapseListener (ExpandableListView.OnGroupCollapseListener onGroupCollapseListener)	设置菜单组关闭的事件处理
public void setOnGroupExpandListener (ExpandableListView.OnGroupExpandListener onGroupExpandListener)	设置菜单组打开的事件处理
public void setOnItemClickListener (AdapterView.OnItemClickListener l)	设置选项单击的事件处理

每一个可以扩展的列表项的旁边都有一个指示符（箭头）用来说明该列表项目前的状态，可以使用方法：setChildIndicator(Drawable)和 setGroupIndicator(Drawable)（或者相应的 XML 文件的属性）去设置这些指示符的样式。当然也可以使用默认的指示符。

```
android.R.layout.simple_expandable_list_item_1,
android.R.layout.simple_expandable_list_item_2
```

注意：在 XML 布局文件中，一般不对 ExpandableListView 的 android:layout_height 属性使用 wrap_content 值，否则可能报错。

2. ExpandableListView 组件的适配器

与 ListView 一样 ExpandableListView 也需要一个适配器作为桥梁来取得数据，与 ListView 不同的是，ExpandableListView 是一个垂直滚动显示两级列表项的视图，它可以有两层：每一层都能够独立展开并显示其子项。这些子项来自于与该视图关联的适配器。

BaseExpandableListAdapter 就是一个用在 ExpandableListView 组件的适配器，此类继承 android.widget.BaseExpandableListAdapter 类。BaseExpandableListAdapter 的子类所需覆写的方法见表 5-8。

表 5-8 BaseExpandableListAdapter 的子类所需覆写的方法

方法	描述
public Object getChild(int groupPosition, int childPosition)	获得指定组中的指定索引的子项数据
public long getChildId(int groupPosition, int childPosition)	获得指定子项数据的 ID
public View getChildView(int groupPosition, int childPosition,boolean isLastChild, View convertView, ViewGroup parent)	获得指定子项的 View 组件

续表

方法	描述
public int getChildrenCount(int groupPosition)	取得指定组中所有子项的个数
public Object getGroup(int groupPosition)	取得指定组数据
public int getGroupCount()	取得所有组的个数
public long getGroupId(int groupPosition)	取得指定索引的组 ID
public View getGroupView(int groupPosition, boolean isExpanded, View convertView, ViewGroup parent)	获得指定组的 View 组件
public boolean hasStableIds()	如果返回 true,表示子项和组的 ID 始终表示一个固定的组件对象
public boolean isChildSelectable(int groupPosition, int childPosition)	判断指定的子选项是否被选中

一般适用于 ExpandableListView 的 Adapter 都要继承 BaseExpandableListAdapter 这个类,并且必须重载 getGroupView 和 getChildView 方法。

```
public abstract View getGroupView (int groupPosition, boolean isExpanded, View
    convertView, ViewGroup parent)
```

功能:取得用于显示给定分组的视图。这个方法仅返回分组的视图对象,要想获取子元素的视图对象,就需要调用 getChildView(int, int, boolean, View, ViewGroup)。

参数:groupPosition 决定返回视图的组位置;isExpanded 该组是展开状态还是收起状态;convertView 视图对象;parent 该视图最终从属的父视图。

返回值:指定位置相应的组视图。

```
public abstract View getChildView (int groupPosition, intchildPosition,
    boolean isLastChild, View convertView, ViewGroup parent)
```

功能:取得显示给定分组给定子位置的数据用的视图。

参数:groupPosition 包含要取得子视图的分组位置;childPosition 分组中子视图(要返回的视图)的位置;isLastChild 该视图是否为组中的最后一个视图;convertView 视图对象;parent 该视图最终从属的父视图。

返回值:指定位置相应的子视图。

3. ExpandableListView 组件的事件处理

ExpandableListView 提供的监听接口见表 5-9。

表 5-9 ExpandableListView 提供的监听接口

监听接口名称	作用	定义方法
ExpandableListView.OnChildClickListener	单击子选项	public boolean onChildClick (ExpandableListView parent, View v,int groupPosition, int childPosition, long id)
ExpandableListView.OnGroupClickListener	单击分组项	public boolean onGroupClick(ExpandableListView parent, View v, int groupPosition, long id)
ExpandableListView.OnGroupCollapseListener	分组关闭	public void onGroupCollapse(int groupPosition)
ExpandableListView.OnGroupExpandListener	分组打开	public void onGroupExpand(int groupPosition)

5.3.2 ExpandableListView 组件实例

ExpandableListView 组件在编程时关键要掌握适配器 BaseExpandableListAdapter 类提供的方法。

实例 5-4:ExpandableListView 组件实例

新建一个项目,项目的命名为:exam5_4,包名称为:org.hnist.demo,利用 ExpandableListView 组件显示各班人员信息所含图片的信息。

1) 修改 activity_main.xml 文件，代码如下：

```xml
<?xml version="1.0" encoding="utf-8"?>
<LinearLayout xmlns:android="http://schemas.android.com/apk/res/android"
    android:layout_width="fill_parent"
    android:layout_height="fill_parent"
    android:orientation="vertical" >
<TextView
    android:layout_width="fill_parent"
    android:layout_height="wrap_content"
    android:textSize="18dp"
    android:gravity="center"
    android:text="12级信工专业学生花名册" />
<ExpandableListView
    android:id="@+id/list"
    android:layout_width="fill_parent"
    android:layout_height="fill_parent"
    android:background="#ffffff" />
</LinearLayout>
```

2) 修改 MainActivity.java 文件，代码如下：

```java
package org.hnist.demo;
import android.app.Activity;
import android.graphics.Color;
import android.os.Bundle;
import android.view.Gravity;
import android.view.View;
import android.view.ViewGroup;
import android.widget.AbsListView;
import android.widget.BaseExpandableListAdapter;//导入BaseExpandableListAdapter
import android.widget.ExpandableListAdapter;  //导入ExpandableListAdapter
import android.widget.ExpandableListView;     //导入ExpandableListView
import android.widget.ExpandableListView.OnChildClickListener;
import android.widget.ImageView;
import android.widget.LinearLayout;
import android.widget.TextView;
import android.widget.Toast;
public class ExpandableList extends Activity{
 protected void onCreate(Bundle savedInstanceState) {
     super.onCreate(savedInstanceState);
     setContentView(R.layout.activity_main);
     //定义BaseExpandableListAdapter对象
     final ExpandableListAdapter adapter = new BaseExpandableListAdapter() {
         //设置各分组的图片
         int[] logos = new int[] { R.drawable.c11, R.drawable.c12,R.drawable.c13};
         //设置各分组的显示文字
         private String[] generalsTypes = new String[] { "1班", "2班", "3班" };
         //子视图显示文字
         private String[][] generals = new String[][] {
             { "陈飞翔", "张晓", "陈俊珊", "郭小嘉", "司马德", "杨大志" },
             { "陈康健", "王俊华", "刘致力", "田晓菲", "黄小英", "赵科健" },
```

```java
            { "王小蒙", "王俊飞", "孙启智", "姜大志", "李丽华" }};
    //子视图图片,对应每个人的图片
    public int[][] generallogos = new int[][] {
            { R.drawable.a11, R.drawable.a12,R.drawable.a13, R.drawable.
              a14,R.drawable.a15, R.drawable.a16 },
            { R.drawable.a21, R.drawable.a22,R.drawable.a23, R.drawable.
              a24,R.drawable.a25, R.drawable.a26 },
            { R.drawable.a31, R.drawable.a32, R.drawable.a33,R.drawable.
              a34, R.drawable.a36} };
//自己定义一个获得文字信息的方法
TextView getTextView() {
    //定义布局参数
    AbsListView.LayoutParams lp = new AbsListView.LayoutParams(
        ViewGroup.LayoutParams.FILL_PARENT, 64);
    TextView textView = new TextView(ExpandableList.this); //创建TextView
    textView.setLayoutParams(lp);                          //设置布局参数
    textView.setGravity(Gravity.CENTER_VERTICAL);          //居中显示
    textView.setPadding(36, 0, 0, 0);                      //与边界的距离
    textView.setTextSize(16);                              //设置文字大小
    textView.setTextColor(Color.BLACK);                    //设置文字颜色
    return textView;         }                             //返回TextView组件
//重写ExpandableListAdapter中的各个方法
    public int getGroupCount() {                           //取得分组的个数
        return generalsTypes.length;}
    public Object getGroup(int groupPosition) {   //取得指定分组
        return generalsTypes[groupPosition]; }
    public long getGroupId(int groupPosition) {   //取得指定分组的ID
        return groupPosition;       }
    public int getChildrenCount(int groupPosition) { //取得子项的个数
        return generals[groupPosition].length;    }
    public Object getChild(int groupPosition, int childPosition) {
        //取得指定子项
        return generals[groupPosition][childPosition];    }
    public long getChildId(int groupPosition, int childPosition) {
//取得指定子项ID
        return childPosition;              }
    public boolean hasStableIds() {  //子项和分组的ID始终表示一个固定的组件对象
        return true;        }
//覆写getGroupView方法
    public View getGroupView(int groupPosition, boolean isExpanded,
    View convertView, ViewGroup parent) {     //取得分组显示组件
        LinearLayout ll = new LinearLayout(ExpandableList.this);
                    //定义布局管理器
        ll.setOrientation(0);            //设置布局管理器的参数
        ImageView logo = new ImageView(ExpandableList.this);
        //定义ImageView
        logo.setImageResource(logos[groupPosition]);//获得显示组图片
        logo.setPadding(20, 0, 0, 0);             //设置边界
        ll.addView(logo);                         //添加ImageView
        TextView textView = getTextView();        //定义TextView组件
        textView.setTextColor(Color.BLACK);       //文字颜色
```

```
        textView.setText(getGroup(groupPosition).toString());//设置显示文字
        ll.addView(textView);                    //添加 TextView 组件
        return ll;             }
//覆写 getChildView 方法
public View getChildView(int groupPosition, int childPosition,
    boolean isLastChild, View convertView, ViewGroup parent)
    {       //取得子项组件
    LinearLayout ll = new LinearLayout(ExpandableList.this); //定
        义布局管理器
    ll.setOrientation(0);                    //设置布局管理器的参数
ImageView generallogo=new ImageView(ExpandableList.this);//定义 ImageView
    //获得显示子项图片
    generallogo.setImageResource(generallogos[groupPosition]
        [childPosition]);
    ll.addView(generallogo);                 //添加 ImageView
    TextView textView = getTextView();       //定义 TextView 组件
    textView.setText(getChild(groupPosition, childPosition).toString());
    //显示文字
    ll.addView(textView);                    //添加 TextView 组件
    return ll;             }
    public boolean isChildSelectable(int groupPosition,int childPosition) {
        return tru  }};                       //判断指定的子选项是否被选中
ExpandableListView expandableListView = (ExpandableListView) findViewById
    (R.id.list);          //获得 ExpandableListView 组件
expandableListView.setAdapter(adapter); //获得数据
//设置 item 单击事件的监听器
expandableListView.setOnChildClickListener(new OnChildClickListener() {
    public boolean onChildClick(ExpandableListView parent, View v,int
        groupPosition, int childPosition, long id) {    //单击该子项后的操作
        Toast.makeText(ExpandableList.this,"您刚才单击了" +
            adapter.getChild(groupPosition,childPosition),Toast.LENGTH_
                    SHORT).show();
        return false;} }); }}
```

保存文件并运行该项目,结果如图 5.5 所示。单击展开或收缩,会将列表项展开或收缩,单击某个选项会触发事件,如图 5.6 所示。

图 5.5 实例 5_4 运行结果

图 5.6 单击某个选项后的显示

5.4 进度条组件 ProgressBar

进度条组件 ProgressBar 是在某些操作的进度发展情况指示器，为用户呈现操作的进度，操作完成时，进度条会被填满。进度条能直观地帮助用户了解等待一定时间的操作所需的时间。ProgressBar 的层次关系如下：

```
java.lang.Object
    android.view.View
        android.widget.ProgressBar
```

要在 Android 程序中使用 ProgressBar 组件必须在程序中使用下面的语句：

```
import android.widget.ProgressBar;        //导入 widget.ProgressBar 类
```

5.4.1 ProgressBar 组件基础知识

1. ProgressBar 组件常见的属性和方法

ProgressBar 组件是 View 组件的子类，所以第 3 章介绍的 View 类属性和方法它都具备，除此之外，还有表 5-10 所列出的属性，表 5-11 所列出的常见方法。

表 5-10 ProgressBar 组件属性

属性	描述
android:progressBarStyle	默认进度条样式，不确定模式
android:progressBarStyleHorizontal	水平进度条样式
android:progressBarStyleLarge	大号进度条样式，也是不确定进度模式
android:progressBarStyleSmall	小号进度条样式，也是不确定进度模式
android:progress	定义默认进度条的范围
android:max	设置最大进度值
android:secondaryProgress	设置次要进度条

表 5-11 ProgressBar 组件常用方法

方法	描述
public ProgressBar(Context context)	创建一个 ProgressBar 实例
public synchronized int getMax ()	返回这个进度条的范围的上限
public synchronized int getProgress()	返回进度
public synchronized int getSecondaryProgress()	返回次要进度
public final synchronized void incrementProgressBy(int diff)	指定增加的进度
public synchronized boolean isIndeterminate()	指示进度条是否是不确定模式
public synchronized void setIndeterminate(boolean indeterminate)	设置不确定模式
public void setVisibility (int v)	设置该进度条是否可视
public synchronized void setMax(int max)	设置这个进度条范围的上限
public synchronized void setProgress(int progress)	设置进度

2. 进度条的分类

Android 系统提供的进度条有对话框进度条、标题进度条和水平进度条。其中对话框进度条在后面再介绍。进度条的样式也分为有条状的水平进度条和圆形转动的圆形进度条。

圆形进度条一般表示一个运转的过程，例如发送短信，连接网络等，表示一个过程正在执行中。一般只要在 XML 布局中定义就可以了：

```
<ProgressBar
  android:id="@+id/PBar"
  android:layout_width="wrap_content"
  android:layout_height="wrap_content"
  android:layout_gravity="center_vertical"/>
```

此时，没有设置它的风格，那么它就是圆形的，一直会旋转的进度条，如果希望是一个超大号圆形 ProgressBar，只需要设置一个 style 风格属性，该 ProgressBar 就有了一个风格，超大号圆形 ProgressBar 的风格是：style="?android:attr/progressBarStyleLarge"，小号 ProgressBar 对应的风格是：style="?android:attr/progressBarStyleSmall"，标题型 ProgressBar 对应的风格是：style="?android:attr/progressBarStyleSmallTitle"，水平进度条对应的风格是：style="?android:attr/progressBarStyleHorizontal"。

3．标题栏进度条创建步骤

1）调用 Activity 程序的 requestWindowFeatures()方法，获得进度条；

例如：requestWindowFeature(Window.FEATURE_PROGRESS); //请求一个窗口进度条特性风格

2）调用 Activity 程序的 setProgressBarIndeterminateVisibility()方法，显示进度条对话框；

例如：setProgressBarVisibility(true);   //设置进度条可视

3）然后设置进度值。

例如：setProgress(myProgressBar.getProgress() * 100); //设置标题栏中前景的一个进度条进度值
setSecondaryProgress(myProgressBar.getSecondaryProgress() * 100);
                                              //设置标题栏中后面的一个进度条进度值

4．水平进度条创建步骤

1）在布局文件中声明 ProgressBar；

```
<ProgressBar
  android:id="@+id/PBar"
  android:layout_width="wrap_content"
  android:layout_height="wrap_content"
  android:layout_gravity="center_vertical"
  style="?android:attr/progressBarStyleHorizontal"   //设置为水平进度条
  android:max="100"                                  //最大进度值为100
  android:progress="50"                              //初始化的进度值
  android:secondaryProgress="70" />                  //初始化次要进度值
```

2）在 Activity 中获得 ProgressBar 实例；

```
private ProgressBar myProgressBar;                   //定义 ProgressBar
myProgressBar = (ProgressBar) findViewById(R.id.PBar); //获得 ProgressBar 实例
```

3）调用 ProgressBar 的 incrementProgressBy()方法增加和减少进度。

```
myProgressBar.incrementProgressBy(5);//ProgressBar   //进度值增加5
myProgressBar.incrementSecondaryProgressBy(5);//ProgressBar//次要进度条进度值增加5
```

5．事件监听

onSizeChanged(int w, int h, int oldw, int oldh)：当进度值改变时引发此事件。

5.4.2 ProgressBar 组件实例

实例 5-5：ProgressBar 组件实例

新建一个项目，项目的命名为：exam5_5，包名称为：org.hnist.demo，建立各种样式的进度条。
1）修改 activity_main.xml 文件，代码如下：

```xml
<RelativeLayout
<?xml version="1.0" encoding="utf-8"?>
<LinearLayout
    xmlns:android="http://schemas.android.com/apk/res/android"
    android:orientation="vertical"
    android:layout_width="match_parent"
    android:layout_height="wrap_content">
    <TextView
        android:layout_width="wrap_content"
        android:layout_height="wrap_content"
        android:textSize="18dp"
        android:text="大号圆形进度条" />
    <ProgressBar
        android:layout_width="wrap_content"
        android:layout_height="wrap_content"
        android:layout_gravity="center_vertical"
        style="?android:attr/progressBarStyleLarge"/>
    <TextView
        android:layout_width="wrap_content"
        android:layout_height="wrap_content"
        android:textSize="18dp"
        android:text="小号圆形进度条" />
    <ProgressBar
        android:layout_width="wrap_content"
        android:layout_height="wrap_content"
        android:layout_gravity="center_vertical"
        style="?android:attr/progressBarStyleSmall"/>
    <TextView
        android:layout_width="wrap_content"
        android:layout_height="wrap_content"
        android:textSize="18dp"
        android:text="水平进度条" />
    <ProgressBar
        android:id="@+id/progress_horizontal"
        style="?android:attr/progressBarStyleHorizontal"
        android:layout_width="300dp"
        android:layout_height="wrap_content"
        android:layout_gravity="center_vertical"
        android:max="200"
        android:progress="50"
        android:secondaryProgress="75" />
    <LinearLayout
        android:orientation="horizontal"
        android:layout_width="match_parent"
        android:layout_height="wrap_content">
        <Button
            android:id="@+id/decrease"
            android:layout_width="wrap_content"
```

```
            android:layout_height="wrap_content"
            android:text="减少" />
    <Button
        android:id="@+id/increase"
        android:layout_width="wrap_content"
        android:layout_height="wrap_content"
        android:text="增加" />
    </LinearLayout>
</LinearLayout>
```

2）修改 MainActivity.java 文件，代码如下：

```
package org.hnist.demo;
package org.hnist.demo;
import android.app.Activity;
import android.os.Bundle;
import android.view.View;
import android.view.Window;
import android.widget.Button;
import android.widget.ProgressBar;                    //导入widget.ProgressBar 类
public class MainActivity extends Activity {
 protected void onCreate(Bundle savedInstanceState) {
  super.onCreate(savedInstanceState);
    requestWindowFeature(Window.FEATURE_PROGRESS); //创建标题栏进度条
        setContentView(R.layout.activity_main);
        setProgressBarVisibility(true);              //让创建标题栏进度条可见
        final ProgressBar progressHorizontal = (ProgressBar) findViewById
            (R.id.progress_horizontal);              //获得水平进度条
        setProgress(progressHorizontal.getProgress() * 100); //设置标题栏进度
                                                             条进度值

//设置标题栏次要进度条进度值
setSecondaryProgress(progressHorizontal.getSecondaryProgress() * 100);
Button button = (Button) findViewById(R.id.increase);
button.setOnClickListener(new Button.OnClickListener() {
    public void onClick(View v) {
        progressHorizontal.incrementProgressBy(5);  //每次单击"增加"按钮增加5
        setProgress(100 * progressHorizontal.getProgress()); } });
button = (Button) findViewById(R.id.decrease);
button.setOnClickListener(new Button.OnClickListener() {
    public void onClick(View v) {
        progressHorizontal.incrementProgressBy(-5);//每次单击"减少"按钮减少5
        setProgress(100 * progressHorizontal.getProgress()); }  });  }}
```

保存文件并运行该项目，结果如图 5.7 所示。添加 4 个 ProgressBar 组件，单击按钮进度条会发生变化。

5.5 拖动条组件 SeekBar

拖动条（SeekBar）组件与 ProgressBar 组件水平形式的显示进度条类似，最大的区别在于，拖动条可以由用户自己进行手工调节，例如当用户需要调整音乐的播放速度或者要调整音量时都会使用到拖动条 SeekBar 类。其层次关系如下：

图 5.7 实例 5_5 运行结果

```
java.lang.Object
    android.view.View
```

```
android.widget.ProgressBar
    android.widget.AbsSeekBar
        android.widget.SeekBar
```

要在 Android 程序中使用 SeekBar 组件，必须要在程序中使用下面的语句：

```
import android.widget.SeekBar;                    //导入 widget.SeekBar 类
```

5.5.1 SeekBar 组件基础知识

1. SeekBar 组件常见的属性和方法

因为 SeekBar 组件是 ProgressBar 组件的子类，所以 ProgressBar 组件所有的属性和方法 SeekBar 组件都继承了，此外该组件还有如表 5-12 所示的方法。

表 5-12 SeekBar 组件常用方法

方法	描述
public SeekBar(Context context)	创建 SeekBar 类的对象
public void setOnSeekBarChangeListener (SeekBar.OnSeekBarChangeListener l)	设置改变监听操作
public synchronized void setMax(int max)	设置增长的最大值
public synchronized void setProgress(int progress)	设置进度值
public synchronized void setSeconddaryProgress (int progress)	设置第二拖动条的数值
public synchronized int getProgress()	返回进度

2. 拖动条的事件

实现 SeekBar.OnSeekBarChangeListener 接口。需要监听三个事件：数值改变（onProgressChanged）；开始拖动（onStartTrackingTouch）；停止拖动（onStopTrackingTouch）。

onStartTrackingTouch 开始拖动时触发，与 onProgressChanged 区别是停止拖动前只触发一次而 onProgressChanged 只要在拖动，则会重复触发。

5.5.2 SeekBar 组件实例

实例 5-6：ProgressBar 组件实例

新建一个项目，项目的命名为：exam5_6，包名称为：org.hnist.demo，建立拖动条，当拖动时会触发相应的事件。

1）修改 activity_main.xml 文件，代码如下：

```
<LinearLayout xmlns:android="http://schemas.android.com/apk/res/android"
    xmlns:tools="http://schemas.android.com/tools"
    android:layout_width="match_parent"
    android:layout_height="match_parent"
    android:orientation="vertical">
    <TextView
        android:id="@+id/myTextView"
        android:layout_width="wrap_content"
        android:layout_height="wrap_content"
        android:textSize="16sp" />
    <SeekBar
        android:id="@+id/mySeekBar"
        android:layout_width="fill_parent"
```

```
        android:layout_height="wrap_content" />
</LinearLayout>
```

2）修改 MainActivity.java 文件，代码如下：

```
package org.hnist.demo;
import android.app.Activity;
import android.os.Bundle;
import android.widget.SeekBar;
import android.widget.TextView;
import android.widget.SeekBar.OnSeekBarChangeListener;
public class MainActivity extends Activity {
    private SeekBar seek;
    private TextView myTextView;
    protected void onCreate(Bundle savedInstanceState) {
        super.onCreate(savedInstanceState);
        setContentView(R.layout.activity_main);
        myTextView = (TextView) findViewById(R.id.myTextView);
        seek = (SeekBar) findViewById(R.id.mySeekBar);
        seek.setProgress(60);           //初始化
        seek.setOnSeekBarChangeListener(seekListener);
        myTextView.setText("当前值为: -" + 60);       }
  //设置监听
    private OnSeekBarChangeListener seekListener = new OnSeekBarChangeListener(){
       //数值改变时触发操作
        public void onProgressChanged(SeekBar seekBar, int progress,boolean
          fromUser) {
   MainActivity.this.myTextView.append("正在拖动,当前值:"+seekBar.get
          Progress()+"\n");}
   //开始拖动时候触发的操作
        public void onStartTrackingTouch(SeekBar seekBar) {
   MainActivity.this.myTextView.append("开始拖动,当前值:"+seekBar.getProgress()
          +"\n"); }
   //停止拖动时候，触发操作
        public void onStopTrackingTouch(SeekBar seekBar) {
   MainActivity.this.myTextView.append("停止拖动,当前值:"+seekBar.getProgress()
          +"\n");}};};
```

保存文件并运行该项目，结果如图 5.8 所示，拖动 SeekBar，会触发相应的事件。

5.6 星级评分条组件 RatingBar

星级评分条组件 RatingBar 一般就是用来做评分用的，用星型来显示等级评定，例如网站的满意度调查，有时也可以把它当做一个水平的进度条来用。RatingBar 的层次关系如下：

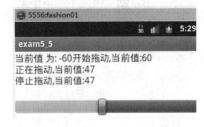

图 5.8 实例 5_6 运行结果

```
java.lang.Object
    android.view.View
        android.widget.ProgressBar
            android.widget.AbsSeekBar
                android.widget.RatingBar
```

要在 Android 程序中使用 RatingBar 组件必须要在程序中使用下面的语句：

```
import android.widget.RatingBar;                    //导入 widget.RatingBar 类
```

5.6.1 RatingBar 组件基础

RatingBar 组件常见的属性和方法如下：

由上面的层次关系发现 RatingBar 是 ProgressBar 的子类，所以 ProgressBar 所有的属性和方法 RatingBar 都继承了，此外还有如表 5-13 所示的属性和表 5-14 所示的方法。

表 5-13 RatingBar 组件常见的属性

属性	描述
android:isIndicator	RatingBar 是否是一个指示器（用户无法进行更改）
android:numStars	显示的星型数量，必须是一个整形值，像"100"
android:rating	默认的评分，必须是浮点类型，像"1.2"
android:stepSize	评分的步长，必须是浮点类型，像"1.2"

表 5-14 RatingBar 组件常用方法

方法	描述
public RatingBar(Context context)	创建 RatingBar 对象
public int getNumStars()	取得评分数量
public float getRating()	取得当前评分值
public float getStepSize()	取得设置的步长
public boolean isIndicator()	判断是否可以操作
public void setIsIndicator(boolean isIndicator)	是否可以操作
public synchronized void setMax(int max)	设置评分等级的范围
public void setNumStars(int numStars)	设置评分星的个数
public void setOnRatingBarChangeListener(RatingBar. OnRatingBarChangeListener listener)	设置操作监听
public void setRating(float rating)	设置当前的评分值
public void setStepSize(float stepSize)	设置每次增长的步长
public RatingBar.OnRatingBarChangeListener getOnRatingBarChangeListener ()	监听器（可能为空）监听评分改变事件

RatingBar 默认情况下利用星星个数来进行评分，如果要把 RatingBar 进行美化，可以将自定义的图片替换为系统默认的图片。

5.6.2 RatingBar 组件实例

实例 5-7：RatingBar 组件实例

新建一个项目，项目的命名为：exam5_7，包名称为：org.hnist.demo，建立 RatingBar，当改变 RatingBar 的值时会触发相应的事件。

1）修改 activity_main.xml 文件，代码如下：

```xml
<?xml version="1.0" encoding="utf-8"?>
<LinearLayout xmlns:android="http://schemas.android.com/apk/res/android"
    android:orientation="vertical"
    android:layout_width="fill_parent"
    android:layout_height="fill_parent"      >
```

```
<RatingBar
    android:id="@+id/myratingBar"
    android:layout_height="wrap_content"
    android:layout_width="wrap_content" />
</LinearLayout>
```

2）修改 MainActivity.java 文件，代码如下：

```
package org.hnist.demo;
import android.app.Activity;
import android.os.Bundle;
import android.widget.RatingBar;
import android.widget.Toast;
public class MainActivity extends Activity {
    public void onCreate(Bundle savedInstanceState) {
        super.onCreate(savedInstanceState);
        setContentView(R.layout.activity_main);
        RatingBar ratingBar = (RatingBar) findViewById(R.id.myratingBar);
        ratingBar.setNumStars(5);           //定义星级
        ratingBar.setRating(4);             //设置默认星级
        //监听星级改变并显示改变值
        ratingBar.setOnRatingBarChangeListener(new RatingBar.OnRating
          BarChangeListener() {
            public void onRatingChanged(RatingBar ratingBar,
                float rating, boolean fromUser) {
                float pp=(int)((rating*20/100)*100);
                Toast.makeText(MainActivity.this,  "您的满意度为：" + String.
                valueOf(pp)+"%",  Toast.LENGTH_LONG).show();}});}}
```

保存文件并运行该项目，结果如图 5.9 所示，修改 RatingBar 时会触发相应的事件。

5.7 自动完成文本框 AutoCompleteTextView

在上网查询资料的时候常常会遇到这样的情况，在输入框输入几个字后，后面会自动出现一些文字信息供选择，Android 系统中的自动完成文本框（AutoCompleteTextView）也可以实现这样的功能。

图 5.9 实例 5_7 运行结果

自动完成文本框（AutoCompleteTextView）能够对用户键入的文本进行有效的扩充提示，不需要用户输入整个内容。这个功能的实现要依靠 android.widget.AutoCompleteTextView 类完成，其层次关系如下：

```
java.lang.Object
    android.view.View
        android.widget.TextView
            android.widget.EditText
                android.widget.AutoCompleteTextView
```

要在 Android 程序中使用 AutoCompleteTextView 组件，必须在程序开始时使用下面的语句：

```
import android.widget.AutoCompleteTextView; //导入 widget.AutoCompleteTextView 类
```

5.7.1 AutoCompleteTextView 组件基础

AutoCompleteTextView 组件常见的属性和方法如下：

由上面的层次关系发现 AutoCompleteTextView 是 EditText 的子类，所以 EditText 所有的属性和方法 AutoCompleteTextView 都继承了，此外还有如表 5-15 所示的属性和表 5-16 所示的方法。

表 5-15　AutoCompleteTextView 组件常见的属性

属性	描述
android:completionHint	设置显示下拉列表的提示题目
android:completionThreshold	至少输入几个字符，它才会具有自动提示的功能
android:dropDownAnchor	后接一个 View 的 ID，会在这个 View 下弹出自动提示
android:dropDownHeight	设置下拉列表的高度
android:dropDownWidth	设置下拉列表的宽度
android:popupBackground	设置下拉列表的背景

表 5-16　AutoCompleteTextView 组件常用方法

方法	描述
public void clearListSelection()	清除所有的下拉列表项
public ListAdapter getAdapter()	取得数据集
public void setAdapter(T adapter)	设置数据集
public void setCompletionHint(CharSequence)	设置出现下拉列表的提示标题
public void setThreshold(int)	至少输入几个字符才会显示提示
public void setDropHeight(int)	设置下拉列表的高度
public void setDropWidth(int)	设置下拉列表的宽度
public void setDropDownbackgroundResource(int)	设置下拉列表的背景
public void setOnClickListener(View.OnClickListener listener)	设置单击事件
public void setOnItemClickListener(AdapterView.OnItemClickListener l)	在选项上设置单击事件
public void setOnItemSelectedListener(AdapterView.OnItemSelectedListener l)	选项选中时的单击事件

AutoCompleteTextView 组件可以很好地帮助用户进行文本信息的输入，它可以和一个字符串数组或 List 对象绑定，当用户输入两个及以上字符时，系统将在 AutoCompleteTextView 组件下方列出字符串数组中所有以输入字符开头的字符串。

AutoCompleteTextView 组件使用的关键是需要使用 AutoCompleteTextView 类的 setAdapter 方法指定一个 Adapter 对象，例如：this.Auto.setAdapter(adapter); 其中 adapter 是一个数据集，可以是字符串数组或 List 对象。

5.7.2 AutoCompleteTextView 组件实例

实例 5-8：AutoCompleteTextView 组件实例

新建一个项目，项目的命名为：exam5_8，包名称为：org.hnist.demo，利用 AutoCompleteTextView 组件，实现文本提示功能。

1）修改 activity_main.xml 文件，代码如下：

```xml
<?xml version="1.0" encoding="utf-8"?>
<LinearLayout
    xmlns:android="http://schemas.android.com/apk/res/android"
```

```
android:orientation="vertical"
android:layout_width="fill_parent"
android:layout_height="fill_parent">
<AutoCompleteTextView
    android:id="@+id/auto"
    android:layout_width="fill_parent"
    android:layout_height="wrap_content" />
</LinearLayout>
```

2）修改 MainActivity.java 文件，代码如下：

```
package org.hnist.demo;
import android.app.Activity;
import android.os.Bundle;
import android.widget.ArrayAdapter;
import android.widget.AutoCompleteTextView;
public class MainActivity extends Activity {
private static final String autoString[] = new String[] { "湖南", "湖南理工",
    "湖南理工学院", "湖南省岳阳市", "湖南的辣椒" };        //设置数据
private AutoCompleteTextView Auto = null;              //定义文本提示组件
public void onCreate(Bundle savedInstanceState) {
    super.onCreate(savedInstanceState);
    super.setContentView(R.layout.activity_main);
    ArrayAdapter<String> adapter = new ArrayAdapter<String>(this,android.R.
        layout.simple_dropdown_item_1line, autoString);   //定义数据集
    this.Auto=(AutoCompleteTextView)super.findViewById(R.id.auto);//取得文
        本提示组件
    this.Auto.setAdapter(adapter);    }}                  //设置数据集
```

保存文件并运行该项目，输入文字会弹出相应的提示，如图 5.10 所示。

还可以使用 MultiAutoCompleteTextView 组件来完成连续输入的功能，读者可以查阅相关 Android API 资料进行学习。

图 5.10 实例 5_8 运行结果

5.8 对话框组件 Dialog

在程序设计时经常需要在界面上弹出一些对话框，用来提示用户输入信息或者让用户做出选择进行一些简单的交互操作，这就是对话框的功能。在 Android 平台的开发之中，所有的对话框是使用 Dialog 类来实现的，其层次关系如下：

```
java.lang.Object
    android.app.Dialog
```

Dialog 类定义的常用方法见表 5-17。

表 5-17　Dialog 类定义的常用方法

方法	描述
public void setTitle(CharSequence title)	设置对话框的显示标题
public void show()	显示对话框
public void hide()	隐藏对话框
public boolean isShowing()	判断对话框是否显示
public void setContentView(View view)	设置组件
public void setContentView(int layoutResID)	设置组件的 ID

续表

方法	描述
public void dismiss()	隐藏对话框
public void closeOptionsMenu()	关闭选项菜单
public void setDismissMessage(Message msg)	设置隐藏对话框时的消息
public void setCancelable(boolean flag)	设置是否可以取消
public void setCancelMessage(Message msg)	设置对话框取消时的消息
public void cancel()	取消对话框,与 dismiss()方法类似
public Window getWindow()	取得 Window 对象
public void setOnShowListener (DialogInterface.OnShowListener listener)	对话框打开时触发事件
public void setOnDismissListener (DialogInterface.OnDismissListener listener)	对话框隐藏时触发事件
public void setOnCancelListener (DialogInterface.OnCancelListener listener)	对话框取消时触发事件

Android 提供的常见对话框有: Alertialog 用于实现警告对话框,ProgressDialog 用于实现带进度条的对话框,DatePickerDialog 用于实现日期选择对话框和 TimePickerDialog 用于实现时间选择对话框。其中 DatePickerDialog 和 TimePickerDialog 见本书 4.9 节。

5.8.1 警告对话框: AlertDialog

警告对话框是在程序运行时,弹出一个对话框显示一条警告信息,提示用户的后续操作,它是项目中经常出现的一种对话框,Android 系统提供了 AlertDialog 类来实现警告框,其层次关系如下:

```
java.lang.Object
    android.app.Dialog
        android.app.AlertDialog
```

要在 Android 程序中使用 AlertDialog 组件必须在程序开始时使用下面的语句:

```
import android.app.AlertDialog;            //导入 app.AlertDialog 类
```

要想实例化 AlertDialog 组件,可以通过 AlertDialog.Builder 类来实现,其常用方法如表 5-18 所示。

表 5-18 AlertDialog.Builder 类定义的常用方法

方法	描述
public AlertDialog.Builder(Context context)	创建 AlertDialog.Builder 对象
public AlertDialog.Builder setMessage (int messageId)	设置显示信息的资源 ID
public AlertDialog.Builder setMessage (CharSequence message)	设置显示信息的字符串
public AlertDialog.Builder setView(View view)	设置显示的 View 组件
public AlertDialog.Builder setSingleChoiceItems (CharSequence[] items, int checkedItem, DialogInterface.OnClickListener listener)	设置对话框显示一个单选的 List,指定默认选项,同时设置监听处理操作
public AlertDialog.Builder setSingleChoiceItems (ListAdapter adapter, int checkedItem, DialogInterface.OnClickListener listener)	设置对话框显示一个单选的 List,指定默认选项,同时设置监听处理操作
public AlertDialog.Builder setMultiChoiceItems (CharSequence[] items, boolean[] checkedItems, DialogInterface.OnMultiChoiceClickListener listener)	设置对话框显示一个复选的 List,同时设置监听处理操作
public AlertDialog.Builder setPositiveButton (CharSequence text, DialogInterface.OnClickListener listener)	为对话框添加一个确认按钮,同时设置监听操作
public AlertDialog.Builder setPositiveButton (int textId, DialogInterface.OnClickListener listener)	为对话框添加一个确认按钮,显示内容由资源文件指定,并设置监听操作
public AlertDialog.Builder setNegativeButton (CharSequence text, DialogInterface.OnClickListener listener)	为对话框设置一个取消按钮,并设置监听操作
public AlertDialog.Builder setNegativeButton (int textId, DialogInterface.OnClickListener listener)	为对话框设置一个取消按钮,显示内容由资源文件指定,并设置监听操作

续表

方法	描述
public AlertDialog.Builder setNeutralButton(CharSequence text, DialogInterface.OnClickListener listener)	设置一个按钮,并设置监听操作
public AlertDialog.Builder setNeutralButton(int textId, DialogInterface.OnClickListener listener)	设置一个按钮,显示内容由资源文件指定,并设置监听操作
public AlertDialog.Builder setItems(CharSequence[] items, DialogInterface.OnClickListener listener)	将信息内容设置为列表项,同时设置监听操作
public AlertDialog.Builder setItems(int itemsId, DialogInterface.OnClickListener listener)	将信息内容设为列表项,列表项内容由资源文件指定,同时设置监听操作
public AlertDialog create()	创建 AlertDialog 的实例化对象
public AlertDialog.Builder setIcon(Drawable icon)	设置显示的图标文件
public AlertDialog.Builder setIcon(int iconId)	设置要显示图标来源的 ID

比较常见的警告框有：只带提示信息的简单对话框、带按钮的警告框、列表式警告框、带复选框或单选框式警告框、带按钮和输入框的警告框。

使用 AlertDialog.Builder 类创建警告框的一般步骤：

1）通过 AlertDialog.Builder(Context)获取一个构造器 Builder；例如：

```
AlertDialog.Builder builder=new AlertDialog.Builder(this);
```

2）使用这个 Builder 类的公共方法来定义警告对话框的所有属性；例如：

```
AlertDialog.Builder builder=new AlertDialog.Builder(MainActivity.this);
    //实例化对象
builder.setTitle("请选择您喜爱的交通工具");        //设置显示标题
    ……
```

3）通过 Builder.Create()来创建 AlertDialog 对象；例如：

```
AlertDialog alert=builder.create();   //通过 Create()来创建 AlertDialog 对象
```

4）直接调用 Builder.Show()显示对话框。例如：

```
alert.show();                    //调用 Show()显示
```

有时不调用 Builder.Create()方法，而是在设置好警告对话框的所有属性后直接调用 show()方法显示 AlertDialog。

在警告对话框中增加按钮时，需设置 DialogInterface.OnClickListener 事件监听接口对象，此接口主要负责进行对话框中按钮的事件处理，此接口定义如下：

```
public interface DialogInterface.OnClickListener {
    public abstract void onClick (DialogInterface dialog, int which) ;}
```

对话框事件处理接口见表 5-19。

表 5-19 对话框事件处理接口

接口名称	描述
DialogInterface.OnClickListener	对话框单击事件处理接口
DialogInterface.OnCancelListener	对话框取消事件处理接口
DialogInterface.OnDismissListener	对话框隐藏事件处理接口
DialogInterface.OnKeyListener	对话框键盘事件处理接口
DialogInterface.OnMultiChoiceClickListener	对话框多选事件处理接口
DialogInterface.OnShowListener	对话框显示事件处理接口

5.8.2 AlertDialog 组件实例

实例 5-9：AlertDialog 组件实例

新建一个项目，项目的命名为：exam5_9，包名称为：org.hnist.demo，利用 AlertDialog 组件，实现不同的警告框。

1）修改 activity_main.xml 文件，代码如下：

```xml
<?xml version="1.0" encoding="utf-8"?>
<LinearLayout
    xmlns:android="http://schemas.android.com/apk/res/android"
    android:id="@+id/MyLayout"
    android:orientation="vertical"
    android:layout_width="fill_parent"
    android:layout_height="fill_parent">
<Button
    android:id="@+id/butAlert"
    android:layout_width="fill_parent"
    android:layout_height="wrap_content"
    android:text="带按钮的警告框"
    android:gravity="center"/>
<Button
    android:id="@+id/listAlert"
    android:text="列表式的警告框"
    android:layout_width="fill_parent"
    android:layout_height="wrap_content"/>
<Button
    android:id="@+id/singleAlert"
    android:text="带单选功能的警告框"
    android:layout_width="fill_parent"
    android:layout_height="wrap_content"/>
<Button
    android:id="@+id/checkAlert"
    android:text="带复选功能的警告框"
    android:layout_width="fill_parent"
    android:layout_height="wrap_content"/>
</LinearLayout>
```

2）修改 MainActivity.java 文件，代码如下：

```java
package org.hnist.demo;
import android.app.Activity;
package org.hnist.demo;
import android.app.Activity;
import android.app.AlertDialog;
import android.app.Dialog;
import android.content.DialogInterface;
import android.os.Bundle;
import android.view.View;
import android.view.View.OnClickListener;
import android.widget.Button;
import android.widget.Toast;
```

```java
public class MainActivity extends Activity {
    private Button butAlert = null ;
    private Button listAlert = null ;
    private Button singleAlert = null ;
    private Button checkAlert = null ;
public void onCreate(Bundle savedInstanceState) {
        super.onCreate(savedInstanceState);
        super.setContentView(R.layout.activity_main);
        this.butAlert = (Button) super.findViewById(R.id.butAlert) ;
        this.listAlert = (Button) super.findViewById(R.id.listAlert) ;
        this.singleAlert = (Button) super.findViewById(R.id.singleAlert) ;
        this.checkAlert = (Button) super.findViewById(R.id.checkAlert) ;
        this.butAlert.setOnClickListener(new butAlertImpl());    //设置单击事件
        this.listAlert.setOnClickListener(new listAlertImpl());  //设置单击事件
        this.singleAlert.setOnClickListener(new singleAlertImpl());
                                                                 //设置单击事件
        this.checkAlert.setOnClickListener(new editAlertImpl()); }
                                                                 //设置单击事件
private class butAlertImpl implements OnClickListener {
        public void onClick(View v) {
            Dialog dialog = new AlertDialog.Builder(MainActivity.this)
                //实例化对象
                .setIcon(R.drawable.alert)                       //设置显示图片
                .setTitle("确定删除？")                          //设置显示标题
                .setMessage("您确定要删除这个文件吗？")          //设置显示内容
                .setPositiveButton("确定",                       //增加一个确定按钮
                    new DialogInterface.OnClickListener() {      //设置操作监听
                        public void onClick(DialogInterface dialog,
                            int whichButton) { } //单击事件
                    }).setNeutralButton("查看文件",              //增加普通按钮
                    new DialogInterface.OnClickListener() {      //设置监听操作
                        public void onClick(DialogInterface dialog,int
                            whichButton) {}}).setNegativeButton("取消",
                //增加取消按钮
                    new DialogInterface.OnClickListener() {//设置操作监听
                        public void onClick(DialogInterface dialog,
                            int whichButton) {    //单击事件
                    }}).create();                                //创建Dialog
            dialog.show(); }}                                    //显示对话框
private class editAlertImpl implements OnClickListener {
    boolean[] flags=new boolean[]{false,false,false,false,false}; //初始复选情况
    public void onClick(View v) {
        final String[] items={"登山","跑步","篮球","羽毛球","乒乓球"}; //定义数组
        AlertDialog.Builder builder = new AlertDialog.Builder(MainActivity.
            this);  builder.setTitle("请选择您喜爱的体育运动");  //设置显示标题
        builder.setMultiChoiceItems(items, flags, new DialogInterface.
        OnMultiChoice ClickListener(){//设置要显示的复选项，第2项默认选中，触发事件
          public void onClick(DialogInterface dialog, int which, boolean
             isChecked) {
                 flags[which]=isChecked;
                 String result = "您选择的是：";
                 for (int i = 0; i < flags.length; i++) {  //确定已选择的选项
                     if(flags[i]){
                         result=result+items[i]+"、"; } }
                Toast.makeText(getApplicationContext(), result, Toast.LENGTH_
                    SHORT).show();}});                //显示选择的选项
```

```
            //添加一个确定按钮
        builder.setPositiveButton(" 确定 ", new DialogInterface.OnClickListener(){
            public void onClick(DialogInterface dialog, int which) { } });
        AlertDialog alert = builder.create();        //创建 Dialog
        alert.show(); }}                             //显示对话框
private class singleAlertImpl implements OnClickListener {
    public void onClick(View v) {
    final CharSequence[] items={"A.1+1=2","B.1+1=3","C.1+2=2","D.1+3=2"};
        //定义数组
    AlertDialog.Builder builder = new AlertDialog.Builder(MainActivity.this);
        //实例化对象
    builder.setTitle("请选择您认为正确的答案");              //设置显示标题
    //设置要显示的单选项,第 2 项默认选中,触发事件
    builder.setSingleChoiceItems(items,1,new DialogInterface.OnClickListener(){
        public void onClick(DialogInterface dialog, int item){ //事件的具体操作
            Toast.makeText(getApplicationContext(), "您选择的是:"+items
                [item], Toast.LENGTH_SHORT).show();}});//显示选择的选项
        //添加一个确定按钮
        builder.setPositiveButton(" 确 定 ", new DialogInterface.
            OnClickListener(){
         public void onClick(DialogInterface dialog, int which) { } });
            AlertDialog alert = builder.create();         //创建 Dialog
            alert.show(); }}                              //显示对话框
private class listAlertImpl implements OnClickListener {
    public void onClick(View v) {
    final CharSequence[] items = { "飞机", "轮船", "火车" };  //定义数组
    AlertDialog.Builder builder=new AlertDialog.Builder(MainActivity.this);
                                                //实例化对象
    builder.setTitle("请选择您喜爱的交通工具");            //设置显示标题
        //设置对话框要显示的 list,触发事件
    builder.setItems(items, new DialogInterface.OnClickListener(){
        public void onClick(DialogInterface dialog, int item){
                                                //事件的具体操作
        Toast.makeText(getApplicationContext(), "您选择的是:"+items[item],
            Toast.LENGTH_SHORT).show();} });       //显示选择的选项
        AlertDialog alert = builder.create();       //创建 Dialog
        alert.show(); }} }                          //调用 Show()显示
```

保存文件并运行该项目,出现如图 5.10 所示的界面,单击各按钮显示不同的警告框。如图 5.11~图 5.15 所示。

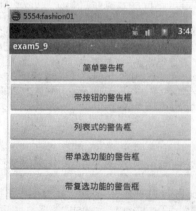

图 5.10 实例 5-9 运行结果

图 5.11 简单警告框

第 5 章 Android 常用高级组件

图 5.12 带按钮的警告框

图 5.13 带列表项的警告框

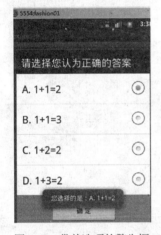

图 5.14 带单选项的警告框

图 5.15 带复选项的警告框

5.8.3 自定义对话框

前面介绍的对话框，界面比较简单，在程序设计的时候往往需要显示一些复杂的界面，要求定制一些对话框，例如：登录提示的对话框。可以这样来实现，先定义一个布局文件定义显示需要显示的组件，然后再将这个布局文件显示包含到对话框之中，这个包含需要 Android 的 LayoutInflater 类的支持。LayoutInflater 类主要的方法有两个：

```
public static LayoutInflater from(Context context)：创建 LayoutInflater 对象
public View inflate(int resource, ViewGroup root)：将布局文件转变为 View 对象
```

实例 5-10：各种属性的运用

新建一个项目，项目的命名为：exam5_10，包名称为：org.hnist.demo，利用 LayoutInflater 类建立一个登录提示的对话框。

1) 在 res\layout 目录下，建立一个 login.xml 文件，代码如下：

```
<?xml version="1.0" encoding="utf-8"?>
<TableLayout
```

```xml
    xmlns:android="http://schemas.android.com/apk/res/android"
    android:id="@+id/LoginLayout"
    android:layout_width="fill_parent"
    android:layout_height="fill_parent">
    <TableRow>
        <TextView
            android:text="用户名："
            android:layout_marginLeft="18dip"
            android:textSize="10pt"
            android:layout_width="wrap_content"
            android:layout_height="wrap_content"/>
        <EditText
            android:width="65pt"
            android:layout_height="wrap_content"/>
    </TableRow>
    <TableRow>
        <TextView
            android:text="密    码："
            android:layout_marginLeft="18dip"
            android:textSize="10pt"
            android:layout_width="wrap_content"
            android:layout_height="wrap_content"/>
        <EditText
            android:password="true"
            android:width="65pt"
            android:layout_height="wrap_content"/>
    </TableRow>
</TableLayout>
```

2）修改 activity_main.xml 文件，代码如下：

```xml
<?xml version="1.0" encoding="utf-8"?>
<LinearLayout
    xmlns:android="http://schemas.android.com/apk/res/android"
    android:id="@+id/MainLayout"
    android:orientation="horizontal"
    android:layout_width="fill_parent"
    android:layout_height="fill_parent">
    <Button
        android:id="@+id/loginBut"
        android:text="自定义的警告框"
        android:layout_width="fill_parent"
        android:layout_height="wrap_content"/>
</LinearLayout>
```

3）修改 MainActivity.java 文件，代码如下：

```java
package org.hnist.demo;
package org.hnist.demo;
import android.app.Activity;
import android.app.AlertDialog;
import android.app.Dialog;
import android.content.DialogInterface;
```

```java
import android.os.Bundle;
import android.view.LayoutInflater;
import android.view.View;
import android.view.View.OnClickListener;
import android.widget.Button;
public class MainActivity extends Activity {
    private Button loginbut = null ;
    public void onCreate(Bundle savedInstanceState) {
        super.onCreate(savedInstanceState);
        super.setContentView(R.layout.activity_main);
        this.loginbut = (Button) super.findViewById(R.id.loginBut) ;
        this.loginbut.setOnClickListener(new OnClickListenerImpl()) ;}  //设置单击事件
    private class OnClickListenerImpl implements OnClickListener {
        public void onClick(View v) {
            LayoutInflater flater=LayoutInflater.from(MainActivity.this);  //创建
                LayoutInflater 对象
            View dialogView = flater.inflate(R.layout.login,null); //将布局文件
                转换为View
            Dialog dialog = new AlertDialog.Builder(MainActivity.this)    //创
                建Dialog
                    .setIcon(R.drawable.c12)                 //设置显示图片
                    .setTitle("用户登录界面")                    //设置显示标题
                    .setView(dialogView)                     //设置显示组件
                    .setPositiveButton("确定",                //设置确定按钮
                        new DialogInterface.OnClickListener() {//单击确定触发的事件
                            public void onClick(DialogInterface dialog, int
                                whichButton) {}})
                    .setNegativeButton("取消",                //设置取消按钮
                        new DialogInterface.OnClickListener() {//单击取消触
                                                                发的事件
                            public void onClick(DialogInterface dialog, int
                                whichButton) {}}).create();  //创建对话框
            dialog.show();  }} }                             //显示对话框
```

保存文件并运行该项目,结果弹出一个按钮,单击该按钮,弹出如图 5.16 所示界面。

5.8.4 带进度条的对话框 ProgressDialog

在进行复杂操作时往往需要有一段操作时间,这时会弹出一个等待的对话框,用户可以通过进度处理对话框显示进度的相关情况,进度处理对话框(ProgressDialog)可以实现这样的功能。层次关系如下所示:

```
java.lang.Object
    android.app.Dialog
        android.app.AlertDialog
            android.app.ProgressDialog
```

图5.16 自定义警告对话框

要在 Android 程序中使用 ProgressDialog 组件必须要在程序中使用下面的语句:

```
        import android.widget.ProgressDialog;    //导入widget.ProgressDialog类
```

从上面可以看出 ProgressDialog 是 AlertDialog 子类,所以 AlertDialog 的属性和方法 ProgressDialog 都具备,此外,还有表 5-20 所列出的常用方法。

表 5-20 ProgressDialog 常用的方法

方法	描述
public static final int STYLE_HORIZONTAL	常量,水平进度显示风格
public static final int STYLE_SPINNER	常量,环形进度显示风格
public ProgressDialog(Context context)	创建进度对话框
public void setMessage(CharSequence message)	设置显示信息
public int getMax()	获得进度条的最大增长值
public int getProgress()	获得当前进度
public int getSecondaryProgress()	获得次要进度条
public void onStart()	启动进度框
public void setProgressStyle (int style)	设置进度条的显示风格
public static ProgressDialog show (Context context, CharSequence title, CharSequence message)	直接创建进度对话框,指定对话框的标题及信息
public void incrementProgressBy(int diff)	设置进度条每次增长的值
public void setMax(int max)	设置进度条的最大增长值
public void setProgress(int value)	设置当前进度

实例 5-11:ProgressDialog 组件实例

新建一个项目,项目的命名为:exam5_11,包名称为:org.hnist.demo,利用 ProgressDialog 类分别建立圆形和水平进度提示对话框。

1)修改 activity_main.xml 文件,代码如下:

```
<RelativeLayout
<?xml version="1.0" encoding="utf-8"?>
<LinearLayout
xmlns:android="http://schemas.android.com/apk/res/android"
android:id="@+id/MainLayout"
android:orientation="vertical"
android:layout_width="fill_parent"
android:layout_height="fill_parent">
<Button
    android:id="@+id/circleBut"
    android:text="圆形进度条"
    android:layout_width="fill_parent"
    android:layout_height="wrap_content"/>
<Button
    android:id="@+id/lineBut"
    android:text="水平进度条"
    android:layout_width="fill_parent"
    android:layout_height="wrap_content"/>
</LinearLayout>
```

2)修改 MainActivity.java 文件,代码如下:

```
package org.hnist.demo;
import android.app.Activity;
import android.app.ProgressDialog;              //导入app.ProgressDialog类
```

```java
import android.content.DialogInterface;
import android.os.Bundle;
import android.view.View;
import android.widget.Button;
public class MainActivity extends Activity {
private Button circleBut,lineBut;
int count= 0;
ProgressDialog ProDialog;                          //声明进度条对话框
    public void onCreate(Bundle savedInstanceState) {
        super.onCreate(savedInstanceState);
        setContentView(R.layout.activity_main);
        circleBut = (Button)this.findViewById(R.id.circleBut);
        lineBut = (Button)this.findViewById(R.id.lineBut);
        circleBut.setOnClickListener(new Button.OnClickListener(){ //单击事件
          public void onClick(View v) {
            ProDialog = new ProgressDialog(MainActivity.this);   //定义进度对话框
            //设置进度条风格，风格为圆形
            ProDialog.setProgressStyle(ProgressDialog.STYLE_SPINNER);
            ProDialog.setTitle("网络连接");                    //设置标题
            ProDialog.setMessage("正在连接网络，请等待......"); //设置提示信息
            ProDialog.setIndeterminate(false);            //设置进度条是否不明确
            ProDialog.setCancelable(true);                //设置是否可按返回键取消
            ProDialog.setButton("确定", new DialogInterface.OnClickListener()
                        {//设置按钮
            public void onClick(DialogInterface dialog, int which) {
            ProDialog.cancel();}});                  //单击"确定"按钮取消对话框
            ProDialog.show();} });
            lineBut.setOnClickListener(new Button.OnClickListener() {
                //单击事件
            public void onClick(View v) {
            count= 0;                                     //计数
            ProDialog = new ProgressDialog(MainActivity.this);   //定义进度对话框
            //设置进度条风格，风格为水平的
            ProDialog.setProgressStyle(ProgressDialog.STYLE_HORIZONTAL);
            ProDialog.setTitle("程序安装");             //设置标题
            ProDialog.setMessage("程序正在安装，请等待......"); //设置提示信息
            ProDialog.setIndeterminate(false);            //设置进度条是否不明确
            ProDialog.setCancelable(true);                //设置是否可按返回键取消
            ProDialog.show();                             //显示进度条对话框
            new Thread(){                                 //定义线程对象
               public void run(){                         //线程主体方法
                  while(count<=100){                      //从1到100循环
                     ProDialog.setProgress(count++);      //由线程来控制进度条
                     try {Thread.sleep(100);}             //休眠0.1秒
                       catch (InterruptedException e) {
                          e.printStackTrace();}}
            ProDialog.cancel();}}.start();} });  }}       //线程启动
```

　　保存文件并运行该项目，弹出两个按钮，分别单击弹出两个按钮，如图5.17和图5.18所示。在水平进度条对话框中用了一个线程，模拟进度情况。

图 5.17 圆形进度对话框

图 5.18 水平进度对话框

5.9 图片切换组件 ImageSwitcher

在 Windows 平台上要查看多张图片，可以通过 Window 图片和传真查看器在"下一张"和"上一张"之间切换，在 Android 平台可以通过 ImageSwitcher 组件达到效果。ImageSwitcher 组件的主要功能是完成图片的切换显示，其层次关系如下：

```
java.lang.Object
    android.view.View
        android.view.ViewGroup
            android.widget.FrameLayout
                android.widget.ViewAnimator
                    android.widget.ViewSwitcher
                        android.widget.ImageSwitcher
```

要在 Android 程序中使用 ImageSwitcher 组件，必须要在程序中使用下面的语句：

```
import android.widget.ImageSwitcher;    //导入 widget.ImageSwitcher 类
```

从上面可以看出 ImageSwitche 是 View 的子类，所以 View 类的属性和方法 ImageSwitcher 都具备，此外，还有表 5-21 所列出的常用方法。

表 5-21 ImageSwitcher 类的常用方法

方法	描述
public ImageSwitcher(Context context)	创建 ImageSwitcher 对象
public void setFactory(ViewSwitcher.ViewFactory factory)	设置 ViewFactory 对象，用于完成两个图片切换时 ViewSwitcher 的转换操作
public void setImageResource(int resid)	设置显示的图片 ID
public void setInAnimation(Animation inAnimation)	图片读进 ImageSwitcher 时的动画效果
public void setOutAnimation(Animation outAnimation)	图片从 ImageSwitcher 消失时的动画效果
public void setImageDrawable (Drawable drawable)	绘制图片
public void setImageURI (Uri uri)	设置图片地址，可读取网络图片

利用 ImageSwitcher 浏览图片的基本步骤如下：
1）定义需要显示的图片，例如：

```java
private int[] images=new int[]{R.drawable.a11,R.drawable.a12,R.drawable.13,
    R.drawable.a14,R.drawable.a15 };   //图片数据
```

2）调用 setFactory 方法，返回一个 View 对象，例如：

```java
private class ViewFactoryImp implements ViewFactory{
  public View makeView() {   //覆写 makeView()方法
    ImageView img=new ImageView(MainActivity.this); //定义 ImageView 组件
    img.setBackgroundColor(0xFFFFFF);   //设置背景
    //按图片居中显示，当图片长/宽超过 View 的长/宽，则截取图片的居中部分显示
    img.setScaleType(ImageView.ScaleType.CENTER);
    img.setLayoutParams(new ImageSwitcher.LayoutParams(
    LayoutParams.FILL_PARENT,LayoutParams.FILL_PARENT));
    return img;}}}
```

3）利用 setImageResource 方法指定图片来源，例如：

```java
this.imageswitch.setImageResource(this.images[this.count]);
```

4）利用 setInAnimation 方法指定图片读入到这个 ImageSwitcher 中时实现的动画效果，一般动画效果是从 Android.R 系统文件里读取的。

例如：

```java
this.imageswitch.setInAnimation(AnimationUtils.loadAnimation(this,android
    .R.anim.fade_in));   //进入时候的动画效果
this.imageswitch.setOutAnimation( AnimationUtils.loadAnimation(this,andro
    id.R.anim.fade_out));} //离开时候的动画效果
```

5）显示指定图片，例如：

```java
MainActivity.this.imageswitch.setImageResource(MainActivity.this.images[-
    -MainActivity.this.count]);}//获得指定图片，并将图片计数减 1
```

5.9.1 ImageSwitcher 组件实例

实例 5-12：ImageSwitcher 组件实例

新建一个项目，项目的命名为：exam5_12，包名称为：org.hnist.demo，利用 ImageSwitcher 组件实现图片的浏览。

1）修改 activity_main.xml 文件，代码如下：

```xml
<LinearLayout xmlns:android="http://schemas.android.com/apk/res/android"
    xmlns:tools="http://schemas.android.com/tools"
    android:layout_width="match_parent"
    android:layout_height="match_parent"
    android:orientation="vertical"
    tools:context=".MainActivity" >
    <ImageSwitcher
        android:id="@+id/myImageS"
        android:layout_width="fill_parent"
        android:layout_height="wrap_content"/>
  <LinearLayout
    android:layout_width="match_parent"
    android:layout_height="match_parent"
    android:orientation="horizontal" >
    <Button
```

```xml
        android:id="@+id/previous"
        android:layout_width="wrap_content"
        android:layout_height="wrap_content"
        android:text="上一张"/>
    <Button
        android:id="@+id/next"
        android:layout_width="wrap_content"
        android:layout_height="wrap_content"
        android:text="下一张"/>
    </LinearLayout>
</LinearLayout>
```

2）修改 MainActivity.java 文件，代码如下：

```java
package org.hnist.demo;
import android.os.Bundle;
import android.app.Activity;
import android.view.View;
import android.view.View.OnClickListener;
import android.view.animation.AnimationUtils;
import android.widget.Button;
import android.widget.ImageSwitcher;            //导入 widget.ImageSwitcher 类
import android.widget.ImageView;
import android.widget.LinearLayout.LayoutParams;
import android.widget.Toast;
import android.widget.ViewSwitcher.ViewFactory; //导入 ViewSwitcher.ViewFactory 类
public class MainActivity extends Activity {
private ImageSwitcher imageswitch=null;      //定义 ImageSwitcher 组件
private Button previous=null;
private Button next=null;
private int count=0;                         //用于计数的变量
private int[] images=new int[]{R.drawable.a11,R.drawable.a12,R.drawable.13,
    R.drawable.a14,R.drawable.a15 };   //图片数据
protected void onCreate(Bundle savedInstanceState) {
super.onCreate(savedInstanceState);
setContentView(R.layout.activity_main);
//获得 ImageSwitcher 组件
this.imageswitch=(ImageSwitcher) super.findViewById(R.id.myImageS);
this.previous=(Button) super.findViewById(R.id.previous);
this.next=(Button) super.findViewById(R.id.next);
this.previous.setOnClickListener(new OnPreButClickImp()); //为上一张按钮添加方法
this.next.setOnClickListener(new OnNextButClickImp());    //为下一张按钮添加方法
this.imageswitch.setFactory(new ViewFactoryImp());        //设置转换工厂
this.imageswitch.setImageResource(this.images[this.count]);
this.imageswitch.setInAnimation(AnimationUtils.loadAnimation(this,android
    .R.anim.fade_in));                    //进入时候的动画效果
this.imageswitch.setOutAnimation( AnimationUtils.loadAnimation(this,andro
    id.R.anim.fade_out));}                //离开时候的动画效果
public class OnPreButClickImp implements OnClickListener{//上一张按钮触发的事件
    public void onClick(View v) {
        if(MainActivity.this.count!=0){
            MainActivity.this.imageswitch.setImageResource(MainActivity.
                this.images[--MainActivity.this.count]);}//获得指定图片，并
                    将图片计数减 1
        else{Toast.makeText(MainActivity.this, "已经是第一张，不能往前了！",
```

```
        Toast.LENGTH_SHORT).show();}}}
public class OnNextButClickImp implements OnClickListener{//下一张按钮触发的事件
    public void onClick(View v) {
        if(MainActivity.this.count!=MainActivity.this.images.length-1){
            MainActivity.this.imageswitch.setImageResource(MainActivity.this.
                images[++MainActivity.this.count]);} //获得指定图片,并将图片计数加1
        else{
Toast.makeText(MainActivity.this,"已经是最后一张!不能往后了!",Toast.LENGTH_
    SHORT).show();}}}
private class ViewFactoryImp implements ViewFactory{
    public View makeView() {   //覆写makeView()方法
        ImageView img=new ImageView(MainActivity.this); //定义ImageView组件
        img.setBackgroundColor(0xFFFFFF); //设置背景
        //按图片居中显示,当图片长/宽超过View的长/宽,则截取图片的居中部分显示
        img.setScaleType(ImageView.ScaleType.CENTER);
        img.setLayoutParams(new ImageSwitcher.LayoutParams(
            LayoutParams.FILL_PARENT,LayoutParams.FILL_PARENT));
        return img;}}}
```

保存文件并运行该项目,结果如图 5.19 所示,单击"下一张"按钮,能实现图片的浏览功能。

5.10 画廊组件 Gallery

如果在手机上要显示很多图片,可以将这些图片形成一个画廊,拖拉画廊中的图片实现浏览,在 Android 系统中可以通过 Gallery 组件实现画廊功能,其层次关系如下所示:

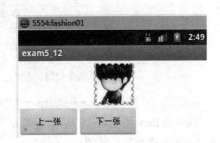

图 5.19 利用 ImageSwitcher 组件浏览图片

```
java.lang.Object
    android.view.View
        android.view.ViewGroup
            android.widget.AdapterView<T extends android.widget.Adapter>
                android.widget.AbsSpinner
                    android.widget.Gallery
```

要在 Android 程序中使用 Gallery 组件必须要在程序中使用下面的语句。

```
import android.widget.Gallery;              //导入widget.Gallery类
```

从上面可以看出 Gallery 是 View 的子类,所以 View 类的属性和方法 Gallery 都具备,此外,还有表 5-22 所列出的常用属性和表 5-23 所列出的常用方法。

表 5-22 Gallery 类中的常用属性

属性	描述
android:gravity	指定在对象的 X 和 Y 轴上如何放置内容
android:spacing	设置图片之间的间距
android:unselectedAlpha	设置未选中的条目的透明度(Alpha)
android:animationDuration	设置布局变化时动画转换所需的时间

表 5-23　Gallery 类中的常用方法

方法	描述
public Gallery(Context context)	创建 Gallery 对象
public void setSpacing(int spacing)	设置两个图片之间的显示间距
public void setAdapter(SpinnerAdapter adapter)	设置图片集
public void setGravity(int gravity)	设置图片的对齐方式
public void setAnimationDuration(int animationDurationMillis)	设置动画转换时间
public void setSpacing (int spacing)	设置 Gallery 中项的间距
public void setUnselectedAlpha(float unselectedAlpha)	设置 Gallery 中未选中项的透明度（alpha）值
public void setOnItemClickListener (AdapterView.OnItemClickListener listener)	设置选项单击事件

Gallery 使用 setAdapter()方法设置要显示的图片集合，要定义适配器，而对于这种适配器操作有两种可选方式。

方式一：用一个自定义的类直接继承 android.widget.BaseAdapter 类实现。

采用这种方式需要覆写 BaseAdapter 类的几个方法，例如：

```java
public abstract int getCount()                //取得图片集中图片的个数
public abstract Object getItem(int position)  //取得一个指定位置的图片对象
public abstract long getItemId(int position)  //取得一个指定位置的对象 ID
public abstract View getView(int position, View convertView, ViewGroup parent)
    //取得一个指定位置的视图显示
```

方式二：直接使用 SimpleAdapter 类完成，通过自定义的显示模板显示所有图片资源。

利用 Gallery 浏览图片的基本步骤如下：

1）定义需要显示的图片，例如：

```java
private Integer[] myImages = { R.drawable.a11, R.drawable.a12,
    R.drawable.a13, R.drawable.a14, R.drawable.a15, R.drawable.a16 };
```

2）利用 BaseAdapter 或 SimpleAdapter 设置图片集，例如：

```java
public class ImageAdapter extends BaseAdapter{//定义ImageAdapter 类继承BaseAdapter
……
```

3）通过 setImageResource 方法来设置需要显示的图片，例如：

```java
myimg.setImageResource(myImages[position]); //设定图片给 imageView 对象
```

实例 5-13：Gallery 组件实例

新建一个项目，项目的命名为：exam5_13，包名称为：org.hnist.demo，利用 Gallery 组件实现图片的浏览。

1）修改 activity_main.xml 文件，代码如下：

```xml
<?xml version="1.0" encoding="utf-8"?>
<LinearLayout
    xmlns:android="http://schemas.android.com/apk/res/android"
    xmlns:tools="http://schemas.android.com/tools"
    android:layout_width="match_parent"
    android:layout_height="match_parent"
    android:orientation="vertical"
    tools:context=".MainActivity" >
```

```xml
<Gallery
  android:id="@+id/gallery"
  android:layout_width="fill_parent"
  android:layout_height="wrap_content"/>
</LinearLayout>
```

2）在 res/values 下新建一个 attrs.xml 文件，代码如下：

```xml
<?xml version="1.0" encoding="utf-8"?>
<resources>
  <declare-styleable name="Gallery">
    <attr name="android:galleryItemBackground" />
  </declare-styleable>
</resources>
```

这是一个引用自制 layout 元素的用法，必须在 res/values 下面添加一个 attrs.xml，并在其中定义 <declare-styleable>选项，目的是自定义 layout 的背景风格，并且通过 TypeArray 的特性，让相同的 Layout 元素可以重复用于每一张图片。

3）修改 MainActivity.java 文件，代码如下：

```java
package org.hnist.demo;
import android.app.Activity;
import android.os.Bundle;
import android.content.Context;
import android.content.res.TypedArray;
import android.view.View;
import android.view.ViewGroup;
import android.widget.AdapterView;
import android.widget.BaseAdapter;
import android.widget.Gallery;
import android.widget.ImageView;
import android.widget.Toast;
import android.widget.AdapterView.OnItemClickListener;
public class MainActivity extends Activity {
 public void onCreate(Bundle savedInstanceState) {
    super.onCreate(savedInstanceState);
    setContentView(R.layout.activity_main);
    Gallery gallery = (Gallery) findViewById(R.id.gallery); //获得Gallery
                                                        对象gallery
    gallery.setAdapter(new ImageAdapter(this));      //新增ImageAdapter,设定
                                                        给gallery
gallery.setOnItemClickListener(new OnItemClickListener() { //设置监听
    public void onItemClick(AdapterView parent, View v, int position, long id) {
        Toast.makeText(MainActivity.this, "这是第" + position+ "张图片",
          Toast.LENGTH_SHORT).show(); } }); }
public class ImageAdapter extends BaseAdapter{//定义ImageAdapter类继承BaseAdapter
    int mGalleryItemBackground;
    private Context mContext;
    public ImageAdapter(Context c) {        //构建ImageAdapter
    mContext = c;
    TypedArray args=obtainStyledAttributes(R.styleable.Gallery);//使用指定
                                                        Gallery 属性
```

```
        mGalleryItemBackground=args.getResourceId(R.styleable.Gallery_android_
            galleryItemBackground, 0);     //取得 Gallery 属性的 Index id
        args.recycle(); } //让对象的 styleable 属性能够反复使用
        public int getCount() { return myImages.length; }   //重写方法 getCount,
                                                             传回图片数目
        public Object getItem(int position){return position;}  //重写方法 getItem,
                                                             传回 position
        public long getItemId(int position){return position;}  //重写 getItemId,
                                                             传回 position
     //重写方法 getView,传回一 View 对象
        public View getView(int position, View convertView, ViewGroup parent) {
          ImageView myimg = new ImageView(mContext);   //定义 ImageView 对象
          myimg.setImageResource(myImages[position]);  //设定图片给 imageView 对象
          myimg.setScaleType(ImageView.ScaleType.FIT_XY);  //重新设定图片的宽高/
          myimg.setLayoutParams(new Gallery.LayoutParams(140,90));//设 Layout 宽高
          myimg.setBackgroundResource(mGalleryItemBackground);//设定 Gallery 背景图
            return myimg; }     //传回 imageView
     //设置图片数据来源
        private Integer[] myImages = { R.drawable.a11, R.drawable.a12, R.
            drawable.a13, R.drawable.a14, R.drawable.a15, R.drawable.a16,
            R.drawable.a17}; } }
```

保存文件并运行该项目,结果如图 5.20 所示,拖动画廊能实现图片的浏览,单击某张图片,能触发相应的事件。

实例 5-14:Gallery 组件与 ImageSwitcher 组件综合实例

新建一个项目,项目的命名为:exam5_14,包名称为:org.hnist.demo,利用 Gallery 组件和 ImageSwitcher 组件实现图片的浏览。

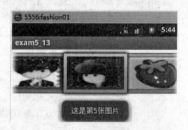

图 5.20 利用 Gallery 组件浏览图片

1) 修改 activity_main.xml 文件,代码如下:

```xml
<?xml version="1.0" encoding="utf-8"?>
<LinearLayout
  xmlns:android="http://schemas.android.com/apk/res/android"
  android:id="@+id/Layout"
  android:orientation="vertical"
  android:layout_width="fill_parent"
  android:layout_height="fill_parent"
  android:gravity="top">
<ImageSwitcher                         增加一个 ImageSwitcher 组件
    android:id="@+id/ImageSwitcher"    命名为 ImageSwitcher
    android:layout_width="fill_parent"
    android:layout_height="wrap_content"/>
<Gallery                               增加一个 Gallery 组件
    android:id="@+id/Gallery"          命名为 Gallery
    android:gravity="center_vertical"  垂直居中显示
    android:spacing="2px"              图片之间相距 2px
    android:layout_width="fill_parent"
    android:layout_height="wrap_content"/>
</LinearLayout>
```

2）在 res/values 下新建一个 islayout.xml 文件，代码如下：

```xml
<?xml version="1.0" encoding="utf-8"?>
<LinearLayout
    xmlns:android="http://schemas.android.com/apk/res/android"
    android:orientation="horizontal"
    android:layout_width="wrap_content"
    android:layout_height="wrap_content"
    android:background="#00FFFF00">
    <ImageView
        android:id="@+id/myimg"
        android:layout_width="wrap_content"
        android:layout_height="wrap_content"
        android:scaleType="center"/>
</LinearLayout>
```

3）修改 MainActivity.java 文件，代码如下：

```java
package org.hnist.demo;
import java.lang.reflect.Field;
import java.util.ArrayList;
import java.util.HashMap;
import java.util.List;
import java.util.Map;
import android.app.Activity;
import android.os.Bundle;
import android.view.View;
import android.widget.AdapterView;
import android.widget.AdapterView.OnItemClickListener;
import android.widget.Gallery;
import android.widget.ImageSwitcher;
import android.widget.ImageView;
import android.widget.LinearLayout.LayoutParams;
import android.widget.SimpleAdapter;
import android.widget.ViewSwitcher.ViewFactory;
public class MainActivity extends Activity {
    private Gallery gallery = null;                          //定义 Gallery 对象
    private List<Map<String,Integer>> list = new ArrayList<Map<String,Integer>>() ;
    private SimpleAdapter simpleAdapter = null;              //定义适配器
    private ImageSwitcher ImageSwitcher = null ;             //定义 ImageSwitcher 对象
    public void onCreate(Bundle savedInstanceState) {
        super.onCreate(savedInstanceState);
        super.setContentView(R.layout.activity_main);
        this.initAdapter() ;                                 //初始化适配器
        this.gallery = (Gallery) super.findViewById(R.id.Gallery) ;//取得Gallery组件
        //取得 ImageSwitcher 组件
        this.ImageSwitcher = (ImageSwitcher) super.findViewById(R.id.ImageSwitcher);
        this.ImageSwitcher.setFactory(new ViewFactoryImpl()) ;//设置图片工厂
        this.gallery.setAdapter(this.simpleAdapter);         //设置图片集
        this.gallery.setOnItemClickListener(new OnItemClickListenerImpl()) ;}
            //设置单击事件
    private class OnItemClickListenerImpl implements OnItemClickListener {
        public void onItemClick(AdapterView<?> parent, View view, int position,
            long id) {
```

```
                Map<String Integer>map=(Map<String, Integer>)MainActivity.this.
                    simpleAdapter.
            getItem(position);       //取出Map
        MainActivity.this.ImageSwitcher.setImageResource(map.get("myimg"));}}
            //设置显示图片
    public void initAdapter(){                                  //初始化适配器
        Field[] fields = R.drawable.class.getDeclaredFields();
        for (int x = 0; x < fields.length; x++) {
            if (fields[x].getName().startsWith("a")){           //所有a命名的图片
                Map<String,Integer> map = new HashMap<String,Integer>() ;//定义Map
                try {
                    map.put("myimg", fields[x].getInt(R.drawable.class)) ;
                        //设置图片资源
                } catch (Exception e) { }
                this.list.add(map) ;}    }                      //保存Map
        this.simpleAdapter = new SimpleAdapter(this,            //实例化SimpleAdapter
                this.list,                                      //要包装的数据集合
                R.layout.is_layout,                             //要使用的显示模板
                new String[] { "myimg" },                       //定义要显示的Map的Key
                new int[] {R.id.myimg });       }               //与模板中的组件匹配
    private class ViewFactoryImpl implements ViewFactory {      //定义视图工厂类
        public View makeView() {
            ImageView img = new ImageView(MainActivity.this);//实例化图片显示
            img.setBackgroundColor(0xFFFFFFFF);                 //设置背景颜色
            img.setScaleType(ImageView.ScaleType.CENTER);       //居中显示
            img.setLayoutParams(new ImageSwitcher.LayoutParams(//自适应图片大小
                LayoutParams.FILL_PARENT, LayoutParams.FILL_PARENT));//定义组件
            return img; }}    }
```

图 5.21 利用 Gallery 和 ImageSwitcher 浏览图片

保存文件并运行该项目，结果如图 5.21 所示，拖动下面的画廊能实现图片的浏览，单击某张图片，能将这张图片显示在 ImageSwitcher 里。

上面的 Gallery 中的适配器采用的是 SimpleAdapter，也可以用 BaseAdapter，通过覆写几个方法来实现。有人可能会问，如果想要上面的图片大一些，如何操作呢？有两个办法：一是在 ImageSwitcher 里显示图片时候图片放大，这样可能会导致失真，参考实例 exam5-14-1 给出的代码；再一个就是准备大图和小图集合，小图集合给 Gallery 用，大图给 ImageSwitcher 用，注意对应图的顺序不要错了。

还有一个网格视图组件 GridView，其操作与 Gallery 组件类似，这里不再赘述，感兴趣的用户可以参考 API 文档学习。

5.11 选项卡组件 TabHost

Tab 选项卡是界面设计时经常使用的界面组件，可以实现多个分页之间的快速切换，每个分页可以显示不同的内容，在 Android 平台提供了 TabHost 组件实现 Tab 选项卡的功能，选项卡组件的主要功能是可以进行应用程序分类管理。每个选项卡称为一个 Tab，而包含这多个选项卡的容器称为 TabHost，TabHost 类的层次关系如下：

```
java.lang.Object
    android.view.View
        android.view.ViewGroup
            android.widget.FrameLayout
                android.widget.TabHost
```

TabHost 是整个 Tab 的容器，包括两部分，TabWidget 和 FrameLayout。其中 TabWidget 是每个 Tab 的选项，FrameLayout 则是 tab 内容。

要在 Android 程序中使用 TabHost 组件，必须要在程序中使用下面的语句：

```
import android.widget.TabHost;                    //导入 widget.TabHost 类
```

5.11.1 TabHost 组件基础

1．TabHost 组件常见的方法

TabHost 组件有如表 5-24 所示的常用方法。

表 5-24 TabHost 组件常用方法

方法	描述
public TabHost(Context context)	创建 TabHost 类对象
public void addTab(TabHost.TabSpec tabSpec)	增加一个 Tab
public TabHost.TabSpec newTabSpec(String tag)	创建一个 TabHost.TabSpec 对象
public void clearAllTabs ()	清除所有关联到当前 TabHost 的选项卡
public View getCurrentView()	取得当前的 View 对象
public int getCurrentTab()	获取当前选项卡的 ID
public String getCurrentTabTag()	获得当前选项卡的 Tag 选项内容
public View getCurrentTabView()	获取当前选项卡的视图 view
public TabHost.TabSpec newTabSpec(String tag)	获取一个 TabHost.TabSpec，并关联到当前 TabHost
public void setup()	建立 TabHost 对象
public void setCurrentTab(int index)	设置当前显示的 Tab 编号
public void setCurrentTabByTag(String tag)	设置当前显示的 Tab 名称
public FrameLayout getTabContentView()	获取保存选项卡内容
public void setOnTabChangedListener (TabHost.OnTabChangeListener l)	设置选项改变时触发
public void setup()	使用 findViewById()加载 TabHost，在新增一个选项卡之前，需要调用它

如果要实现选项卡的显示界面，有两种实现途径：

1）直接让一个 Activity 程序继承 TabActivity 类；

2）利用 findViewById()方法取得 TagHost 组件，要进行一些必要的配置。

如果使用 findViewById()方法取得 TagHost 组件，那么在新增一个选项卡之前，需要调用 setup()方法。在 TabActivity 里使用 getTabHost()方法获取 TabHost 组件，就不需要调用 setup()方法。

2．直接让一个 Activity 程序继承 TabActivity 类的方式实现选项卡功能

这种方法相对比较简单，直接用 TabActivity 类提供的方法来实现操作，TabActivity 类提供的常用方法如表 5-25 所示。

可以直接用 getTabHost()方法获得一个 TabHost 类的对象，不能使用 findViewById()方法进行 TabHost 对象的实例化，要通过 LayoutInflater 类完成布局管理器中定义组件的实例化操作。

表 5-25　TabActivity 类的常用方法

方法	描述
public TabActivity()	无参构造，方便子类继承时调用
public TabHost getTabHost()	取得 TabHost 类的对象
public TabWidget getTabWidget()	取得 TabWidget 类的对象
public void setDefaultTab(String tag)	设置默认选中的选项
public void setDefaultTab(int index)	设置默认选中的选项

　　LayoutInflater 类的作用类似于前面介绍过的 findViewById()，不同点是 LayoutInflater 是找 layout 下的布局管理文件，并且将它实例化，而 findViewById()是找布局管理文件下的具体组件（如：Button、EditText 等）。LayoutInflater 类常用的方法如表 5-26 所示。

表 5-26　LayoutInflater 类的常用方法

方法	描述
LayoutInflater(Context context)	创建一个LayoutInflater对象
public View inflate (int resource, ViewGroup root)	设置所需要的布局管理器的资源 ID，设置组件的容器及是否包含设置组件的参数
public View inflate(int resource, ViewGroup root, boolean attachToRoot)	设置所需要的布局管理器的资源 ID，设置组件的容器及是否包含设置组件的参数
public static LayoutInflater from(Context context)	从指定的容器之中获得 LayoutInflater 对象

　　其中，resource：View 的 layout 的 ID；root：如果返回 null，则将此 View 作为根，此时就可以应用此 View 中的其他控件了，如果返回非 null，则将默认的 layout 作为 View 的根；attachToRoot：如果为 true 就将此解析的 xml 作为 View 根，为 false 则为默认的 xml 为根视图 View。

　　Android 程序想要创建一个界面的时候，一般的做法是新建一个类，继承 Activity 基类，然后在 onCreate 里面使用 setContentView 方法来载入一个在 XML 里定义好的界面。

　　其实在 Activity 程序里面就是使用了 LayoutInflater 类来载入 XML 文件指定界面的，通过 getSystemService(Context.LAYOUT_INFLATER_SERVICE)方法可以获得一个 LayoutInflater 对象，然后使用 inflate 方法来载入 layout 的 XML 文件，对于一个没有被载入或者想要动态载入的界面，需要使用 inflate 来载入。对于一个已经载入的界面，可以使用这个界面调用 findViewById 方法来获得其中的组件了。

　　getSystemService()是 Android 很重要的一个 API，它是 Activity 的一个方法，根据传入的 NAME 来取得对应的 Object，然后转换成相应的服务对象。常见系统服务见表 5-27。

表 5-27　getSystemService 返回的系统服务对象

传入的 Name	返回的对象	描述
WINDOW_SERVICE	WindowManager	管理打开的窗口程序
LAYOUT_INFLATER_SERVICE	LayoutInflater	取得 xml 里定义的 view
ACTIVITY_SERVICE	ActivityManager	管理应用程序的系统状态
POWER_SERVICE	PowerManger	电源的服务
ALARM_SERVICE	AlarmManager	闹钟的服务
NOTIFICATION_SERVICE	NotificationManager	状态栏的服务
KEYGUARD_SERVICE	KeyguardManager	键盘锁的服务
LOCATION_SERVICE	LocationManager	位置的服务，如 GPS
SEARCH_SERVICE	SearchManager	搜索的服务
VEBRATOR_SERVICE	Vebrator	手机震动的服务

传入的 Name	返回的对象	描述
CONNECTIVITY_SERVICE	Connectivity	网络连接的服务
WIFI_SERVICE	WifiManager	Wi-Fi 服务
TELEPHONY_SERVICE	TeleponyManager	电话服务

如果要增加一个选项的使用的方法 addTab(TabHost.TabSpec tabSpec)，如果有多个选项就要增加多个 TabHost.TabSpec 的对象，TabHost.TabSpec 类是 TabHost 定义的内部类，如果要想取得此类的实例化对象依靠 TabHost 类中的 newTabSpec()方法完成。

每个选项卡都有一个选项卡指示符、内容和 tag 选项，TabHost.TabSpec 类可以设置这些参数。TabHost.TabSpec 类定义的常用方法见表 5-28。

表 5-28 TabHost.TabSpec 类定义的常用方法

方法	描述
public TabHost.TabSpec setContent(int viewId)	设置要显示的组件 ID
public TabHost.TabSpec setContent(Intent intent)	指定一个加载 activity 的 Intent 对象作为选项卡的内容
public TabHost.TabSpec setContent(TabHost. TabContentFactory contentFactory)	指定 TabHost.TabContentFactory 用于创建选项卡的内容
public TabHost.TabSpec setIndicator(View view)	指定一个视图作为选项卡的指示符
public TabHost.TabSpec setIndicator(CharSequence label)	设置一个选项
public TabHost.TabSpec setIndicator(CharSequence label, Drawable icon)	为选项卡指示符指定一个选项和图标
public String getTag ()	获取 tag 选项字符串

继承 TabActivity 类的方式实现选项卡功能的步骤：
1）设计所有分页的界面布局；
2）建立一个类继承 TabActivity 类；例如：

```
public class MainActivity extends TabActivity {……
```

3）通过方法获得 TabHost 对象，例如：

```
TabHost th = getTabHost();
```

4）将指定布局管理文件实例化；

```
LayoutInflater.from(this).inflate(R.layout.activity_main,th.getTabContent
    View(), true);
```

5）设置选项卡的标题和内容。例如：

```
th.addTab(th.newTabSpec("tab1").setIndicator("文件").setContent(R.id.file));
```

虽然这种方法很简单，但是建议不要采用这种形式实现选项卡功能，因为 TabActivity 类从 API 级别 13 开始已经废弃不用了，移植到手机上时可能会因为版本的问题出现错误。

3．在布局管理器之中定义 TabHost 组件实现选项卡功能

如果使用布局管理器的方式实现选项卡功能，必须先了解 android.widget.TabWidget 类，其层次关系如下：

```
java.lang.Object
    android.view.View
        android.view.ViewGroup
            android.widget.LinearLayout
                android.widget.TabWidget
```

如果要想通过配置实现选项卡功能，则对配置文件的编写上也有要求：
1）所有用于选项配置的文件，必须以"<TabHost>"为根节点；

2）为了保证选项页和选项内容显示正常可以采用一个布局管理器进行布局；

3）定义一个"<TagWidget>"的选项，用于表示整个选项容器，另外在定义此组件的时候要引入"tabs"的组件，表示允许加入多个选项页；

4）由于 TabHost 是 FrameLayout 的子类，所以要想定义选项页必须使用 FrameLayout 布局，而后在此布局中定义所需要的选项页组件，而且框架布局上必须引用 tabcontent 组件(android:id="@android:id/tabcontent")，这与 getTabContentView()功能类似。

表 5-29 TabWidget 类常用属性

方法	描述
android:divider	可绘制对象，被绘制在选项卡窗口间充当分割物
android:tabStripEnabled	确定是否在选项卡绘制
android:tabStripLeft	用来绘制选项卡下面的分割线左边部分的可视化对象
android:tabStripRight	用来绘制选项卡下面的分割线右边部分的可视化对象

表 5-30 TabWidget 类常用方法

方法	描述
public TabWidget(Context context)	创建 TabWidget 实例
public void addView(View child)	向 TabWidget 增加组件
public int getTabCount()	返回选项卡的数量
public void setEnabled(boolean enabled)	配置是否启用
public void focusCurrentTab(int index)	设置当前选项卡并且让其获得焦点
public void setCurrentTab (int index)	设置当前选项卡

5.11.2 TabHost 组件实例

实例 5-15：TabHost 组件实例

新建一个项目，项目的命名为：exam5_15，包名称为：org.hnist.demo，直接让一个 Activity 程序继承 TabActivity 类的方式实现选项卡功能。

1）修改 activity_main.xml 文件，代码如下：

```xml
<?xml version="1.0" encoding="utf-8"?>
<TabHost xmlns:android="http://schemas.android.com/apk/res/android"
    android:layout_width="fill_parent"
    android:layout_height="fill_parent">
<LinearLayout
    android:id="@+id/file"
    android:layout_width="fill_parent"
    android:layout_height="fill_parent"
    android:orientation="vertical"
    android:gravity="center_horizontal">
    <Button
        android:id="@+id/open"
        android:layout_width="150dp"
        android:layout_height="wrap_content"
        android:text="打开文件"/>
    <Button
        android:id="@+id/save"
        android:layout_width="150dp"
        android:layout_height="wrap_content"
```

```xml
            android:text="保存文件" />
        <Button
            android:id="@+id/saveAs"
            android:layout_width="150dp"
            android:layout_height="wrap_content"
            android:text="文件另存为" />
</LinearLayout>
<LinearLayout
    android:id="@+id/edit"
    android:layout_width="fill_parent"
    android:layout_height="fill_parent"
    android:orientation="vertical"
    android:gravity="center_horizontal">
    <TextView
        android:id="@+id/copy"
        android:layout_width="fill_parent"
        android:layout_height="wrap_content"
        android:text="复制"/>
    <TextView
        android:id="@+id/paste"
        android:layout_width="fill_parent"
        android:layout_height="wrap_content"
        android:text="粘贴"/>
</LinearLayout>
<LinearLayout
    android:id="@+id/seek"
    android:layout_width="fill_parent"
    android:layout_height="fill_parent"
    android:orientation="vertical"
    android:gravity="center_horizontal">
  <EditText
            android:id="@+id/edit1"
            android:layout_width="wrap_content"
            android:layout_height="wrap_content"
            android:text="请输入检索关键字..."
            android:textSize="18sp"/>
<Button
            android:id="@+id/seekbut"
            android:layout_width="wrap_content"
            android:layout_height="wrap_content"
            android:text="搜索"/>
</LinearLayout>
<LinearLayout
    android:id="@+id/time"
    android:layout_width="fill_parent"
    android:layout_height="fill_parent"
    android:orientation="vertical"
    android:gravity="center_horizontal">
    <TimePicker
        android:id="@+id/seekbut"
        android:layout_width="wrap_content"
        android:layout_height="wrap_content"
        android:text="设置时间"/>
</LinearLayout>
</TabHost>
```

2. 修改 MainActivity.java 文件，代码如下：

```java
package org.hnist.demo;
import android.app.AlertDialog;
import android.app.Dialog;
import android.app.TabActivity;
import android.content.DialogInterface;
import android.os.Bundle;
import android.view.LayoutInflater;
import android.widget.TabHost;
import android.widget.TabHost.OnTabChangeListener;
public class MainActivity extends TabActivity {       //继承 TabActivity 类
  public void onCreate(Bundle savedInstanceState) {
    super.onCreate(savedInstanceState);
    TabHost th = getTabHost();                    //获得 TabHost 对象
    LayoutInflater.from(this).                    //获得 LayoutInflater 对象
    inflate(R.layout.activity_main,               //定义要转换的布局管理器
    th.getTabContentView(),                       //指定选项增加的容器
    true);                                        //实例化布局管理器中的组件
    //增加一个选项   定义 TabSpec   设置选项文字   设置显示组件
    th.addTab(th.newTabSpec("文件").setIndicator("文件").setContent(R.id.file));
    th.addTab(th.newTabSpec("编辑").setIndicator("编辑").setContent(R.id.edit));
    th.addTab(th.newTabSpec("查看").setIndicator("查看").setContent(R.id.seek));
    th.addTab(th.newTabSpec("设置时间").setIndicator("设置时间").setContent
        (R.id.time));
        //选项切换事件处理
    th.setOnTabChangedListener(new OnTabChangeListener() {//设置监听
        public void onTabChanged(String tabId) {        //如果选项发生变化触发事件
            Dialog dialog = new AlertDialog.Builder(MainActivity.this) //定义对话框
                .setTitle("提示")
                .setMessage("当前选中："+tabId+"选项")
                .setPositiveButton("确定",new DialogInterface.OnClickListener(){
                    public void onClick(DialogInterface dialog, int whichButton)
                    { dialog.cancel();} }).create();            //创建按钮
dialog.show(); } }); }}
```

保存文件并运行该项目，结果如图 5.22 所示，单击某个选项，例如 "文件" 选项，会触发事件，如图 5.23 所示。

图 5.22 利用 TabHost 实现选项卡功能

图 5.23 单击 "文件" 选项触发的事件

Android 系统提供的组件还有很多，新的组件还在不断地开发出来，限于篇幅其他组件不再赘述，感兴趣的用户可以参考 Android SDK API 文档资料学习其他组件的使用。

本章小结

本章着重介绍了 Android 系统提供的常见高级组件，例如：列表显示、进度条、对话框、画廊组件、选项卡组件等，这些组件都有自己相应的属性、方法和事件触发处理机制。

习题

1. ListView 组件中有哪些事件？写出代码。
2. ProgressBar 组件与 ProgressDialog 组件的区别与联系有哪些？
3. 利用 DatePickerDialog 和 TimePickerDialog 将当前时间设置为 2013-10-15 上午 12:30。
4. AlertDialog 有哪些种类？
5. AutoCompleteTextView 组件实现自动提示功能的步骤有哪些？
6. 参照实例 5-13，编写程序利用 GridView 组件显示一组图片。
7. 假设 ImageSwitcher 组件中显示的图片比本身大，应该在哪里修改什么语句？
8. Gallery 组件要用到适配器，有哪些适配器可用？该如何操作？
9. 用两种方法设置 TabHost 组件含有 2 个选项。

第 6 章 Android 组件之间的通信

学习目标：
- 了解使用 Intent 进行组件通信的原理
- 掌握使用 Intent 启动 Activity 的方法
- 掌握获取 Activity 返回值的方法
- 掌握 Message、Handler、Looper 类的使用及消息的传递
- 掌握 Service 的定义及使用
- 了解系统提供的 Service 程序
- 掌握发送和接收广播消息的方法

前面的章节学习了一些组件和事件的处理机制，所涉及的程序都是在 Activity 程序中进行的，在一个项目中，往往会包含多个 Activity 程序，这多个 Activity 程序之间如何解决控制权的转换，如何进行数据的通信，主要依据 Intent 组件来实现。

6.1 Android 四大组件简介

Android 四大基本组件分别是 Activity、Service（服务）、Content Provider（内容提供者），BroadcastReceiver（广播接收者）。并不是每一个 Android 应用程序都需要这四种组件，例如前面介绍的程序都是在一个 Activity 程序中进行的。下面对这四个组件进行一些简单的说明，让大家有一个整体的认识。

Activity：Activity 是活动的意思，一个 Activity 通常表现为一个可视化的用户界面，是 Android 程序与用户交互的窗口，也是 Android 组件中最基本、最复杂的一个组件。从外部来看，一个 Activity 占据当前的窗口，响应所有窗口事件，具备控件、菜单等界面元素。从内部逻辑来看，Activity 需要为了保持各个界面的状态，还需要管理生命周期和一些转跳逻辑。对于开发者而言，需要派生一个 Activity 的子类，进行编码实现各种功能方法。

Service：Service 是服务的意思，服务是运行在后台的一个组件，它就像一个没有界面的 Activity。它的很多方面与 Activity 类似，例如：封装有一个完整的功能逻辑实现，接受上层指令，完成相关的事件，定义好需要接受的 Intent 提供同步和异步的接口等。服务不提供用户界面，例如在后台下载文件、播放音乐，在你播放音乐的同时还可以做其他事情，不会妨碍用户与其他活动的交互。另一个组件（比如 Activity）可以启动一个服务，并运行或者绑定到它。

BroadcastReceiver：BroadcastReceiver 是广播接收者的意思，它不执行任何任务。广播是一种广泛运用在应用程序之间传输信息的机制，而 BroadcastReceiver 是对发送出来的广播进行过滤接收并响应的一类组件。

BroadcastReceiver 不包含任何用户界面。然而它们可以启动一个 Activity 以响应接收到的信息，或者通过 NotificationManager 通知用户。可以通过多种方式使用户知道有新的通知产生，手机震动、闹钟等。

在 Android 中还有一个很重要的概念就是 Intent，Intent 是一个对动作和行为的抽象描述，负责组件之间、程序之间进行消息传递。而 BroadcastReceiver 组件提供了一种把 Intent 作为一个消息广播出去，由所有对其感兴趣的程序对其作出反应的机制。

ContentProvider：是 ContentProvider 内容提供者的意思，作为应用程序之间唯一的共享数据的途径，ContentProvider 主要的功能就是存储并检索数据以及向其他应用程序提供访问数据的接口。

在 Android 中，每个应用程序都是用自己的用户 ID 并在自己的进程中运行。这样做的好处是，可以有效地保护系统及应用程序，避免被其他应用程序所影响，每个进程都拥有独立的进程地址空间和虚拟空间。Android 的数据都是属于应用程序自身，其他的应用是不能直接进行操作的。如果要实现不同应用之间的数据共享，就要用到 ContentProvider 组件，将在后面的第 8 章详细介绍 ContentProvider 组件。

在 Android 中，Intent 作为连接组件的纽带，除了 ContentProvider 是通过 Content Resolver 激活外，其他 3 种组件 Activity、Service 和 BroadcastReceiver 都是由 Intent 激活的，Intent 在不同的组件之间传递消息，将一个组件的请求意图传给另一个组件。

6.2 Intent 简介

Intent 译成中文就是"意图"，Intent 组件在 Android 中是一个十分重要的组件，它是连接不同应用的桥梁和纽带，也是让组件级复用（Activity 和 Service）成为可能的一个重要因素。Intent 组件的主要作用是运行在相同或不同应用程序的 Activity，Service，Broadcase receiver 间，进行切换和数据的传递。Intent 组件常用的方法如表 6-1 所示。

表 6-1 Intent 常用的方法

方法	描述
public void startActivity(Intent intent)	启动一个 Activity，并通过 Intent 传送数据
public void startActivityForResult(Intent intent, int requestCode)	启动并接收另一个 Activity 程序回传数据，当 requestCode 大于 0 才可以触发 onActivityResult()
public Intent getIntent()	返回启动当前 Activity 程序的 Intent
protected void onActivityResult(int requestCode, int resultCode, Intent data)	当需要接收 Intent 回传数据的时候覆写此方法对回传操作进行处理
public void finish()	调用此方法会返回之前的 Activity 程序，并自动调用 onActivityResult()方法
public final Cursor managedQuery (Uri uri, String[] projection, String selection, String[] selectionArgs, String sortOrder)	处理返回的 Cursor 结果集

要在 Android 程序中使用 Intent 组件必须要在程序中使用下面的语句：

```
import android.content.Intent;        //导入 content.Intent 类
```

6.2.1 利用 Intent 启动 Activity

一个 Activity 类似于 Web 开发中的一个页面，在 Web 设计中经常会用到超级链接来实现从一个页面跳转到另外的一个页面，在 Android 系统中，应用程序一般都有多个 Activity，这多个 Activity 就需要通信，Intent 组件可以在这多个 Activity 之间传递要操作的信息，也可以启动其他的 Activity 程序。

启动 Activity 方式有两种：显示启动和隐式启动。显式启动，必须在 Intent 中指明启动的 Activity 所在的类。隐式启动，Android 系统根据 Intent 的动作和数据来决定启动哪一个 Activity，也就是说在隐式启动时，Intent 中只包含需要执行的动作和所包含的数据，而无须指明具体启动哪一个 Activity，选择权由 Android 系统和最终用户来决定。

使用 Intent 显式启动 Activity 的基本步骤：
1）创建一个 Intent；
2）指定当前的应用程序上下文以及要启动的 Activity；例如：

```
    Intent it = new Intent(Send.this, Receive.class);//实例化 Intent，指定当前的
        应用程序以及要启动的 Activity
```

3）把创建好的这个 Intent 作为参数传递给 startActivity()或者 startActivityForResult()方法，这两个方法的区别是，是否接收另一个 Activity 程序回传数据，startActivityForResult 会有数据的回传，当 requestCode 大于 0 才可以触发 onActivityResult()。例如：

```
    Send.this.startActivity(it);                    //启动 Activity，不会接收数据的回传
    Send.this.startActivityForResult(it, 1);  //启动 Activity，会接收数据的回传
```

4）在 AndroidManifest.xml 文件中注册两个 Activity，应使用<activity>标签，嵌套在<application>标签内部。

例如：

```
    <activity android:name="Send" android:label="@string/app_title">
        <intent-filter>
            <action android:name="android.intent.action.MAIN" />
            <category android:name="android.intent.category.LAUNCHER" />
        </intent-filter>
    </activity>
    <activity android:name="Receive" android:label="@string/receive_name" />
</application>
```

<application>节点下共有两个<activity>节点，分别代表应用程序中所使用的两个 Activity，Send 和 Receive。其中 Send 是程序的主入口，Receive 是通过 Send 启动的。

6.2.2　利用 Intent 在 Activity 之间传递数据

1．利用 Intent 的 startActivity 传递数据

startActivity 可以把数据传到指定的地方，但是不可以获得从接收方反馈的数据。

传递数据方的 Activity 中的关键代码：

```
    Intent it = new Intent(Send.this, Receive.class); //实例化 Intent，接收方是 Receive
    String info1=name.getText().toString(); //取得文本框输入的内容
    it.putExtra("sendinfo", info1) ;        //将 sendinfo 赋值为字符串 info1 的值，传出
    Send.this.startActivity(it);            //启动 Activity，不会接收数据的回传
```

在接收数据方的 Activity 中的关键代码：

```
    Intent it = super.getIntent() ;                         //取得启动此程序的 Intent
    String info = it.getStringExtra("sendinfo") ;           //取得传来的 sendinfo 值
```

2．利用 Intent 的 startActivityForResult 传递数据

startActivityForResult 可以把数据传到指定的地方，还可以把接收方的数据传过来。

传递数据方的 Activity 中的关键代码：

```
    Intent it = new Intent(Send.this, Receive.class); //实例化 Intent，接收方是 Receive
    String info1=name.getText().toString();           //取得文本框输入的内容
    it.putExtra("sendinfo", info1) ;        //将 sendinfo 赋值为字符串 info1 的值，传出
    Send.this.startActivityForResult(it, 1); //启动 Activity，会接收数据的回传，1
        是一个 requestCode 的值，它大于 0 才可以触发 onActivityResult()
```

重载 onActivityResult 方法，用来接收传过来的数据。

第6章 Android 组件之间的通信

如果想要接收回传的数据，需要 Activity 常量的支持，Activity 提供的操作常量如下：

RESULT_OK：表示操作正常的状态码

RESULT_CANCELED：表示操作取消的状态码

RESULT_FIRST_USER：表示用户自定义的操作状态码

```
protected void onActivityResult(int requestCode, int resultCode, Intent data) {
    switch (resultCode) {                           //判断操作类型
    case RESULT_OK:                                 //成功操作
        msg.setText("返回的内容是：" + data.getStringExtra("returninfo"));//
            取得回传的 returninfo 值，并显示出来
        break;
    case RESULT_CANCELED:                           //取消操作
        msg.setText("操作取消。");
        break ;
    default:
        break;          } }}
```

在接收数据方的 Activity 中的关键代码：

```
Intent it = super.getIntent() ;                    //取得启动此程序的 Intent
String info = it.getStringExtra("sendinfo") ;      //取得传送来的 sendinfo 值
……
Receive.this.getIntent().putExtra("returninfo", retu) ;//返回信息变量名为
    returninfo，它的值为 retu，retu 为一个字符串
Receive.this.setResult(RESULT_OK, Receive.this.getIntent()) ; //设置返回数据
    的状态，RESULT_OK 与 Send.java 中的 onActivityResult()里判断的对应
Receive.this.finish() ;                            //结束 Intent
```

6.2.3 Intent 组件传递数据实例

实例 6-1：Intent 组件传递数据实例

新建建一个项目，项目的命名为：exam6_1，包名称为：org.hnist.demo，利用 Intent 的 startActivityForResult 可以在 Activity 之间进行数据的传递和回传。

1）在 res 文件夹下选择 layout，右键单击 activity_main.xml，选择"Refactor"，再选择"Rename"，如图 6.1 所示。

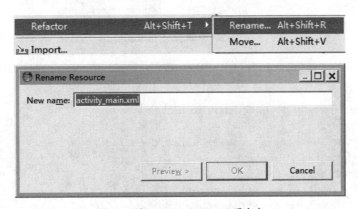

图 6.1　给 activity_main.xml 重命名

将图 6.1 中的 activity_main.xml 改为 send_main.xml，单击"OK"按钮。将 send_main.xml 作如下的修改：

```xml
<?xml version="1.0" encoding="utf-8"?>
<LinearLayout
 xmlns:android="http://schemas.android.com/apk/res/android"
 android:id="@+id/MyLayout"
 android:orientation="vertical"
 android:layout_width="fill_parent"
 android:layout_height="fill_parent">
<EditText
    android:id="@+id/myedit"
    android:layout_width="wrap_content"
    android:layout_height="wrap_content"
    android:selectAllOnFocus="true"
    android:text="输入您的姓名"/>
<Button
    android:id="@+id/mybut"
    android:layout_width="wrap_content"
    android:layout_height="wrap_content"
    android:text="传送数据到 Receive"/>
<TextView
    android:id="@+id/msg"
    android:layout_width="wrap_content"
    android:layout_height="wrap_content"/>
</LinearLayout>
```

2）鼠标右键单击"send_main.xml"，选择"Copy"，再右键单击"layout"文件夹，选择"Paste"，如图 6.2 所示。

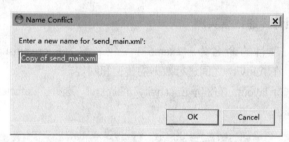

图 6.2 复制 send_main.xml

在图 6.2 中输入 receive_main.xml，单击"OK"按钮，建立一个 receive_main.xml 布局文件，做如下修改：

```xml
<?xml version="1.0" encoding="utf-8"?>
<LinearLayout
 xmlns:android="http://schemas.android.com/apk/res/android"
 android:id="@+id/MyLayout"
 android:orientation="vertical"
 android:layout_width="fill_parent"
 android:layout_height="fill_parent">
<TextView
    android:id="@+id/show"
```

```
        android:layout_width="wrap_content"
        android:layout_height="wrap_content"/>
<Button
    android:id="@+id/retu"
    android:layout_width="wrap_content"
    android:layout_height="wrap_content"
    android:text="返回数据到 Send。"/>
</LinearLayout>
```

3）在 src 文件夹下选择 org.hnist.demo，右键单击 MainActivity.java，选择"Refactor"，再选择"Rename"，如图 6.3 所示。

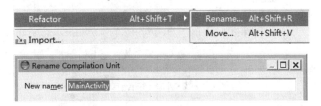

图 6.3 给 MainActivity.java 重命名

将 MainActivity 改为 Send，单击"Finish"按钮，建立一个 Send.java 文件，做如下修改：

```java
package org.hnist.demo;
import android.app.Activity;
import android.content.Intent;                    //导入 content.Intent 类
import android.os.Bundle;
import android.view.View;
import android.view.View.OnClickListener;
import android.widget.Button;
import android.widget.TextView;
import android.widget.EditText;
public class Send extends Activity {
 private Button mybut = null ;
 private TextView msg = null ;
 private EditText name = null ;
 public void onCreate(Bundle savedInstanceState) {
     super.onCreate(savedInstanceState);
     super.setContentView(R.layout.send_main);        //默认布局管理器
     this.mybut = (Button) super.findViewById(R.id.mybut) ;
     this.msg = (TextView) super.findViewById(R.id.msg) ;
     this.name = (EditText) super.findViewById(R.id.myedit) ;
     this.mybut.setOnClickListener(newOnClickListenerImpl());}//定义单击事件
 private class OnClickListenerImpl implements OnClickListener {
     public void onClick(View view) {
         Intent it = new Intent(Send.this, Receive.class);  //实例化 Intent
         String info1=name.getText().toString();            //取得文本框输入的内容
         it.putExtra("sendinfo", info1) ;                   //设置附加信息
         Send.this.startActivityForResult(it, 1); }}        //启动 Activity
 protected void onActivityResult(int requestCode, int resultCode, Intent data) {
     switch (resultCode) {                                  //判断操作类型
     case RESULT_OK:                                        //成功操作
         msg.setText("返回的内容是: " + data.getStringExtra("returninfo"));
```

```
            break;
        case RESULT_CANCELED:                              //取消操作
            msg.setText("操作取消。");
            break ;
        default:
            break;         }    }}
```

4）鼠标右键单击 Send.java，选择"Copy"，右键单击"src"文件夹，选择"Paste"，如图 6.4 所示。

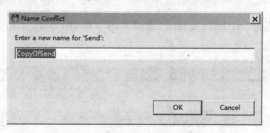

图 6.4　复制 Send.java 文件

在图 6.4 中输入 receive.java，单击"OK"按钮，建立一个 receive.java 文件，做如下修改：

```
package org.hnist.intentdemo;
import android.app.Activity;
import android.content.Intent;                     //导入 content.Intent 类
import android.os.Bundle;
import android.view.View;
import android.view.View.OnClickListener;
import android.widget.Button;
import android.widget.TextView;
public class Receive extends Activity {
private TextView show = null ;
private Button rebut = null ;
@Override
public void onCreate(Bundle savedInstanceState) {
    super.onCreate(savedInstanceState);
    super.setContentView(R.layout.receive_main); //调用默认布局管理器
    this.show = (TextView) super.findViewById(R.id.show) ;
    this.rebut = (Button) super.findViewById(R.id.retu) ;
    Intent it = super.getIntent() ;              //取得启动此程序的 Intent
    String info = it.getStringExtra("sendinfo") ;   //取得设置的附加信息
    this.show.setText(info) ;                    //设置文本显示信息
    this.rebut.setOnClickListener(new OnClickListenerImpl()) ; } //设置监听
private class OnClickListenerImpl implements OnClickListener {
    public void onClick(View view) {
        String retu="我收到你发来的信息:"+show.getText().toString();;
        Receive.this.getIntent().putExtra("returninfo", retu) ; //返回信息
        //设置返回数据的状态
        Receive.this.setResult(RESULT_OK, Receive.this.getIntent()) ;
        Receive.this.finish() ;    }    }}           //结束 Intent
```

5）修改 AndroidManifest.xml 文件，添加一个 activity。

```
<?xml version="1.0" encoding="utf-8"?>
<manifest xmlns:android="http://schemas.android.com/apk/res/android"
    ……
```

```xml
        <activity android:name="Send" android:label="@string/send_name">
            <intent-filter>
                <action android:name="android.intent.action.MAIN" />
                <category android:name="android.intent.category.LAUNCHER" />
            </intent-filter>
        </activity>
        <activity android:name="Receive" android:label="@string/receive_name" />
    </application>
</manifest>
```

保存所有文件，运行该项目，显示一个文本编辑框和一个按钮，如图 6.5 所示。在文本框中输入内容，如图 6.6 所示，单击按钮，会将输入的信息在另一个 Activity 中显示，如图 6.7 所示，单击其中的返回按钮，会将指定信息会传到主程序中，如图 6.8 所示。

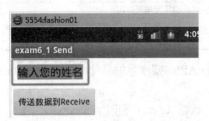

图 6.5　实例 6-1 运行界面

图 6.6　输入传送信息

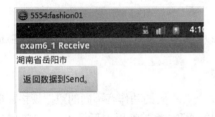

图 6.7　跳转到 Receive 程序界面

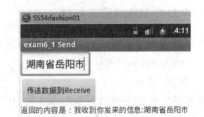

图 6.8　返回 Send 程序界面

6.3　深入了解 Intent

Android 中提供了 Intent 组件来协助应用程序之间的交互与通信，Intent 负责对应用中一次操作的动作、动作涉及数据、附加数据进行描述，根据此 Intent 的描述，Android 会找到对应的组件，将 Intent 传递给调用的组件，并完成组件的调用。

前面的例子说明了 Intent 能够实现 Activity 之间的切换和数据的传递，事实上这仅仅是利用 Intent 实现了对附加信息的传递，Intent 的功能还有很多，例如：打开网页、打电话、发短信、E-mail 等，采用 Intent 都可以轻松实现。

6.3.1　Intent 的构成

Intent 数据结构两个最重要的部分是动作和动作对应的数据。典型的动作类型有：MAIN、VIEW、DAIL、EDIT 等，而动作对应的数据一般以 URI 的形式进行表示。

1. Intent 的动作（Action）

在 Intent 中，Action 就是希望触发的动作，当你指明了一个 Action，执行者就会依照这个动作的指示,接受相关输入,表现对应行为,产生相应的输出。可以通过 setAction()方法进行设置 Action,通过 getAction()方法读取。在 Android 系统之中已经为用户准备好了一些表示 Action 操作的常量,如表 6-2 所示。

表 6-2 常用的 Action 常量

Action 名称	AndroidManifest.xml 配置名称	描述
ACTION_MAIN	android.intent.action.MAIN	作为一个程序的入口
ACTION_VIEW	android.intent.action.VIEW	用于数据的显示
ACTION_DIAL	android.intent.action.DIAL	调用电话拨号程序
ACTION_EDIT	android.intent.action.EDIT	用于编辑给定的数据
ACTION_PICK	android.intent.action.PICK	从特定的一组数据之中进行数据选择操作
ACTION_RUN	android.intent.action.RUN	运行数据
ACTION_SEND	android.intent.action.SEND	调用发送短信程序
ACTION_GET_CONTENT	android.intent.action.GET_CONTENT	根据指定 Type 来选择打开操作内容的 Intent
ACTION_CHOOSER	android.intent.action.CHOOSER	创建文件操作选择器

除了上述的常用常量外,更多的动作请参考 Intent 类的 API,要注意的是使用这些常量的时候可能要在 AndroidManifest.xml 文件中做相应的配置,才会生效。

动作很大程度上决定了剩下的 intent 如何构建,特别是数据(data)和附加(extras)字段,应该尽可能明确指定动作,并紧密关联到其他 intent 字段。

2. Intent 动作对应的数据

一共分为以下六种数据:数据(Data)、数据类型(Type)、操作类别(Category)、附加信息(Extras)、组件(Component)、标志(Flags)。

1)数据(Data)

数据是指作用于动作的数据的 URI 和数据的 MIME 类型。不同的动作有不同的数据规格。例如,如果动作是 ACTION_VIEW,数据字段是一个 http:URI,接收活动将被调用去下载和显示 URI 指向的数据。不同的动作对应不同的 Data,常见的 Data 如表 6-3 所示。

表 6-3 常见的动作与数据对应表

操作类型	Data(Uri)格式	范例
浏览网页	http://网页地址	http://www.hnist.cn
拨打电话	tel:电话号码	tel:07308648870
发送短信	smsto:短信接收人号码	smsto: 13207304568
查找 SD 卡文件	file:///sdcard/文件或目录	file:///sdcard/aa.jpg
显示地图	geo:坐标,坐标	geo:35.89, 29.6

如果要设置数据,就要借助 android.net.Uri 类完成,参照 API 了解其方法和属性。

2)数据类型(Type)

指定要传送数据的 MIME 类型,可以直接通过 setType()方法进行设置,getType()读取类型。一般 Intent 的数据类型能够根据数据本身进行判定,但是通过 setType()方法设置属性,可以强制采用显式指定的类型而不再进行推导。常见的几种 MIME 类型见表 6-4。

表 6-4 常见的几种 MIME 类型

作用	MIME 类型
发送短信	vnd.android-dir/mms-sms
设置图片	image/png
普通文本	text/plain
设置音乐	audio/mp3

3)操作类别(Category)

有时通过 Action,配合 Data 或 Type,很多时候可以准确地表达出一个完整的意图了,但也会需要加一些约束在里面才能够更精准。比如，如果你虽然很喜欢做俯卧撑，但一次做三个还只是在特殊的时候才会发生，那么你可能会说：每次吃撑了的时候，我都想做三个俯卧撑。吃撑了，这就对应着 Intent 的 Category 的范畴，它给所发生的意图附加一个约束。在 Android 中，一个实例是：所有应用的主 Activity 都需要一个 Category 为 CATEGORY_LAUNCHER，Action 为 ACTION_MAIN 的 Intent。

这个选项指定了将要执行的这个 action 的其他一些额外的约束，可以通过 addCategory()方法设置多个类别，removeCategory()方法删除一个之前添加的种类，getCategories()方法获取 Intent 对象中的所有种类。常见的几种 Gategory 见表 6-5。

表 6-5 常见的几种 Category

Category 名称	AndroidManifest.xml 配置名称	描述
CATEGORY_LAUNCHER	android.intent.category.LAUNCHER	表示此程序显示在应用程序列表中
CATEGORY_HOME	android.intent.category.HOME	显示主桌面，即开机时的第一个界面
CATEGORY_PREFERENCE	android.intent.category.PREFERENCE	运行后将出现一个选择面板
CATEGORY_BROWSABLE	android.intent.category.BROWSABLE	显示一张图片、Email 信息
CATEGORY_DEFAULT	android.intent.category.DEFAULT	设置一个操作的默认执行
CATEGORY_OPENABLE	android.intent.category.OPENABLE	当 Action 设置为 GET_CONTEN 时用于打开指定的 Uri
CATEGORY_GADGET .		设置活动可以嵌入另一个活动

4）附加信息（Extras）

传递的是一组键值对，可以使用 putExtra()方法进行设置，主要的功能是传递数据（Uri）所需要的一些额外的操作信息。使用 extras 可以为组件提供扩展信息，比如，如果要执行"发送电子邮件"这个动作，可以将电子邮件的标题、正文等保存在 extras 里，传给电子邮件发送组件。常见的几种 Extras 见表 6-6。

表 6-6 常见的几种 Extras

操作数据	附加信息	作用
短信操作	sms_body	表示要发送短信的内容
彩信操作	Intent.EXTRA_STREAM	设置发送彩信的内容
指定接收人	Intent.EXTRA_BCC	指定接收 E-mail 或信息的接收人
Email 收件人	Intent.EXTRA_EMAIL	用于指定 E-mail 的接收者，接收一个数组
Email 标题	Intent.EXTRA_SUBJECT	用于指定 E-mail 邮件的标题
Email 内容	Intent.EXTRA_TEXT	用于设置邮件内容

5）组件（Component）

指明了将要处理的 Activity 程序，通常 Android 会根据 Intent 中包含的其他属性的信息，比如 action、data/type、category 进行查找，最终找到一个与之匹配的目标组件。但是，如果 component 这个属性有指定的话，将直接使用它指定的组件，而不再执行上述查找过程。所有的组件信息都被封装在一个 ComponentName 对象之中，这些组件都必须在 AndroidManifest.xml 文件中的"<application>"中注册。

6）标志（Flags）

在 android.content.Intent 中一共定义了 20 种不同的 Flags，Flags 是一个整形数，由一系列的标志位构成，这些标志，是用来指明运行模式的。例如：指示 Android 系统如何去启动一个活动和启动之后如何对待它。所有这些标志都定义在 Intent 类中。可以通过 addFlags()方法进行增加。

6.3.2 Intent 常用用法示例

在 Android 系统之中提供了多种 Intent 动作，可以在 Intent 中指定程序要执行的动作（比如：view,edit,dial），以及程序执行该动作时所需要的数据，然后调用 startActivity()，Android 系统会自动寻找最符合指定要求的应用程序，并按要求执行该程序。

Intent 常用的一些操作如下：

1）从 google 搜索内容。

```
Intent intent = new Intent();                              //实例化 Intent
intent.setAction(Intent.ACTION_WEB_SEARCH);                //指定动作
intent.putExtra(SearchManager.QUERY,"searchString")        //设置数据
startActivity(intent);                                     //启动这个应用
```

2）浏览网页。

```
Uri uri =Uri.parse("http://www.hnist.cn");                      //定义 uri 数据
Intent it = new Intent(Intent.ACTION_VIEW,uri);//实例化 Intent，指定动作和数据
startActivity(it);                                              //启动这个应用
```

3）显示地图。

```
Uri uri = Uri.parse("geo:38.899533,-77.036476");
Intent it = new Intent(Intent.Action_VIEW,uri);
startActivity(it);
```

4）路径规划。

```
Uri uri =Uri.parse("http://maps.google.com/maps?f=dsaddr=startLat%20startLng&
    daddr=endLat%20endLng&hl=en");
Intent it = new Intent(Intent.ACTION_VIEW, uri);
startActivity(it);
```

5）拨打电话。

```
Uri uri =Uri.parse("tel:07308748870");
Intent it = new Intent(Intent.ACTION_DIAL,uri);
startActivity(it);
```

6）调用发短信的程序。

```
Intent it = new Intent(Intent.ACTION_VIEW);
it.putExtra("sms_body", "TheSMS text");
it.setType("vnd.android-dir/mms-sms");
startActivity(it);
```

7）发送短信。

```
Uri uri =Uri.parse("smsto:13207304569");
Intent it = new Intent(Intent.ACTION_SENDTO, uri);
it.putExtra("sms_body", "TheSMS text");
startActivity(it);
String body="this is sms demo";
Intent mmsintent = new Intent(Intent.ACTION_SENDTO, Uri.fromParts("smsto",
    number, null));
mmsintent.putExtra(Messaging.KEY_ACTION_SENDTO_MESSAGE_BODY,body);
mmsintent.putExtra(Messaging.KEY_ACTION_SENDTO_COMPOSE_MODE,true);
mmsintent.putExtra(Messaging.KEY_ACTION_SENDTO_EXIT_ON_SENT,true);
startActivity(mmsintent);
```

8）发送彩信。

```
Uri uri =Uri.parse("content://media/external/images/media/23");
Intent it = new Intent(Intent.ACTION_SEND);
```

第6章 Android 组件之间的通信

```
    it.putExtra("sms_body","some text");
    it.putExtra(Intent.EXTRA_STREAM, uri);
    it.setType("image/png");
    startActivity(it);
    StringBuilder sb = new StringBuilder();
    sb.append("file://");
    sb.append(fd.getAbsoluteFile());
    Intent intent = new Intent(Intent.ACTION_SENDTO, Uri.fromParts("mmsto", number,
        null));
    //Below extra datas are all optional.
    intent.putExtra(Messaging.KEY_ACTION_SENDTO_MESSAGE_SUBJECT,subject);
    intent.putExtra(Messaging.KEY_ACTION_SENDTO_MESSAGE_BODY,body);
    intent.putExtra(Messaging.KEY_ACTION_SENDTO_CONTENT_URI,sb.toString());
    intent.putExtra(Messaging.KEY_ACTION_SENDTO_COMPOSE_MODE,composeMode);
    intent.putExtra(Messaging.KEY_ACTION_SENDTO_EXIT_ON_SENT,exitOnSent);
    startActivity(intent);
```

9）发送 E-mail。

```
    Uri uri =Uri.parse("mailto:xxx@abc.com");
    Intent it = new Intent(Intent.ACTION_SENDTO, uri);
    startActivity(it);
```

或：

```
    Intent it = new Intent(Intent.ACTION_SEND);
    it.putExtra(Intent.EXTRA_EMAIL,"me@abc.com");
    it.putExtra(Intent.EXTRA_TEXT, "The email body text");
    it.setType("text/plain");
    startActivity(Intent.createChooser(it,"Choose Email Client"));
```

或：

```
    Intent it=new Intent(Intent.ACTION_SEND);
    String[] tos={"me@abc.com"};
    String[]ccs={"you@abc.com"};
    it.putExtra(Intent.EXTRA_EMAIL, tos);
    it.putExtra(Intent.EXTRA_CC, ccs);
    it.putExtra(Intent.EXTRA_TEXT, "Theemail body text");
    it.putExtra(Intent.EXTRA_SUBJECT, "The email subject text");
    it.setType("message/rfc822");
    startActivity(Intent.createChooser(it,"Choose Email Client"));
```

发送附件：

```
    Intent it = new Intent(Intent.ACTION_SEND);
    it.putExtra(Intent.EXTRA_SUBJECT, "Theemail subject text");
    it.putExtra(Intent.EXTRA_STREAM,"file:///sdcard/mysong.mp3");
    sendIntent.setType("audio/mp3");
    startActivity(Intent.createChooser(it,"Choose Email Client"));
```

10）播放多媒体。

```
    Intent it = new Intent(Intent.ACTION_VIEW);
    Uri uri =Uri.parse("file:///sdcard/song.mp3");
    it.setDataAndType(uri,"audio/mp3");
    startActivity(it);
```

或：

```
Uri uri =Uri.withAppendedPath(MediaStore.Audio.Media.INTERNAL_CONTENT_URI,"1");
Intent it = new Intent(Intent.ACTION_VIEW,uri);
startActivity(it);
```

11）uninstall apk。

```
Uri uri =Uri.fromParts("package", strPackageName, null);
Intent it = newIntent(Intent.ACTION_DELETE, uri);
startActivity(it);
```

12）install apk。

```
Uri installUri = Uri.fromParts("package","xxx", null);
returnIt = newIntent(Intent.ACTION_PACKAGE_ADDED, installUri);
```

13）打开照相机。

方式一：

```
Intent i = new Intent(Intent.ACTION_CAMERA_BUTTON, null);
this.sendBroadcast(i);
```

方式二：

```
long dateTaken = System.currentTimeMillis();
String name = createName(dateTaken) + ".jpg";
fileName = folder + name;
ContentValues values = new ContentValues();
values.put(Images.Media.TITLE, fileName);
values.put("_data", fileName);
values.put(Images.Media.PICASA_ID, fileName);
values.put(Images.Media.DISPLAY_NAME, fileName);
values.put(Images.Media.DESCRIPTION, fileName);
values.put(Images.ImageColumns.BUCKET_DISPLAY_NAME, fileName);
Uri photoUri = getContentResolver().insert(MediaStore.Images.Media. EXTERNAL_
            CONTENT_URI,values);
Intent inttPhoto = new Intent(MediaStore.ACTION_IMAGE_CAPTURE);
inttPhoto.putExtra(MediaStore.EXTRA_OUTPUT, photoUri);
startActivityForResult(inttPhoto, 10);
```

14）从 gallery 选取图片。

```
Intent i = new Intent();
i.setType("image/*");
i.setAction(Intent.ACTION_GET_CONTENT);
startActivityForResult(i, 11);
```

15）打开录音机。

```
Intent mi = new Intent(Media.RECORD_SOUND_ACTION);
startActivity(mi);
```

16）显示应用详细列表。

```
Uri uri =Uri.parse("market://details?id=app_id");
//Uri uri =Uri.parse("market://details?id=<packagename>");
Intent it = new Intent(Intent.ACTION_VIEW,uri);
startActivity(it);
```

17）寻找应用。

```
Uri uri =Uri.parse("market://search?q=pname:pkg_name");
Intent it = new Intent(Intent.ACTION_VIEW,uri);
startActivity(it);
//where pkg_name is the full package pathfor an application
```

18）打开联系人列表。

方式一：

```
Intent i = new Intent();
i.setAction(Intent.ACTION_GET_CONTENT);
i.setType("vnd.android.cursor.item/phone");
startActivityForResult(i, REQUEST_TEXT);
```

方式二：

```
Uri uri = Uri.parse("content://contacts/people");
Intent it = new Intent(Intent.ACTION_PICK, uri);
startActivityForResult(it, REQUEST_TEXT);
```

19）打开另一程序。

```
Intent i = new Intent();
ComponentName cn = newComponentName("com.yellowbook.android2", "com.yellowbook.
    android2.AndroidSearch");
i.setComponent(cn);
i.setAction("android.intent.action.MAIN");
startActivityForResult(i, RESULT_OK);
```

20）调用系统编辑添加联系人（高版本 SDK 有效）。

```
Intent it = newIntent(Intent.ACTION_INSERT_OR_EDIT);
it.setType("vnd.android.cursor.item/contact");
//it.setType(Contacts.CONTENT_ITEM_TYPE);
it.putExtra("name","myName");
it.putExtra(android.provider.Contacts.Intents.Insert.COMPANY, "organization");
it.putExtra(android.provider.Contacts.Intents.Insert.EMAIL,"email");
it.putExtra(android.provider.Contacts.Intents.Insert.PHONE,"homePhone");
it.putExtra(android.provider.Contacts.Intents.Insert.SECONDARY_PHONE,
    "mobilePhone");
it.putExtra(android.provider.Contacts.Intents.Insert.TERTIARY_PHONE,
    "workPhone");
 it.putExtra(android.provider.Contacts.Intents.Insert.JOB_TITLE,"title");
 startActivity(it);
```

21）调用系统编辑添加联系人。

```
Intent intent = newIntent(Intent.ACTION_INSERT_OR_EDIT);
intent.setType(People.CONTENT_ITEM_TYPE);
intent.putExtra(Contacts.Intents.Insert.NAME, "Jackon");
intent.putExtra(Contacts.Intents.Insert.PHONE, "+1234567890");
intent.putExtra(Contacts.Intents.Insert.PHONE_TYPE,Contacts.PhonesColumns
    .TYPE_MOBILE);
intent.putExtra(Contacts.Intents.Insert.EMAIL, "com@com.com");
intent.putExtra(Contacts.Intents.Insert.EMAIL_TYPE,Contacts.ContactMethod
    sColumns.TYPE_WORK);
startActivity(intent);
```

要注意的是有些动作可能要 AndroidManifest.xml 文件中申明其权限，Android 常用权限见表 6-7。

表 6-7 常见的 AndroidManifest.xml 权限设置

设置属性（前缀 android.permission）	描述
SEND_SMS	发短信
READ_SMS	读短信
CALL_PHONE	打电话
SET_TIME_ZONE	设置时间
INTERNET	访问网络
CHANGE_NETWORK_STATE	改变网络状态
CHANGE_WIFI_STATE	改变 WIFI 状态
CAMERA	照相机
SET_WALLPAPER	设置壁纸
VIBRATE	允许震动
ACCESS_FINE_LOCATION	访问位置信息
BLUETOOTH	访问蓝牙设备
READ_CALENDAR	读取日历
READ_CONTACTS	读取联系人
WRITE_CONTACTS	修改联系人
RECORD_AUDIO	录音
MOUNT_UNMOUNT_FILESYSTEMS	在 SD 卡中创建或删除文件
WRITE_EXTERNAL_STORAGE	向 SD 卡中写入数据
READ_PHONE_STATE	读取电话状态

6.3.3 Intent 操作实例

在 Android 系统中设置了 Intent 动作，设置了 URL 以及附加的数据就可以完成 Intent 的操作了。

实例 6-2：Intent 组件操作实例

新建一个项目，项目的命名为：exam6_2，包名称为：org.hnist.demo，利用 Intent 编程实现浏览网页和拨打电话的功能。

1）新建布局管理文件 activity_main.xml，代码如下：

```xml
<?xml version="1.0" encoding="utf-8"?>
<LinearLayout
xmlns:android="http://schemas.android.com/apk/res/android"
android:id="@+id/MyLayout"
android:orientation="vertical"
android:layout_width="fill_parent"
android:layout_height="fill_parent">
<EditText
    android:id="@+id/tel"
    android:layout_width="fill_parent"
    android:layout_height="wrap_content"
    android:selectAllOnFocus="true"
    android:text="输入电话号码或网址"/>
<Button
    android:id="@+id/telbut"
```

```xml
        android:layout_width="fill_parent"
        android:layout_height="wrap_content"
        android:text="拨打电话"/>
<Button
    android:id="@+id/netbut"
    android:layout_width="fill_parent"
    android:layout_height="wrap_content"
    android:text="打开网页"/>
</LinearLayout>
```

2）新建 Activity 文件 MainActivity.java，代码如下：

```java
package org.hnist.demo;
import android.app.Activity;
import android.content.Intent;
import android.net.Uri;
import android.os.Bundle;
import android.view.View;
import android.view.View.OnClickListener;
import android.widget.Button;
import android.widget.EditText;
public class MainActivity extends Activity {
private Button telbut = null ;
private Button netbut = null ;
private EditText tel = null ;
public void onCreate(Bundle savedInstanceState) {
    super.onCreate(savedInstanceState);
    super.setContentView(R.layout.activity_main);
    this.telbut = (Button) super.findViewById(R.id.telbut) ;
    this.netbut = (Button) super.findViewById(R.id.netbut) ;
    this.tel = (EditText) super.findViewById(R.id.tel) ;
    this.netbut.setOnClickListener(new OnClickListenernet());//定义单击事件
    this.telbut.setOnClickListener(new OnClickListenertel());}//定义单击事件
private class OnClickListenernet implements OnClickListener {
    public void onClick(View view) {
        String netaddress = MainActivity.this.tel.getText().toString() ;
        Uri uri = Uri.parse("http://" + netaddress) ;       //指定数据
        Intent it = new Intent() ;                           //实例化 Intent
        it.setAction(Intent.ACTION_VIEW);                    //指定 Action
        it.setData(uri) ;                                    //设置数据
        MainActivity.this.startActivity(it);     }  }        //启动 Activity
private class OnClickListenertel implements OnClickListener {
    public void onClick(View view) {
        String tel = MainActivity.this.tel.getText().toString() ;
        Uri uri = Uri.parse("tel:" + tel) ;                  //指定数据
        Intent it = new Intent() ;                           //实例化 Intent
        it.setAction(Intent.ACTION_CALL);                    //指定 Action
        it.setData(uri) ;                                    //设置数据
        MainActivity.this.startActivity(it); }   }}          //启动 Activity
```

3）要拨打电话，修改 AndroidManifest.xml 文件，才能生效。

```xml
<?xml version="1.0" encoding="utf-8"?>
<manifest xmlns:android="http://schemas.android.com/apk/res/android"
```

```
......
</application>
<uses-permission
        android:name="android.permission.CALL_PHONE"/>
</manifest>
```

保存所有文件,运行该项目,弹出如图 6.9 所示界面,在输入框中输入 "www.hnist.cn",单击 "打开网页" 按钮,结果如图 6.10 所示,返回后在输入框中输入 07308648870,单击 "拨打电话" 按钮,结果如图 6.12 所示。

图 6.9　实例 6-2 运行界面

图 6.10　跳转到指定网站

图 6.11　输入电话号码

图 6.12　拨打电话

这里只给出了打开网页和拨打电话的代码,其他 Intent 操作,读者可以自己参照 6.3.2 节给出的相关代码实现。

6.4　Activity 的生命周期

Activity 是整个 Android 平台的基本组成,在系统中的 Activity 被一个 Activity 栈管理,同一时刻只有最顶上的那个 Activity 是处于运行状态的,当一个新的 Activity 启动时,将被放置到栈顶,成为运行状态，前一个 Activity 保留在栈中，不再放到前台，直到新的 Activity 退出为止。一个 Activity 生命周期包含三个阶段：

1）运行状态（Running State），在屏幕的前台（Activity 栈顶），叫做运行状态，此时 Activity 程序显示在屏幕前台，并且具有焦点，可以和用户的操作动作进行交互，例如：向用户提供信息、捕获用户单击按钮的事件并作处理。

2）暂停状态（Paused State），如果一个 Activity 失去焦点，但是依然可见，叫做暂停状态（Paused）。一个暂停状态的 Activity 依然保持活力（保持所有的状态，成员信息，和窗口管理器保持连接），但是不可以与其进行交互，在系统内存不够的时候将被结束。

3）停止状态（Stopped State），如果一个 Activity 被另外的 Activity 完全覆盖掉，叫做停止状态（Stopped）。它依然保持所有状态和成员信息，但是它不再可见，当系统内存不够的时候将被结束。

如果一个 Activity 是 Paused 或者 Stopped 状态，系统可以将该 Activity 从内存中删除，Android 系统采用两种方式进行删除，要么要求该 Activity 结束，要么直接结束它的进程。当该 Activity 再次显示给用户时，它必须重新开始和重置前面的状态。

当 Activity 程序在不同状态之间进行切换时，可以通过 Activity 类提供的方法来进行操作，常用的方法如表 6-8 所示。

表 6-8 Activity 程序的生命周期控制方法

方法	可关闭否	描述
protected void onCreate(Bundle savedInstanceState)	不可以	当启动新的 Activity 的时候被调用
protected void onRestart()	不可以	当 Activity 对用户即将可见时调用
protected void onStart()	不可以	重新启动 Activity 时调用（此方法是重启留在缓存中的 Activity）
protected void onResume()	不可以	当 Activity 界面可与用户交互时调用
protected void onPause()	可以	当系统要启动一个其他的 Activity 时调用，用于保存当前数据
protected void onStop()	可以	该 Activity 已经不可见时调用
protected void onDestroy()	可以	当 Activity 被 finish 或手机内存不足被销毁的时候调用

在一个 Activity 正常启动的过程中，它们被调用的顺序是 onCreate → onStart → onResume，在 Activity 被另一个 Activity 关掉的时候顺序是 onPause → onStop → onDestroy，这样就是一个完整的生命周期。

onCreate：在这里创建界面，做一些数据的初始化工作。

onStart：此时变成用户可见但是还不能进行交互。

onResume：此时可以和用户进行交互。

onPause：此时是用户可见但不可交互，这个时候程序的优先级降低，有可能被系统收回。在这里保存的数据，应该早在 onResume 里就读出来了，注意：这个方法里做的事情时间要短，因为下一个 Activity 不会等到这个方法完成才启动。

onStop：此时变得不可见，被下一个 Activity 完全覆盖了。

onDestroy：这是 Activity 被关掉前最后一个被调用方法，可能的原因是调用 finish()方法或者是系统为了节省空间将它暂时性关掉。

图 6.13 显示了 Activity 的重要状态转换。

整个的生命周期，从 onCreate()开始到 onDestroy()结束。Activity 在 onCreate()设置所有的"全局"状态，在 onDestory()释放所有的资源。

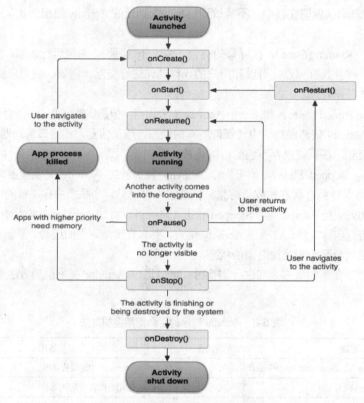

图 6.13　Activity 程序的生命周期图

可以看见的生命周期，从 onStart()开始到 onStop()结束。在这段时间，可以看到 Activity 在屏幕上，尽管有可能不在前台，不能和用户交互。在这两个接口之间，需要保持显示给用户的 UI 数据和资源等。

前台的生命周期，从 onResume()开始到 onPause()结束。在这段时间里，该 Activity 处于所有 Activity 的最前面，和用户进行交互。Activity 可以在 Resumed 和 Paused 状态之间切换。

所有的 Activity 都需要实现 onCreate(Bundle)去初始化设置，大部分 Activity 需要实现 onPause()去提交更改过的数据，当前大部分的 Activity 也需要实现 onFreeze()接口，以便恢复在 onCreate(Bundle)里面设置的状态。

从流程图上可以看出，无论现在是 onPause 状态还是 onStop 状态，当系统的内存不足时，都会将该 Activity 结束，如果发生这种情况一些重要的数据和状态可能还来不及保存，Google 专门提供了一个回调函数，onSaveInstanceState(Bundle)，通过该函数，可以将这些数据在销毁前进行保存，然后供 onCreate()或 onRestoreInstanceState()重新读出。

下面通过一个具体的例子来说明 Activity 的生命周期。两个 Activity 分别命名为 Activity1 和 Activity2，其中 Activity1 是默认启动的，界面有一个按钮，单击它会跳转到 Activity2，Activity2 中也有个按钮，单击它会跳转到 Activity1，如图 6.14 所示。来看看 Activity1 和 Activity2 的生命周期情况。

 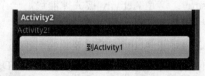

图 6.14　举例说明 Activity 程序的生命周期

1）启动该程序，Activity1 将按顺序调用 onCreate → onStart → onResume 方法。

2）当单击 Activity1 的按钮到 Activity2 时，执行的顺序是 Activity1 的 onPause 方法，这样可以将 Activity1 程序中未完成的数据保存或终止某些操作，接下来是 Activity2 按顺序调用 onCreate → onStart → onResume 方法，这三个方法过后，Activity2 会完全覆盖 Activity1，Activity1 将不可见，因此 Activity1 会调用 onStop 方法。（注意这里如果不覆盖 Activity1 则不会调用 onStop 方法）

3）当单击 Activity2 按钮到 Activity1 时，与 2）执行情况完全类似。

4）如果在 Activity2 中是单击手机上的"返回"按钮到 Activity1 时，首先执行的是 Activity2 的 onPause 方法，因为是返回操作，所以会读取缓存中的 Activity1，执行 onRestart 方法接着是 Activity1 按顺序调用 onStart → onResume 方法。接着是 Activity2 执行 onStop 方法，最后会执行 Activity2 的 onDestroy 销毁，是因为 Android 的缓存是不可逆的，只能后退不能向前。一旦程序关掉了，要再次启动就必须从头来：onCreate → onStart() → onResume()。

6.5 Android 中的消息处理机制

Windows 程序是采用消息驱动的模式，并且有全局的消息循环系统。Google 参考了 Windows 的消息循环机制，在 Android 系统中通过 Looper、Handler 来实现消息处理机制。Android 的消息处理是针对线程的，每个线程都可以有自己的消息队列和消息循环。

6.5.1 消息处理机制基础

要了解 Android 的消息处理机制，有几个概念必须了解。

1）消息类：Message，理解为线程间通信的数据单元，主要功能是进行消息的封装，同时指定消息的操作形式。

2）消息队列类：Message Queue，用来存放通过 Handler 发布的消息，按照先进先出的顺序执行。

3）消息操作类：Handler，它是 Message 的主要处理者，Message 对象封装了所有的消息，而这些消息的处理就需要 Handler 类完成。

4）消息通道类：Looper，是 Message Queue 和 Handler 之间桥梁的角色，循环取出 Message Queue 里面的 Message，交给相应的 Handler 进行处理。

5）线程：UI thread 通常就是 main thread，每一个线程里可含有一个 Looper 对象以及一个 Message Queue 数据结构。在应用程序里可以定义 Handler 的子类别来接收 Looper 所送出的消息。

整个消息处理的大概流程是：

1）包装 Message 对象（指定 Handler、回调函数和携带数据等）；

2）通过 Handler 的 sendMessage() 等方法将 Message 发送出去；

3）在 Handler 的处理方法里面将 Message 添加到 Handler 绑定的 Looper 的 MessageQueue；

4）Looper 的 loop() 方法通过循环不断从 MessageQueue 里面提取 Message 进行处理，并移除处理完毕的 Message；

5）调用 Message 绑定的 Handler 对象的 dispatchMessage() 方法完成对消息的处理。

Looper 本身提供的就是一个消息队列的集合，而每个消息都可以通过 Handler 增加和取出，而操作 Handler 的对象就是主线程（UI Thread）和子线程。

大致可以这样理解：假如一个隧道就是一个消息队列（Looper），那么里面的每一部汽车就是一个一个消息（Message），隧道里的管理者（Handle）保证秩序，告知哪辆车进入队列，离开队列，确保车实现先进先出。

1. 消息类：Message

它的主要功能是进行消息的封装，可以定义一个包含任意类型的描述数据对象，此对象可以发送给 Handler。对象包含两个额外的 int 字段和一个额外的对象字段，这样可以使得在很多情况下不用做分配工作。其常用的变量和方法如表 6-9 所示。

表 6-9 Message 类常用的变量和方法

变量或方法	描述
public int what	变量，用于定义此 Message 属于何种操作
public Object obj	变量，用于定义此 Message 传递的信息数据
public int arg1	变量，传递一些整型数据时使用，一般很少使用
public int arg2	变量，传递一些整型数据时使用，一般很少使用
public Messenger replyTo	指定此 Message 发送到哪个 Messenger 对象
public Runnable getCallback()	获取回调对象，此对象会在 message 处理时执行
public Bundle getData()	获取附加在此事件上的任意数据的 Bundle 对象
public Handler getTarget()	取得操作此消息的 Handler 对象
public long getWhen()	返回此消息的传输时间，以毫秒为单位
public void sendToTarget()	向 Handler 发送此消息
public void setTarget(Handler target)	设置将接收此消息的 Handler 对象
public String toString()	返回一个 Message 对象的描述信息
public static Message obtain(Handler h,int what,int arg1,int arg2,Object obj)	从全局池中分配的一个 Message 对象,可不带参数

其中：参数 h 表示设置的 target 值；what 表示设置的 what 值；arg1 表示设置的 arg1 值；arg2 表示设置的 arg2 值；obj 表示设置的 obj 值。

获取 Message 对象的最好方法是调用 Message.obtain()或者 Handler.obtainMessage()，以节省资源。如果一个 Message 只需要携带简单的 int 型信息，应优先使用 Message.arg1 和 Message.arg2 属性来传递信息,这比用 Bundle 更节省内存,尽可能使用 Message.what 来标识信息,以便用不同方式处理 Message。

要在 Android 的 java 程序中使用 Message 类必须要在程序中使用下面的语句。

```
import android.os.Message;            //导入 os.Message 类
```

2. 消息操作类：Handler

Handler 主要用来与 Looper 沟通，增加新消息到 Message Queue 里，或者接收 Looper 所送来的消息。其常用的方法如表 6-10 所示。

表 6-10 Handler 类常用的方法

方法	描述
public Handler()	创建一个新的 Handler 实例
public Handler(Looper looper)	使用指定的队列创建一个新的 Handler 实例
public Handler(Handler.Callback callback)	通过回调对象创建一个新的 Handler 实例
public final Message obtainMessage(int what, Object obj)	获得一个 Message 对象
public final Message obtainMessage(int what, int arg1, int arg2, Object obj)	获得一个带指定参数的 Message 对象
public void handleMessage(Message msg)	处理消息的方法，子类要覆写此方法
public final boolean hasMessages(int what)	判断是否有指定的 Message
public final boolean hasMessages(int what, Object object)	判断是否有指定的 Message
public final void removeMessages(int what)	删除掉指定的 Message
public final void removeMessages(int what, Object object)	删除掉指定的 Message
public final boolean sendEmptyMessage(int what)	发送一个空消息
public final boolean sendEmptyMessageAtTime(int what, long uptimeMillis)	在指定的日期时间发送消息
public final boolean sendEmptyMessageDelayed(int what, long delayMillis)	等待指定的时间之后发送消息
public final boolean sendMessage(Message msg)	发送消息

Handler 的作用是把消息加入特定的 Looper 消息队列中,并分发和处理该消息队列中的消息。如另一个线程怎样把消息放入主(UI)线程的消息队列,可以通过 Handler 对象,通过调用 Handler 主线程的 sendMessage 接口,把消息队列放入主线程的消息队列,并在该 Handler 的 handleMessage()来处理消息。

要在 Android 程序中使用 Handler 类必须要在程序中使用下面的语句。

```
import android.os.Handler;            //导入 os.Handler 类
```

3. 消息通道:Looper

它主要用来管理线程里的 Message Queue,在一个 Activity 类之中,会自动启动 Looper 对象,在一个用户自定义的类之中,需要用户手工调用 Looper 类中的若干方法之后才可以正常启动 Looper 对象。Looper 类的常用方法见表 6-11。

表 6-11 Looper 类的常用方法

方法	描述
public static final synchronized Looper getMainLooper()	取得主线程
public static final Looper myLooper()	返回当前的线程
public static final void prepare()	初始化 Looper 对象
public static final void prepareMainLooper()	初始化主线程 Looper 对象
public void quit()	消息队列结束时调用
public static void loop()	启动消息队列
public static MessageQueue myQueue()	返回当前的 MessageQueue
public Thread getThread()	获得 looper 分配的线程
public String toString ()	成为字符串

要在 Android 程序中使用 Looper 类必须要在程序中使用下面的语句。

```
import android.os.Looper;             //导入 os.Looper 类
```

6.5.2 一个简单的消息处理实例

实例 6-3:简单的消息处理实例

新建一个项目,项目的命名为:exam6_3,包名称为:org.hnist.demo,利用 Message 类、Looper 和 Handler 类实现同线程内不同组件间的消息传递。

1)新建布局管理文件 activity_main.xml,代码如下:

```xml
<?xml version="1.0" encoding="utf-8"?>
<LinearLayout
    xmlns:android="http://schemas.android.com/apk/res/android"
    android:orientation="vertical"
    android:layout_width="fill_parent"
    android:layout_height="fill_parent">
<TextView
    android:id="@+id/msg"
    android:layout_width="fill_parent"
    android:layout_height="wrap_content" />
<EditText
    android:id="@+id/edit"
```

```xml
        android:layout_width="fill_parent"
        android:layout_height="wrap_content"
        android:text="等待接收数据" />
    <Button
        android:id="@+id/start"
        android:layout_width="fill_parent"
        android:layout_height="wrap_content"
        android:text="传送数据到其他组件" />
</LinearLayout>
```

2）新建 Activity 文件 MainActivity.java，代码如下：

```java
package org.hnist.demo;
import android.app.Activity;
import android.os.Bundle;
import android.os.Handler;              //导入 os.Handler 类
import android.os.Looper;                //导入 os.Looper 类
import android.os.Message;               //导入 os.Message 类
import android.view.View;
import android.view.View.OnClickListener;
import android.widget.Button;
import android.widget.EditText;
import android.widget.TextView;
public class MainActivity extends Activity {
    private TextView showmsg;                        //定义文本显示组件
    private EditText edit;                           //定义按钮组件
    private Button sendbut;                          //定义按钮组件
    private static final int SET = 1 ;               //定义 what 操作码
    public void onCreate(Bundle savedInstanceState) {
        super.onCreate(savedInstanceState);
        super.setContentView(R.layout.activity_main);
        this.showmsg = (TextView) super.findViewById(R.id.msg);   //取得文本组件
        this.sendbut = (Button) super.findViewById(R.id.start);   //取得按钮组件
        this.edit = (EditText) super.findViewById(R.id.edit);     //取得按钮组件
        this.sendbut.setOnClickListener(new OnClickListenerImpl());  }//设置单
                                                                      击事件
    private class OnClickListenerImpl implements OnClickListener {
        public void onClick(View view) {
            switch (view.getId()) {                  //判断操作的组件 ID
            case R.id.start:                         //表示按钮操作
                Looper loop = Looper.myLooper();     //取得当前的线程
                MyHandler hand = new MyHandler(loop); //构造一个 Handler
                hand.removeMessages(0) ;             //清空所有的消息队列
                String data = "单击按钮后传送的数据"; //设置要发送的数据
                Message mymsg = hand.obtainMessage(SET,1,1,data);//获得 Message 对象
                hand.sendMessage(mymsg);             //发送消息
                break;}}}
    private class MyHandler extends Handler {        //定义类继承 Handler
        public MyHandler(Looper looper) {            //接收 Looper
            super(looper);          }                //调用父类构造
        public void handleMessage(Message msg) {     //处理消息
            switch (msg.what) {                      //判断操作形式
            case 1:
```

```
                    MainActivity.this.showmsg.setText(msg.obj.toString());//设置文
                                                                         本内容
                    MainActivity.this.edit.setText(msg.obj.toString());}}}}
                    //设置编辑框内容
```

保存所有文件，运行该项目，弹出如图 6.13 所示界面，单击按钮，结果如图 6.14 所示。

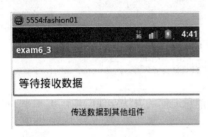

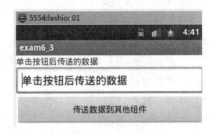

图 6.13　等待传送数据　　　　　　　　图 6.14　传送数据到其他组件

程序启动时，当前线程(即主线程 main thread)已产生了一个 Looper 对象，并且有了一个 MessageQueue 数据结构。

语句 Looper loop = Looper.myLooper();通过调用 Looper 类的 myLooper()函数，以取得目前线程里的 Looper 对象。

语句 MyHandler hand = new MyHandler(loop);构造一个 MyHandler 对象来与 Looper 沟通。Activity 等对象可以由 MyHandler 对象来将消息传给 Looper，然后放入 MessageQueue 里；MyHandler 对象也扮演 Listener 的角色，可接收 Looper 对象所送来的消息。

语句 Message mymsg = hand.obtainMessage(SET,1,1,data);先构造一个 Message 对象，并将数据存入对象里，形成消息。

语句 hand.sendMessage(mymsg);通过定义的 MyHandler 对象 hand 将消息 mymsg 传给 Looper，然后放入 MessageQueue 里。

Looper 对象看到 MessageQueue 里有消息 mymsg，就将它广播出去，hand 对象接收到此信息时，会呼叫其 handleMessage()函数来处理，于是按照要求输出指定的数据到文本组件和编辑框组件上。

6.5.3　线程基础知识

当一个 Android 应用程序启动时，系统会启动一个 Linux 进程（Process），并在此进程中开启一个主线程（Main Thread）。主线程是应用与界面交互的地方一般不能阻塞，否则可能会导致进程的关闭。但是有时会运行一些比较费时的程序可能导致主线程阻塞，后续组件不能执行，Android 系统就会自动地把整个进程都给关闭。因此，需要在进程里面增加线程，让一些比较费时的组件运行在其他线程里面，从而保证主线程畅通，而不至于让系统关闭应用程序的整个进程。

进程概念

进程是表示资源分配的基本单位，是调度运行的基本单位。例如，用户运行自己的程序，系统就创建一个进程，并为它分配资源，包括各种内存空间、磁盘空间、I/O 设备等。然后，把该进程放入进程的就绪队列。进程调度程序选中它，为它分配 CPU 以及其他有关资源，该进程才真正运行。所以，进程是系统中的并发执行的单位。

在 Mac、Windows NT 等采用微内核结构的操作系统中，进程的功能发生了变化：它只是资源分配的单位，而不再是调度运行的单位。在微内核系统中，真正调度运行的基本单位是线程。因此，实现并发功能的单位是线程。

线程概念

线程是进程中执行运算的最小单位,即执行处理机调度的基本单位。如果把进程理解为在一个任务的话,那么线程表示完成该任务的许多可能的子任务之一。线程可以在处理器上独立调度执行,这样,在多处理器环境下就允许几个线程各自在单独处理器上进行。操作系统提供线程就是为了方便而有效地实现这种并发性。

进程和线程的关系:

1) 一个线程只能属于一个进程,而一个进程可以有多个线程,但至少有一个线程。
2) 资源分配给进程,同一进程的所有线程共享该进程的所有资源。
3) 处理机分给线程,即真正在处理机上运行的是线程。
4) 线程在执行过程中,需要协作同步。线程间要利用消息通信的办法实现同步。
5) 二者均可并发执行。

Android 中线程的创建

Android 中线程的创建一般有两个常见方法:

1) 在Android中实现 Runnable 类并复写 Run()方法创建线程,其实该线程和 Android 的 Activity 是同一个线程,而不是单独的线程。

```
Runnable updateThread=new Runnable(){
    public void run(){……}}
```

2) 使用 Android 系统框架提供的 HandlerThread 创建新的线程。这是一个真正的线程。

① 创建一个 MyHandler 继承 Handler 类,并在 MyHandler 的构造函数中使用父类的构造函数来接受线程的 Looper,并覆写 handlerMessage 来接收消息。

```
class MyHandler extends Handler{
    public MyHandler(Looper looper)
        {super(looper); }
    public void handleMessage(Messagemsg){
        super.handleMessage(msg);……}}
```

② 创建一个 HandlerThread 线程对象,并启动该线程。

```
HanderThread myHandlerThread = new HanderThread("ThreadName");
myHandlerThread.Start();
```

③ 实例化 MyHandler 并把 myHandlerThread 线程的 Looper 对象传递过去。

```
MyHandler myHandler = new MyHandler(myHandlerThread.getLooper());
```

④ 创建一个 myHandler 的消息对象,并把消息传递给指定的线程。

```
Message msg = myHandler.obtainMessage();
msg.sendToTarget();
```

Android 主线程与子线程间的消息传递

当子线程要发送 Message 给主线程时,首先要为当前线程创建一个 Handler 类对象,由 Handler 调用相应的方法,将需要发送的 Message 发送到 MessageQueue(消息队列)中,当 Looper 发现 MessageQueue 中有未处理的消息时就会将此消息广播出去,此时主线程的 Handler 接收到此 Message 时,就会调用相应的方法来处理这条信息,完成主界面更新。

当子线程要发送 Message 给主线程时,有以下几个关键步骤:

1) 首先要创建两个 Handler 类对象,一个是主线程,一个是子线程;

例如：

```
    private Handler mainHandler, subHandler;        //定义两个 Handler 对象
```

2）定义子线程，注意在子线程中完成如下操作：初始化 Looper、定义子线程的 Handler 对象、覆写 handlerMessage 方法、发送消息，启动子线程的消息队列；例如：

```
class ChildThread implements Runnable {            //定义子线程类
 public void run() {
    Looper.prepare();                              //初始化 Looper
       this.subHandler = new Handler() {           //子线程的 Handler 对象
          public void handleMessage(Message msg) {//覆写 handlerMessage
             switch (msg.what) {                   //判断 what 操作
             case SETCHILD:                        //主线程发送给子线程的信息
                System.out.println("Main to Child Message : "+ msg.obj);
             Message toMain= this.mainHandler.obtainMessage();//创建 Message
                   toMain.obj="这是子线程发给主线程消息："+ super.getLooper().
                         getThread().getName();//设置发送消息的内容
             toMain.what = SETMAIN;                //设置主线程操作的状态码
             this.mainHandler.sendMessage(toMain); //发送消息
             break;    }}};
    Looper.loop();}}                               //启动该线程的消息队列
```

3）定义主线程，注意在主线程中完成如下操作：定义主线程的 Handler 对象、覆写 handlerMessage 方法、接收消息，启动子线程。例如：

```
this.mainHandler = new Handler() {                 //主线程的 Handler 对象
       public void handleMessage(Message msg) {    //消息处理
          switch (msg.what) {                      //判断 Message 类型
          case SETMAIN:                            //设置主线程的操作类
                MyThreadDemo.this.msg.setText("主线程接收数据："
                      + msg.obj.toString());       //设置文本内容
          break;   }}};
   new Thread(new ChildThread(), "Child Thread").start(); //启动子线程
```

实例 6-4：线程之间消息的传递实例

新建一个项目，项目的命名为：exam6_4，包名称为：org.hnist.demo，利用 Message 类、Looper 和 Handler 类实现子线程传递消息给主线程。

1）新建布局管理文件 activity_main.xml，代码如下：

```
<?xml version="1.0" encoding="utf-8"?>
<LinearLayout
xmlns:android="http://schemas.android.com/apk/res/android"
android:orientation="vertical"
android:layout_width="fill_parent"
android:layout_height="fill_parent">
<TextView
    android:layout_width="fill_parent"
    android:layout_height="wrap_content"
    android:id="@+id/msg"
    android:text="等待子线程发送消息。"/>
<Button
    android:layout_width="fill_parent"
    android:layout_height="wrap_content"
```

```xml
        android:id="@+id/start"
        android:text="接收子线程消息同时发送消息给子线程"/>
</LinearLayout>
```

2) 新建 Activity 文件 MainActivity.java，代码如下：

```java
package org.hnist.demo;
import android.app.Activity;
import android.os.Bundle;
import android.os.Handler;            //导入 os.Handler 类
import android.os.Looper;              //导入 os.Looper 类
import android.os.Message;             //导入 os. Message 类
import android.view.View;
import android.view.View.OnClickListener;
import android.widget.Button;
import android.widget.TextView;
public class MainActivity extends Activity {
public static final int ToMain = 1;        //设置一个 what 标记
public static final int ToSub = 2;         //设置一个 what 标记
private Handler mainHandler, subHandler;   //定义 Handler 对象
private TextView mymsg;                    //文本显示组件
private Button mybut;                      //按钮组件
class subThread implements Runnable {      //创建子线程类
    public void run() {
        Looper.prepare();                  //初始化 Looper
        MainActivity.this.subHandler = new Handler() {//定义子线程的 Handler 对象
            public void handleMessage(Message msg) {//覆写 handlerMessage
                switch (msg.what) {        //判断 what 操作
                case ToSub:                //主线程发送给子线程的信息
                    System.out.println(" Main Child Message : "+ msg.obj);
                        //打印消息
        Message toMain = MainActivity.this.mainHandler.obtainMessage();//创建消息
                    toMain.obj = "This is a message from Sub"; //设置显示文字
                    toMain.what = ToMain;          //设置主线程操作的状态码
                    MainActivity.this.mainHandler.sendMessage(toMain);
                        //发送消息
                    break;  }}};
        Looper.loop();   }}              //启动该线程的消息队列
public void onCreate(Bundle savedInstanceState) {
    super.onCreate(savedInstanceState);
    super.setContentView(R.layout.activity_main);
    this.mymsg = (TextView) super.findViewById(R.id.msg);
    this.mybut = (Button) super.findViewById(R.id.start);
    this.mainHandler = new Handler() {        //定义主线程的 Handler 对象
        public void handleMessage(Message msg) { //消息处理
            switch (msg.what) {                //判断 Message 类型
            case ToMain:                       //设置主线程的操作类
                MainActivity.this.mymsg.setText("主线程接收数据是："+msg.obj.
                    toString());               //设置文本内容
                break;}}};
```

```
    new Thread(new subThread()).start();              //启动子线程
    this.mybut.setOnClickListener(new OnClickListenerImpl()) ;}//单击事件操作
private class OnClickListenerImpl implements OnClickListener {
    public void onClick(View view) {
        if (MainActivity.this.subHandler != null) {   //判断是否已经实例化子
                                                        线程Handler
            Message childMsg = MainActivity.this.subHandler.obtainMessage();
                //创建消息
            childMsg.obj = "This is a message from Main.";    //设置消息内容
            childMsg.what = ToSub;                     //操作码
            MainActivity.this.subHandler.sendMessage(childMsg);}}}//向子线程
                                                            发送消息
protected void onDestroy() {
    super.onDestroy();
    MainActivity.this.subHandler.getLooper().quit();   }}  //结束队列
```

保存所有文件，运行该项目，弹出如图 6.15 所示界面，单击按钮，结果如图 6.16 所示。

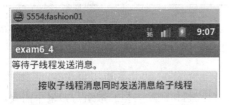
图 6.15　等待接收和传送数据

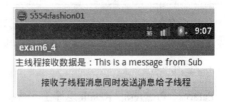
图 6.16　接收传来的数据

主线程也给子线程发送了消息，打开 LogCat 可以发现有一条消息，如图 6.17 所示。

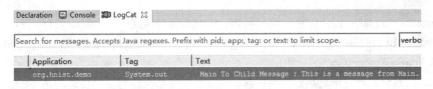
图 6.17　主线程发送给子线程的消息

这个消息为什么不用 TextView 或者其他组件显示出来呢？注意子线程不能更新主线程的组件数据，否则会报错。

6.5.4　异步处理工具类：AsyncTask

前面的学习知道主线程和子线程可以相互传递消息，但是子线程无法直接对主线程里面的组件进行更新，Android 提供了 android.os.AsyncTask 类可以在后台进行操作之后更新主线程的组件，它的层次关系如下：

```
Java.lang.Object
    android.os.AsyncTask<Params, Progress, Result>
```

要在 Android 的 Java 程序中使用 AsyncTask 类必须要在程序中使用下面的语句。

```
import android.os.AsyncTask;                         //导入 os.AsyncTask 类
```

在 AsyncTask 类中有三个泛型参数，在这个类中定义了一些常用方法如表 6-12 所示。

表 6-12 AsyncTask 类的常用方法

方法	描述
public AsyncTask ()	创建一个新的异步任务，这个构造函数必须在 UI 线程上调用
public final boolean cancel(boolean mayInterruptIfRunning)	指定是否取消当前线程操作
public final AsyncTask<Params, Progress, Result> execute(Params... params)	执行 AsyncTask 操作
public final boolean isCancelled()	判断子线程是否被取消
protected final void publishProgress(Progress... values)	更新线程进度
public final Result get(long timeout, TimeUnit unit)	等待计算结束并返回结果，最长等待时间为：timeOut
public final AsyncTask.Status getStatus()	获得任务的当前状态
protected abstract Result doInBackground(Params... params)	在后台完成任务执行，可以调用 publishProgress()方法更新线程进度
protected void onProgressUpdate(Progress... values)	在主线程中执行，用于显示任务的进度
protected void onPreExecute()	在主线程中执行，在 doInBackground()之前执行
protected void onPostExecute(Result result)	在主线程中执行，方法参数为任务执行结果
protected void onCancelled()	主线程中执行，在 cancel()方法之后执行

异步任务的定义是一个在后台线程上运行，其结果是在 UI 线程上发布的计算。异步任务被定义成三个泛型参数(Params, Progress, Result)和四个步骤(begin, doInBackground,processProgress,end)。

1．三个泛型参数

Params、Progress、Result 这个三个参数可为任意类型的数据和任意类型的数组，如果不需要，则用 void 代替。

1）Params：启动时需要的参数类型，例如：HTTP 请求的 URL，对应的方法 doInBackground(Params...parames);

2）Progress：后台执行任务的百分比，例如：进度条需要传递的是 Integer，Progress 对应的方法 onProgressUpdate()和 publishProgress(Progress...progress)，用来反映线程执行的进度，其中 publishProgress 方法必须在 doInBackground 方法中调用；

3）Result：后台执行完毕之后返回的信息，例如：完成数据信息显示传递的是 String。对应的方法 onPostExecute(Result)，后台进程得出的结果，作为参数传递给此方法。

例如：

```
Result doInBackground(){
    A();              //方法 A
  this.publishProgress("state1","I like it");
  B();                //方法 B
  this.publishProgress("state2","for test");
  return result; }
onProgressUpdate(String values) {
    if(values[0].equals("state1"))
        C();          //将 A 读取的数据在 UI 上展现
    else if(values[0].equals("state2"))
     Log.e("value",values[1]);}
```

2．四个步骤和对应的方法：

1）begin 和 onPreExecute()。

任务启动后（通过 execute()方法启动任务），这个步骤用来在 UI 线程中做一些初始化的工作。

2）doInBackground 和 doInBackground()。

当 onPreExecute()方法执行完后，立即在后台线程运行，用来处理一些耗时的计算及其他引起 UI 线程阻塞的操作，处理的结果 Result 返回给 onPostExecute(Result)方法，也可以使用 publishProgress() 和 UI 线程进行交互。

3）processProgress 和 onProgressUpdate()。

每次当在后台线程里调用了 publishProgress()方法后，onProgressUpdate()都会在 UI 线程中执行。在后台线程还未结束时，用来进行 UI 线程和后台线程的交互。

4）end 和 onPostExecute()。

当后台线程执行完毕之后，后台线程将得到的结果传递给 onPostExecute()方法，这个步骤在 UI 线程上展现后台线程执行完毕后最终得到的结果。

上面这四个方法只有 doInBackground()是在后台线程中执行，其他都在 UI 线程中执行。这四个方法都是 protected，必须继承 AsyncTask 类，必须覆写 doInbackground()方法，可能还要覆写 onPostExecute()方法。具体覆写哪些方法根据实际需要决定，如果要在后台进程尚未执行完成需要和 UI 交互，就要覆写 onProgressUpdate()方法，如果只需要等后台进程执行完毕得到结果后再和 UI 交互，则覆写 onPostExecute()方法就行。

3. 使用 AsyncTask 遵循的线程规则

1）这个类的实例必须在 UI 线程中创建。
2）execute()必须在 UI 线程中调用。
3）不要自己动手去调用上面的四个方法。
4）这个任务只能被执行一次，如果尝试多次执行会抛出异常。

实例 6-5：利用 AsyncTask 类进行消息传递的实例

新建一个项目，项目的命名为：exam6_5，包名称为：org.hnist.demo，利用 AsyncTask 类实现线程间消息的传递。

1）新建布局管理文件 activity_main.xml，代码如下：

```xml
<?xml version="1.0" encoding="utf-8"?>
<LinearLayout xmlns:android="http://schemas.android.com/apk/res/android"
    android:layout_width="fill_parent"
    android:layout_height="fill_parent"
    android:orientation="vertical"
    android:layout_margin="10dip"  >
    <TextView
        android:id="@+id/show"
        android:layout_width="fill_parent"
        android:layout_height="wrap_content"
        android:text="这里显示计数" />
    <Button android:id="@+id/Start"
        android:layout_width="wrap_content"
        android:layout_height="wrap_content"
        android:text="开始"/>
    <Button android:id="@+id/Stop"
        android:layout_width="wrap_content"
        android:layout_height="wrap_content"
        android:text="停止"/>
```

```xml
<ProgressBar
    android:id="@+id/progressBar01"
    style="?android:attr/progressBarStyleHorizontal"      //水平进度条
    android:layout_width="fill_parent"
    android:layout_height="wrap_content"
    android:max="100"
    android:progress="30"/>
<ProgressBar
    android:id="@+id/progressBar02"
    style="?android:attr/progressBarStyleLarge"           //大号圆形进度条
    android:layout_width="wrap_content"
    android:layout_height="wrap_content"
    android:max="100"
    android:progress="30"
    android:visibility="gone"/>
</LinearLayout>
```

2）新建 Activity 文件 MainActivity.java，代码如下：

```java
package org.hnist.demo;
package org.hnist.demo;
import android.app.Activity;
import android.os.AsyncTask;                              //导入os.AsyncTask类
import android.os.Bundle;
import android.util.Log;
import android.view.View;
import android.view.View.OnClickListener;
import android.widget.Button;
import android.widget.ProgressBar;
import android.widget.TextView;
public class MainActivity extends Activity {
private TextView show;
private Button startBtn,stopBtn;
private ProgressBar progressBar1,progressBar2;
private AsyncTask<Integer,Integer,Integer> task = null;   //定义AsyncTask
                                                          //对象task
private boolean stop;
private int count = 0;
    public void onCreate(Bundle savedInstanceState) {
        super.onCreate(savedInstanceState);
        setContentView(R.layout.activity_main);
        show = (TextView)MainActivity.this.findViewById(R.id.show);
        startBtn = (Button)MainActivity.this.findViewById(R.id.Start);
        stopBtn = (Button)MainActivity.this.findViewById(R.id.Stop);
        progressBar1 = (ProgressBar)MainActivity.this.findViewById(R.id.progressBar01);
        progressBar2 = (ProgressBar)MainActivity.this.findViewById(R.id.progressBar02);
        startBtn.setOnClickListener(new StartOnClickListener());
        stopBtn.setOnClickListener(new StopOnClickListener()); }
        class StartOnClickListener implements OnClickListener
    {   public void onClick(View v) {
            if(task == null){
                task = new CounterTask();     //实例化task
```

```java
                task.execute(count);    //在应用程序中启动一个子线程，执行doInBackground
                progressBar2.setVisibility(View.VISIBLE); }}}  //让第二个进度条可见
class StopOnClickListener implements OnClickListener
{   public void onClick(View v) {
        if(task != null){
            stop = true;            //停止线程
            task = null;            //清除线程
            progressBar2.setVisibility(View.GONE); }} }  //让第二个进度条消失
//定义 CounterTask 类继承 AsyncTask
class CounterTask extends AsyncTask<Integer,Integer,Integer>{
    protected Integer doInBackground(Integer... params){   //覆写doInbackground()
                                                           方法
        Integer initCounter = params[0];
        Log.v("This is a test","In doInBackground...");  //打印输出，在LogCat
                                                          可以看见
        stop = false;
        while(!stop){
            publishProgress(initCounter);//将中间结果传递到onProgressUpdate中去
            try {
                Thread.sleep(1*200);        //休眠0.2秒
            } catch (InterruptedException e) {
                e.printStackTrace();}
            initCounter += 1 ;
            if(initCounter == 100){
                break;  }}
        return initCounter;}             //将返回结果传递到onPostExecute
    protected void onPostExecute(Integer result) {  //覆写onPostExecute()方法
        super.onPostExecute(result);
        Log.v("Hello!","In onPostExecute...");//打印输出，在LogCat可以看见
        progressBar1.setProgress(result);      //设置第一个进度条的进度值
        progressBar2.setProgress(result);      //设置第二个进度条的进度值
        String text = result.toString();       //将计算结果转换为字符串
        show.setText(text);}                    //更新文本组件的显示内容
    protected void onProgressUpdate(Integer... values) {//覆写onProgressUpdate
                                                         ()方法
        super.onProgressUpdate(values);
        Log.v("Hello!","In onProgressUpdate...");  //打印输出，在LogCat可以看见
        progressBar1.setProgress(values[0]);    //更新进度条的进度值
        progressBar2.setProgress(values[0]);    //更新进度条的进度值
        String text = values[0].toString();     //将计算结果转换为字符串
        show.setText(text); }                   //更新文本组件的显示内容
    protected void onPreExecute() {             //覆写onPreExecute ()方法
        Log.v("Hello","In onPreExecute...");    //打印输出，在LogCat可以看见
        progressBar1.setProgress(0);            //更新进度条的进度值
        progressBar2.setProgress(0);            //更新进度条的进度值
        super.onPreExecute();       } }}        //调用父类的onPreExecute()方法
```

保存所有文件，运行该项目，弹出如图 6.18 所示界面，单击按钮，结果如图 6.19 所示。

分析程序的代码发现 4 个覆写的方法中除了 doInBackground()方法不能更新主线程的组件外，另外的几个方法都能够更新主线程的组件。利用 AsyncTask 类实现线程之间消息的传递比前面介绍的方

法相比要简单些，但是 AsyncTask 不能完全取代子线程，在一些逻辑较为复杂或者需要在后台反复执行的逻辑就可能需要子线程来实现了。

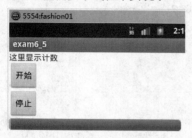

图 6.18　等待接收数据更新组件内容

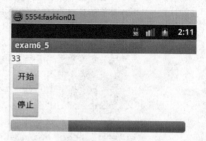

图 6.19　接收传来的数据更新组件内容

6.6　Service

在 Android 系统开发之中，Services 是一个重要的组成部分，它有两个主要目的：后台运行程序和跨进程访问。通过启动一个服务，可以在不显示界面的前提下在后台运行指定的任务，这样可以不影响用户做其他事情。通过 AIDL 服务可以实现不同进程之间的通信，这也是服务的重要用途之一。

6.6.1　Service 基础

Service 主要的功能是为 Activity 程序提供一些必要的支持或者提供一些不需要运行界面的功能。在开发时用户只需要继承 android.app.Service 类就可以完成 Service 程序的开发。

要在 Android 的 java 程序中使用 Service 组件必须要在程序中使用下面的语句。

import android.app.Service;   //导入 app.Service 类

在 Service 之中也有自己的生命周期方法，如表 6-13 所示。

表 6-13　Service 的生命周期控制方法

方法及常量	描述
public static final int START_CONTINUATION_MASK	继续执行 Service
public static final int START_STICKY	用于显式的启动和停止 Service
public abstract IBinder onBind(Intent intent)	设置 Activity 和 Service 之间的绑定
public void onCreate()	当一个 Service 创建时调用
public int onStartCommand(Intent intent, int flags, int startId)	启动 Service，由 startService()方法触发
public void onDestroy()	Service 销毁时调用，由 stopService()方法触发

要实现 Service 操作还需要 Activity 类中一些方法的支持，如表 6-14 所示。

表 6-14　Activity 类中操作 Service 的方法

方法	描述
public ComponentName startService(Intent service)	启动一个 Service
public boolean stopService(Intent name)	停止一个 Service
public boolean bindService(Intent service, ServiceConnection conn, int flags)	与一个 Service 绑定
public void unbindService(ServiceConnection conn)	取消与一个 Service 的绑定

Service 的生命周期控制相对 Activity 的要简单些，Service 的生命周期里一般只需要使用 startService()和 stopService()两个方法来开始和终止服务。

Service 分为两种：用于应用程序内部的本地服务和用于应用程序之间的远程服务。

创建一个 Service 很简单，按照下面的步骤就可以建立一个 Service。

1）编写一个服务类，该类必须从 android.app.Service 继承。

```
public class MyService extends Service { …… }
```

2）服务不能自己启动，要想启动这个服务，可以在 Activity 中调用 startService 或 bindService 方法，想停止服务，需要调用 stopService 方法。

3）在 AndroidManifest.xml 中声明。

在<application>节点下添加如下的代码：

```
<service android:name=". MyService " />    //MyService 为建立的服务类名
```

6.6.2 Service 的启动和停止

1. 用 startService()方法启动 Service

使用 startService()方法启用服务，调用者与服务之间没有关联，即使调用者退出了，服务仍然运行，在服务未被创建时，系统会先调用服务的 onCreate()方法，接着调用 onStart()方法。如果调用 startService()方法前服务已经被创建，多次调用 startService()方法并不会导致多次创建服务，但会导致多次调用 onStart()方法。采用 startService()方法启动的服务，只能调用 stopService()方法结束服务，服务结束时会调用 onDestroy()方法。

采用 startService()方法启动服务和 stopService()方法停止服务的代码如下：

```
……
    Intent it = new Intent(MyServiceDemo.this, MyService.class);
    startService(it);    //启动 Service
……
    Intent it = new Intent(MyServiceDemo.this, MyService.class);
    stopService(it);    //停止 Service
……}
```

注意：这里的 MyServiceDemo 是你的主程序名字，并不是固定不变的值。

2. 用 bindService()方法启动 Service

使用 bindService()方法启用服务，调用者与服务绑定在了一起，调用者一旦退出，服务也就终止，在服务未被创建时，系统会先调用服务的 onCreate()方法，接着调用 onBind()方法。这个时候调用者和服务绑定在一起，调用者退出了，系统就会先调用服务的 onUnbind()方法，接着调用 onDestroy()方法。如果调用 bindService()方法前服务已经被绑定，多次调用 bindService()方法并不会多次调用 onCreate() 和 onBind()方法。如果调用者希望与正在绑定的服务解除绑定，可以调用 unbindService()方法，调用该方法也会导致系统调用服务的 onUnbind()→onDestroy()方法。

采用 bindService()方法启动服务的代码如下：

```
……
private ServiceConnection conn = new ServiceConnection() {    //定义连接名
    //连接到 Service
    public void onServiceConnected(ComponentName name, IBinder service) {……}
    //与 Service 断开连接
    public void onServiceDisconnected(ComponentName name) {……}
……
```

```
            Intent it = new Intent(MyServiceDemo.this, MyService.class);
              //实现绑定
            MyServiceDemo.this.bindService(it, conn, Context.BIND_AUTO_CREATE);
……          //取消 Service 绑定
            MyServiceDemo.this.unbindService(conn);
……          //启动 Service
            Intent it = new Intent(MyServiceDemo.this, MyService.class);
            MyServiceDemo.this.startService(it));
……          //停止 Service
            Intent it = new Intent(MyServiceDemo.this, MyService.class);
            MyServiceDemo.this.stopService(it));
……    }
```

注意：bindService 开启了一个服务后，一定要记得执行停止服务操作。否则系统将会抛出 android.app.ServiceConnectionLeaked 异常。

3. 通过 RPC 机制来实现不同进程间 Service 的调用。

远程 Service 调用，是 Android 系统为了提供进程间通信而提供的实现方式，这种方式采用一种称为远程进程调用技术来实现，远程进程调用就是 RPC（Remote Procedure Call）。

RPC 是指在一个进程里，调用另外一个进程里的服务。Android 通过接口定义语言来生成两个进程间的访问代码。接口定义语言（Android Interface Definition Language，AIDL）是 Android 系统的一种接口描述语言，Android 编译器可以将 AIDL 文件编译成一段 Java 代码，生成相对的接口。

6.6.3 绑定 Service

使用 startService()方法启用服务，调用者与服务之间没有关联，即使调用者退出了，服务仍然运行，直到 Android 系统关闭后服务才停止。使用 bindService()方法可以将一个 Activity 和 Service 进行绑定，调用者与服务绑定在了一起，调用者一旦退出，服务也就终止。

在 Activity 类中专门提供了一个用于绑定 Service 的 bindService()方法，此方法返回的是一个 android.content.ServiceConnection 接口的参数，它主要定义了两个方法，见表 6-15。

表 6-15 ServiceConnection 接口定义的方法

方法	描述
public abstract void onServiceConnected (ComponentName name, IBinder service)	当与一个 Service 建立连接的时候调用
public abstract void onServiceDisconnected(ComponentName name)	当与一个 Service 取消连接的时候调用

ServiceConnection 接口主要的功能是当一个 Activity 程序与 Service 建立连接之后，可以通过 ServiceConnection 接口执行 Service 连接（或取消）处理操作，在 Activity 连接到 Service 程序之后，会触发 Service 类中的 onBind()方法，在此方法中要返回一个 IBinder 接口的对象，IBinder 接口定义的常量和方法如表 6-16 所示。

表 6-16 IBinder 接口的常量和方法

常量及方法	描述
public static final int DUMP_TRANSACTION	IBinder 协议的事务码：清除内部状态
public static final int FIRST_CALL_TRANSACTION	用户指令的第一个事务码可用
public static final int FLAG_ONEWAY	transact()方法单向调用的标志位，表示调用者不会等待从被调用者那里返回的结果，而立即返回

常量及方法	描述
public static final int INTERFACE_TRANSACTION	IBinder 协议的事务码：向事务接收端询问其完整的接口规范
public static final int LAST_CALL_TRANSACTION	用户指令的最后一个事务码可用
public static final int PING_TRANSACTION	IBinder 协议的事务码：pingBinder()
public static final int LIKE_TRANSACTION	IBinder 协议的事务码
public static final int TWEET_TRANSACTION	IBinder 协议的事务码，发送一个信号给目标
public abstract void dump(FileDescriptor fd, String[] args)	向指定的数据流输出对象状态
public abstract String getInterfaceDescriptor()	取得被 Binder 对象所支持的接口名称
public abstract boolean isBinderAlive()	检查 Binder 所在的进程是否活着
public abstract void linkToDeath(IBinder.DeathRecipient recipient, int flags)	如果指定的 Binder 消失，则为通知注册一个新的接收器
public abstract boolean pingBinder()	检查远程对象是否存在
public abstract IInterface queryLocalInterface(String descriptor)	取得对一个接口绑定对象本地实现
public abstract boolean transact(int code, Parcel data, Parcel reply, int flags)	执行一个一般的操作
public abstract boolean unlinkToDeath(IBinder.DeathRecipient recipient, int flags)	删除一个接收通知的接收器

实例 6-6：Service 组件操作实例

新建一个项目，项目的命名为：exam6_6，包名称为：org.hnist.demo，利用 Service 编程实现服务的启动和绑定服务的启动。

1）新建布局管理文件 activity_main.xml，代码如下：

```xml
<?xml version="1.0" encoding="utf-8"?>
<?xml version="1.0" encoding="utf-8"?>
<LinearLayout
    xmlns:android="http://schemas.android.com/apk/res/android"
    android:orientation="vertical"
    android:layout_width="fill_parent"
    android:layout_height="fill_parent">
    <Button
        android:id="@+id/startSrv"
        android:layout_width="150dp"
        android:layout_height="wrap_content"
        android:layout_gravity="center"
        android:text="启动服务" />
    <Button
        android:id="@+id/stopSrv"
        android:layout_width="150dp"
        android:layout_height="wrap_content"
        android:layout_gravity="center"
        android:text="停止服务" />
    <Button
        android:id="@+id/bindSrv"
        android:layout_width="150dp"
        android:layout_height="wrap_content"
        android:layout_gravity="center"
        android:text="绑定服务" />
    <Button
        android:id="@+id/unbindSrv"
```

```xml
        android:layout_width="150dp"
        android:layout_height="wrap_content"
        android:layout_gravity="center"
        android:text="解除绑定服务" />
</LinearLayout>
```

2）编写一个服务类 MyService.java，该类必须继承 android.app.Service，代码如下：

```java
package org.hnist.demo;
import android.app.Service;
import android.content.Intent;
import android.os.Binder;
import android.os.IBinder;
public class MyService extends Service {          //新建类 MyService 继承 Service
    private IBinder myBinder = new Binder(){      //定义 IBinder 接口
        public String getInterfaceDescriptor() {  //取得接口描述信息
            return "MyServiceUtil class.";}} ;    //返回 Service 类的名称
    public IBinder onBind(Intent intent) {        //绑定时触发
        return myBinder;}
    public void onRebind(Intent intent) {         //重新绑定时触发
        super.onRebind(intent); }
    public boolean onUnbind(Intent intent) {      //解除绑定时触发
        return super.onUnbind(intent);    }
    public void onCreate() {                      //创建时触发
        super.onCreate();      }
    public void onDestroy() {                     //销毁时触发
        super.onDestroy();    }
    public int onStartCommand(Intent intent, int flags, int startId) {
        //启动时触发
        return Service.START_CONTINUATION_MASK;  }}
```

3）新建 Activity 文件 MainActivity.java，代码如下：

```java
package org.hnist.demo;
package org.hnist.demo;
import android.app.Activity;
import android.content.ComponentName;
import android.content.Context;
import android.content.Intent;
import android.content.ServiceConnection;
import android.os.Bundle;
import android.os.IBinder;
import android.os.RemoteException;
import android.view.View;
import android.view.View.OnClickListener;
import android.widget.Button;
public class MainActivity extends Activity {
private Button start;
private Button stop;
private Button bind;
private Button unbind;
private ServiceConnection serviceConnection = new ServiceConnection() {
    //连接到 Service
```

```java
        public void onServiceConnected(ComponentName name, IBinder service) {
            try {System.out.println("Service Connect Success");
            } catch (RemoteException e) {
                e.printStackTrace();  }}
        public void onServiceDisconnected(ComponentName name) { //与Service断开连接
            System.out.println("Service Disconnected."); }};
    public void onCreate(Bundle savedInstanceState) {
        super.onCreate(savedInstanceState);
        super.setContentView(R.layout.activity_main);
        this.start = (Button) super.findViewById(R.id.startSrv);
        this.stop = (Button) super.findViewById(R.id.stopSrv);
        this.bind = (Button) super.findViewById(R.id.bindSrv);
        this.unbind = (Button) super.findViewById(R.id.unbindSrv);
        this.start.setOnClickListener(new StartOnClickListenerImpl()) ;
        this.stop.setOnClickListener(new StopOnClickListenerImpl()) ;
        this.bind.setOnClickListener(new BindOnClickListenerImpl()) ;
        this.unbind.setOnClickListener(new UnbindOnClickListenerImpl()); }
    private class StartOnClickListenerImpl implements OnClickListener {
        public void onClick(View v) {
            //启动 Service
            MainActivity.this.startService(new Intent(MainActivity.this, MyService.
                class)); }}
    private class StopOnClickListenerImpl implements OnClickListener {
        public void onClick(View v) {
            //停止 Service
            MainActivity.this.stopService(new Intent(MainActivity.this, MyService.
                class));}}
    private class BindOnClickListenerImpl implements OnClickListener {
        public void onClick(View v) {
            //绑定 Servic
            MainActivity.this.bindService(new Intent(MainActivity.this, MyService.
                class),
MainActivity.this.serviceConnection,Context.BIND_AUTO_CREATE);}}
    private class UnbindOnClickListenerImpl implements OnClickListener {
        public void onClick(View v) {
            //取消 Service 绑定
            MainActivity.this.unbindService(MainActivity.this.serviceConnection);}}}
```

4）在 AndroidManifest.xml 中声明。

在<application>节点下添加如下的代码：

```xml
<service android:name=". MyService " />     //MyService 为建立的服务类名
```

保存所有文件，运行该项目，在模拟手机上按"Menu"然后选择"设置→应用程序→正在运行的服务"，弹出如图 6.20 所示的界面，表示服务并没有启动。

再运行该程序，如图 6.21 所示，单击"启动服务"按钮，再查看正在运行的服务情况，选择"设置→应用程序→正在运行的服务"，出现如图 6.22 所示的界面，表示服务已经启动。

单击 exam6_6，可以查看运行的是 MyService 服务，可以停止该服务，如图 6.23 所示。

图 6.20　启动服务前查看服务的情况

图 6.21　程序运行的主界面

图 6.22　启动服务后查看服务的情况

图 6.23　查看运行的服务

在模拟手机上按"Menu"然后选择"设置→应用程序→管理应用程序",找到"exam6_6"程序,如图 6.24 所示,单击该程序,选择"强行停止",停止该程序,如图 6.25 所示。再来查看正在运行的服务情况,发现服务并没有随着程序的停止和终止。

图 6.24　程序运行的主界面

图 6.25　启动服务后查看服务的情况

再运行该程序,先单击"绑定服务",然后单击"启动服务"按钮,再查看正在运行的服务情况,可以看到服务已经启动,然后用前面的方法,选择"设置→应用程序→管理应用程序",找到"exam6_6"程序,单击该程序,选择"强行停止",停止该程序,再查看正在运行的服务情况,发现服务已经随着程序的停止和终止,说明绑定已经发生作用了。

6.6.4　Service 的生命周期

Service 与 Activity 一样,也有一个从启动到销毁的过程,但 Service 的这个过程比 Activity 简单得多。Service 启动到销毁的过程只会经历如下 3 个阶段:创建服务、开始服务、销毁服务。

Service 生命周期主要包含以下几个方法：onCreate()、onBind(Intent intent)、onStart(Intent intent,int startId)、onDestory()。

对于使用 startService()方法打开的服务，其生命周期为：onCreate→onStart→onDestroy 服务关闭。

对于使用 bindService ()方法打开的服务，其生命周期为：onCreate→onBind→onUnbind→onDestroy 服务关闭。

Service 的生命周期示意图如图 6.26 所示。

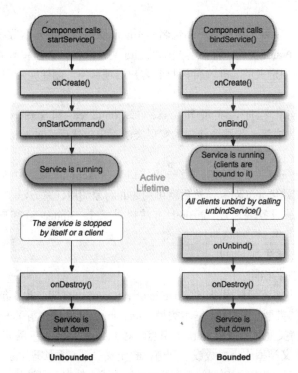

图 6.26 Service 的生命周期示意图

6.6.5 跨进程调用 Service（AIDL 服务）

1．AIDL 服务的概念

AIDL 是 Android Interface Definition Language 的简称，即 Android 接口定义语言，它用来公开服务的接口，使其他的应用程序也可以访问本应用程序提供的服务，将这种可以跨进程访问的服务称为 AIDL 服务。

2．建立 AIDL 服务

建立 AIDL 服务分为服务端和客户端的建设两部分。

建立服务端

1）创建一个 aidl 文件，创建完成后，eclipse 插件自动在 gen 目录下生成同名字的 java 文件。里面包含一个 Stub 抽象类，这个类继承自 android.os.Binder，这个类是实现整个远程调用的核心。

2）然后创建一个类来继承上面说到的那个 Stub 抽象类，实现里面的抽象方法。（这些抽象方法是根据 aidl 文件自动生成的）。

3）创建一个自定义 Service 继承自 Service，实现 onBind 方法，注意此 onBind 方法必须返回第二步创建的那个 Stub 类的子类。

4）创建一个 Activity 程序，此 Activity 程序只要实现把 Service 启动了即可。

5）在 AndroidManifest.xml 文件中配置 AIDL 服务。

这样服务器端就开发完毕，运行后启动了一个可供远程调用的 Service。关键还是通过 onBind 提供一个 Binder 给客户端。

建立客户端

1）在客户端建立一个项目，注意客户端包名称不能和服务器端的包名称相同。

2）客户端也需要一个 aidl 文件，注意客户端的 aidl 文件的包名必须和服务器端的 aidl 包名一致，名字也相同，可将服务端中定义的 aidl 文件拷贝到客户端项目中，创建完后同样会在 gen 下生成一个接口。

3）创建一个 Activity 程序，实现其 onServiceConnected 和 onServiceDisconnected 方法，onServiceConnected 方法生成第一步那个接口的实现类的对象，然后通过 bindService 方法绑定服务，得到 aidl 对象。

客户端关键点是创建 aidl 文件自动生成了一个接口，在 Activity 程序中必须绑定服务程序开启的 Service，在绑定过程中初始化 aidl 对象，然后就可用 aidl 对象调用任意方法了。

整个设计的关键点是服务端通过 onBind 方法提供一个 Binder 给客户端，然后客户端通过 bindService 方法得到 aidl 对象。

编写 Aidl 文件时，需要注意下面几点：

1）接口名和 aidl 文件名相同。

2）接口和方法前不用加访问权限修饰符 public,private,protected 等，也不能用 final,static。

3）AIDL 默认支持的类型包括 Java 基本类型（int、long、boolean 等）和（String、List、Map、CharSequence），使用这些类型时不需要 import 声明。对于 List 和 Map 中的元素类型必须是 AIDL 支持的类型。如果使用自定义类型作为参数或返回值，自定义类型必须实现 Parcelable 接口。

4）自定义类型和 AIDL 生成的其他接口类型在 aidl 描述文件中，应该显式 import，即便在该类和定义的包在同一个包中。

5）AIDL 服务中的 onBind 方法必须返回 AIDL 接口对象，它也是 onServiceConnected 事件方法的第 2 个参数值。

6）客户端 bindService 方法的第 1 个参数是 Intent 对象，该对象构造方法的参数需要指定 AIDL 服务的 ID，也就是在 AndroidManifest.xml 文件中<service>标签的<action>子标签的 android:name 属性的值。

实例 6-7：AIDL 服务操作实例

利用 Service 组件编程实现跨进程调用 Service 端的数据和方法。

服务端的建立

1）新建一个项目，项目的命名为：exam6_7Server，包名称为：org.hnist.demo。

2）右击包文件，选择 NEW→File，弹出如图 6.27 所示界面，输入文件名字 IPerson.aidl。

单击"Finish"按钮完成文件的创建，输入如下内容：

```
package org.hnist.demo;
interface IPerson {
    void setName(String name);
```

```
    void setAge(int age);
    void setEmail(String email);
    String display();}
```

保存文件,如果输入没有错误,会自动在 gen 文件夹下产生一个 IPerson.java 文件,该文件里面包含 IPerson.Stub 类,如图 6.28 所示。

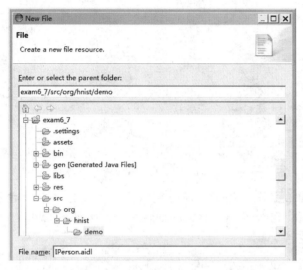

图 6.27 建立 IPerson.aidl 文件

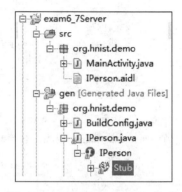

图 6.28 成功建立 IPerson.aidl 文件

3)在该包下建立 IPersonImpl.java 文件,以继承 IPerson.Stub 类,这里是核心部分,在这里定义客户端要调用的数据和方法。示例代码如下:

```
package org.hnist.demo;
import org.hnist.demo.IPerson;
import android.os.RemoteException;
public class IPersonImpl extends IPerson.Stub{     //继承 IPerson.Stub 类
    //声明三个变量
    private String name="张晓飞";
    private int age=21;
    private String email="zxf@163.com";
    public String display() throws RemoteException {//定义display方法显示信息
        return "姓名:"+name+"  年龄:"+age+"   邮箱:"+email; }
    public void setName(String name) throws RemoteException {   //定义方法设置name
        this.name = name;        }
    public void setAge(int age) throws RemoteException {    //定义方法设置age
        this.age = age;}
    public void setEmail(String email) throws RemoteException {   //定义方法设置email
        this.email = email;  }}
```

4)在该包下建立一个 Service,MyRemoteService.java 文件,代码如下:

```
package org.hnist.demo;
import android.app.Service;
import android.content.Intent;
import android.os.Binder;
import android.os.IBinder;
```

```
import org.hnist.demo.IPerson.Stub;
public class MyRemoteService extends Service{        //定义服务MyRemoteService
    private Binder iPerson = new IPersonImpl();      //声明IPerson接口
    public IBinder onBind(Intent intent) {           //实现onBind()方法
        return iPerson                               //返回值iPerson
```

5）在该包下建立一个Activity，MainActivity.java文件，代码如下：

```
package org.hnist.demo;
import android.app.Activity;
import android.content.Intent;
import android.os.Bundle;
public class MainActivity extends Activity {
 public void onCreate(Bundle savedInstanceState) {
     super.onCreate(savedInstanceState);
     Intent intent = new Intent();                   //定义Intent
     intent.setAction("org.hnist.demo.MyRemoteService");//设置Intent Action属性
     startService(intent);    }}                     //绑定服务
```

6）在AndroidManifest.xml文件中配置AIDL服务，声明此Service，注意此service的声明必须包含一个action，此action也用于客户端的调用使用。在</activity>与</application>之间加入如下代码即可。

```
<service android:name="MyRemoteService">
    <intent-filter>
            <action android:name="org.hnist.demo.MyRemoteService"/>
    </intent-filter>
</service>
```

到此，服务端建立完毕。

客户端的建立

1）建立另外一个新的Android项目exam6_7Client，包命名为：org.hnist.exam67C。

2）将服务端中定义的IPerson.aidl文件拷贝到客户端项目中，创建完后同样会在gen下生成一个接口，如图6.29所示。注意在不同的包中。

3）在该包下建立一个Activity，MainActivity.java文件，代码如下：

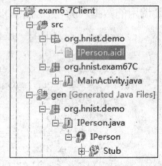

图6.29 客户端成功建立IPerson.aidl文件

```
package org.hnist.exam67C;
import org.hnist.demo.IPerson;
import android.app.Activity;
import android.app.Service;
import android.content.ComponentName;
import android.content.Intent;
import android.content.ServiceConnection;
import android.os.Bundle;
import android.os.IBinder;
import android.view.View;
import android.view.View.OnClickListener;
import android.widget.Button;
import android.widget.TextView;
```

第6章 Android组件之间的通信

```
import android.widget.Toast;
public class MainActivity extends Activity {
    private IPerson iPerson;                                //声明 IPerson 接口
    private Button getBtn;
    private TextView Myinfo;
    private ServiceConnection conn = new ServiceConnection() {//实例化 ServiceConnection
        synchronized public void onServiceConnected(ComponentName name, IBinder
            service) { iPerson = IPerson.Stub.asInterface(service);}   //建立和
            服务的连接获得 IPerson 接口
    public void onServiceDisconnected(ComponentName name) {iPerson=null;}};
        //连接断开
    public void onCreate(Bundle savedInstanceState) {
        super.onCreate(savedInstanceState);
        setContentView(R.layout.activity_main);
        getBtn = (Button) findViewById(R.id.Start);
        Myinfo = (TextView) findViewById(R.id.myinf0);
        Intent intent = new Intent("org.hnist.demo.MyRemoteService"); //实例化 Intent
        bindService(intent, conn, Service.BIND_AUTO_CREATE);//设置 Intent Action 属性
        getBtn.setOnClickListener(new OnClickListener() {            //设置监听事件
            public void onClick(View v) {
                try{
                    iPerson.setName("王小军");  //设置 name 的值,调用服务中的方法
                    String msg = iPerson.display();//设置 msg 的值,调用服务中的方法
                    MainActivity.this.Myinfo.setText(msg);
                }catch (Exception e) {e.printStackTrace();}}});}
    protected void onDestroy(){unbindService(conn); //解除绑定
        super.onDestroy();   }}
```

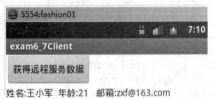

图 6.30 调用服务端的方法和数据

保存所有文件,先运行 exam6_7Server 的项目,以便安装 AIDL 服务,再运行客户端的 exam6_7Client 项目,单击按钮,结果如图 6.30 所示。

来分析一下这个运行结果,客户端的 exam6_7Client 项目中的任何一个文件中都没有 "name:王小军; 年龄: 21 邮箱: zxf@163.com" 这样的数据,为什么会显示这样的结果呢? 这个数据原本是在 exam6_7Server 项目中的 IPersonImpl.java 文件中的。

在 exam6_7Client 项目的 MainActivity.java 文件中有连接服务和获得 IPerson 接口的语句:

```
synchronized public void onServiceConnected(ComponentName name, IBinder
    service) { iPerson = IPerson.Stub.asInterface(service); //获得 IPerson 接口
```

IPerson 接口是在 exam6_7Server 项目中 MyRemoteService.java 服务中语句声明,以便客户端调用,语句如下:

```
private Binder iPerson = new IPersonImpl();             //声明 IPerson 接口
    public IBinder onBind(Intent intent) {              //实现 onBind()方法
        return iPerson;}}                               //返回值
```

这个服务中有语句: return iPerson,返回值是 iPerson 接口,IPersonImpl.java 中定义 IPersonImpl 是 IPerson.Stub 的子类。IPersonImpl.java 文件中有语句定义了 display()方法,因此 iPerson 也有 display()方法。

因此启动 exam6_7Client 项目,会将服务也启动,从而获得返回值 iPerson,这个 iPerson 也具有

IPersonImpl 定义的 4 个方法，通过语句 iPerson.setName("王小军"); 设置 name 的值，语句 String msg = iPerson.display(); 获得 msg 的值，然后再通过 TextView 组件将 msg 值显示出来，从而得到如图 6.30 所示的结果。

6.6.6 Service 系统服务

前面的学习，我们知道可以利用 Service 组件设计一些方法和数据让客户调用，Android 系统就设计了这样的服务，称之为 Service 系统服务。通俗地说就是 Android 系统自身来的一些内置的功能，例如，当手机接到来电时，会响铃，会显示对方的号码，等等。这些服务是能够被应用程序调用的。

在 Android 操作系统中，为了方便用户使用系统服务，在 android.content.Context 类中将所有的系统服务名称以常量的形式进行了绑定，如表 6-17 所示。

表 6-17 Context 类中定义的常见系统服务

常量	描述
public static final String CLIPBOARD_SERVICE	剪贴板服务
public static final String WINDOW_SERVICE	窗口服务
public static final String ALARM_SERVICE	闹铃服务
public static final String AUDIO_SERVICE	音频服务
public static final String NOTIFICATION_SERVICE	Notification 服务
public static final String SEARCH_SERVICE	搜索服务
public static final String POWER_SERVICE	电源管理服务
public static final String WIFI_SERVICE	WIFI 服务
public static final String ACTIVITY_SERVICE	运行程序服务
public static final String TELEPHONY_SERVICE	取得手机网络信息

这里只给出了常见的系统服务名称，更多的系统服务名称可以查阅 Android 相关开发文档获得。

系统服务实际上可以看作是一个对象，通过 Activity 类的 getSystemService()方法可以获得指定的系统服务。getSystemservice()方法只有一个 String 类型的参数，表示系统服务的 ID，这个 ID 在整个 Android 系统中是唯一的。例如：

```
getSystemService(Context.WINDOW_SERVICE);      //获得 WindowManager 服务
getSystemService(Context.WIFI_SERVICE);        //获得 WIFI 服务
getSystemService(Context.CLIPBOARD_SERVICE);   //获得 ClipboardManager 服务
```

获得服务后会返回一个指定类的对象，就可以利用这个类提供的方法做相应的操作了。例如：

```
//获得音频服务
AudioManager AM=(AudioManager) getSystemService(Context.AUDIO_SERVICE);
//获得最大铃音
int max = AM.getStreamMaxVolume(AudioManager.STREAM_VOICE_CALL);
//获得当前铃音
int current= AM.getStreamVolume(AudioManager.STREAM_VOICE_CALL);……
```

使用某个服务前要先了解该服务怎样获得，该服务会返回一个怎样的类，这个类有哪些方法和常量等等，下面以取得手机网络信息为例介绍一个系统服务。

1. 取得手机网络信息

要想取得手机网络信息的服务，可以使用 getSystemService 方法，里面携带的参数是 Context.TELEPHONY_SERVICE，返回的是 android.telephony.TelephonyManager 类型的对象，这个类有表 6-18 所示的常用常量和表 6-19 所示的常用方法。

表 6-18 TelephonyManager 类的常用常量

常量及方法	描述
public static final int NETWORK_TYPE_CDMA	使用 CDMA 网络
public static final int NETWORK_TYPE_GPRS	使用 GPRS 网络
public static final int PHONE_TYPE_CDMA	使用 CDMA 通信
public static final int PHONE_TYPE_GSM	使用 GSM 通信
public static final String ACTION_PHONE_STATE_CHANGED	电话状态改变
public static final int CALL_STATE_IDLE	空闲（无呼入或已挂机）
public static final int CALL_STATE_OFFHOOK	摘机（有呼入）
public static final int CALL_STATE_RINGING	响铃（接听中）
public static final int DATA_CONNECTED	数据连接状态：已连接
public static final int DATA_DISCONNECTED	数据连接状态：断开

表 6-19 TelephonyManager 类的常用方法

常量及方法	描述
public String getNetworkCountryIso()	获取网络的国家 ISO 代码
public String getDeviceSoftwareVersion()	获得软件版本
public String getDeviceId()	取得设备标识
public String getLine1Number()	取得手机号码
public String getNetworkOperatorName()	取得移动提供商的名称
public int getNetworkType()	取得移动网络的连接类型
public int getPhoneType()	取得电话网络类型
public boolean isNetworkRoaming()	判断电话是否处于漫游状态
public void listen(PhoneStateListener listener, int events)	注册电话状态监听器
public String getSimCountryIso()	获取 SIM 卡中国家 ISO 代码
public String getSimOperator()	获得 SIM 卡中移动国家代码（MCC）和移动网络代码（MNC）
public String getSimOperatorName()	获取服务提供商姓名
public String getSimSerialNumber()	SIM 卡序列号

实例 6-8：取得手机网络信息操作实例

新建一个项目，项目的命名为：exam6_8，包名称为：org.hnist.demo，编程列出获得手机的网络信息。

1）新建布局管理文件 activity_main.xml，代码如下：

```xml
<?xml version="1.0" encoding="utf-8"?>
<LinearLayout
    android:id="@+id/LinearLayout01"
    android:layout_width="fill_parent"
    android:layout_height="fill_parent"
    xmlns:android="http://schemas.android.com/apk/res/android">
    <ListView
        android:id="@+id/MyListView"
        android:layout_width="fill_parent"
        android:layout_height="wrap_content" />
</LinearLayout>
```

2）新建 Activity 文件 MainActivity.java，代码如下：

```java
package org.hnist.demo;
import java.util.ArrayList;
import java.util.List;
import android.app.Activity;
import android.content.Context;
import android.os.Bundle;
import android.telephony.TelephonyManager;
import android.widget.ArrayAdapter;
import android.widget.ListAdapter;
import android.widget.ListView;
public class MainActivity extends Activity {
    private ListView infolist = null;
    private TelephonyManager TM = null;              //定义TelephonyManager对象
    private ListAdapter adapter = null;              //适配器组件
    private List<String> ListInfo = new ArrayList<String>(); /
    public void onCreate(Bundle savedInstanceState) {
        super.onCreate(savedInstanceState);
        super.setContentView(R.layout.activity_main);
        this.infolist = (ListView) super.findViewById(R.id.MyListView);
        this.TM = (TelephonyManager) super.getSystemService(Context.TELEPHONY_
            SERVICE);   //取得手机服务
        this.listInfo();   }                         //定义listInfo方法列表显示
    private void listInfo() {                        //执行列表显示操作
        this.ListInfo.add(this.TM.getLine1Number() == null ? "没有手机号码" : "
            手机号码："
                + this.TM.getLine1Number());         //取得手机号码
        this.ListInfo.add(this.TM.getNetworkOperatorName() == null ?
                "没有移动网络服务商" : "移动网络服务商：" +   this.TM.getNetwork-
                    OperatorName());                 //取得移动网络商名称
    //判断网络类型
        if (this.TM.getPhoneType()  == TelephonyManager.NETWORK_TYPE_CDMA) {
          this.ListInfo.add("移动网络类型：CDMA");}
            else if (this.TM.getPhoneType() == TelephonyManager.NETWORK_TYPE_GPRS) {
                this.ListInfo.add("移动网络类型：GPRS");
            } else {   this.ListInfo.add("移动网络类型：未知");}
    //判断电话网络类型
        if (this.TM.getNetworkType() == TelephonyManager.PHONE_TYPE_GSM) {
          this.ListInfo.add("手机网络类型：GSM"); }
            else if (this.TM.getNetworkType() == TelephonyManager.PHONE_TYPE_CDMA) {
                this.ListInfo.add("手机网络类型：CDMA");
            } else {this.ListInfo.add("手机网络类型：UTMS");  }
    //是否是漫游
        this.ListInfo.add("是否漫游：" + (this.TM.isNetworkRoaming() ?"漫游":
            "非漫游"));
        this.ListInfo.add(this.TM.getDeviceId() == null ? "没有设备标识ID" : "
            设备标识ID："+ this.TM.getDeviceId());    //取得设备标识
        this.adapter = new ArrayAdapter<String>(this,   //实例化ArrayAdapter
                android.R.layout.simple_list_item_1,    //每行显示一条数据
                this.ListInfo);                         //定义显示数据
        this.infolist.setAdapter(this.adapter); }}      //设置适配器
```

3）修改 AndroidManifest.xml 文件，增加读取手机状态的权限，添加下面的语句

```
<uses-permission android:name="android.permission.READ_PHONE_STATE">
</uses-permission>
```

保存所有文件，首先打开模拟手机，单击右边的"Menu"按钮，选择"设置→关于手机→状态消息"，如图 6.31 所示。运行该程序，结果如图 6.32 所示。

图 6.31　手机状态信息　　　　　图 6.32　编程获取的手机网络信息

6.7　BroadcastReceiver

BroadcastReceiver 是"广播接收者"的意思，它用来接收来自系统和应用中的广播。在 Android 系统中，广播在很多地方得到体现。例如当网络状态改变时系统会产生一条广播，接收到这条广播就能及时地做出提示和保存数据等操作。Android 中的广播机制设计得非常方便，可以将很多复杂的事情用广播形式告知就可以了，极大地减少了开发工作量。

6.7.1　BroadcastReceiver 基础

广播是一种广泛运用的在应用程序之间传输信息的机制。而 BroadcastReceiver 是对发送出来的广播进行过滤接收并响应的组件。BroadcastReceiver 自身并不实现图形用户界面，当它收到某个通知后，BroadcastReceiver 可以启动 Activity 作为响应，或者通过 NotificationMananger 提醒用户，或者启动 Service 等等。广播属于触发式操作，当有了指定操作之后会自动启动广播。

要实现一个 BroadcastReceiver 方法的基本步骤如下：

1）继承 BroadcastReceiver，并重写 onReceive()方法，例如：

```
public class MyBroadcastReceiver extends BroadcastReceiver {
    public MyBroadcastReceiver(){......}                    //构造方法
    public void onReceive(Context context, Intent intent) {......} }//重写 onReceive()
```

在 onReceive 方法内，可以获取随广播而来的 Intent 中的数据，这里面可以包含很多有用的信息，这里是广播消息核心部分所在！

创建完 BroadcastReceiver 之后，还不能马上工作，还需要为它注册一个指定的广播地址。没有注册广播地址的 BroadcastReceiver 就像一个缺少选台按钮的收音机，虽然功能具备，但也无法收到各电台的信号。

2）为 BroadcastReceiver 注册广播地址。

注册有静态注册和动态注册两种方式。

① 静态注册方式是在 AndroidManifest.xml 的 application 里面定义 receiver 并设置要接收的 action。

```xml
<receiver                                              //定义广播处理
    android:name=".MyBroadcastReceiver"                //广播处理类
    android:enabled="true">                            //启用广播
    <intent-filter>                                    //匹配 Action 操作时广播
        <action android:name="android.intent.action.EDIT" />
    </intent-filter>
</receiver>
```

配置后，只要是 android.intent.action.EDIT 这个地址的广播，MyBroadcastReceiverUtil 都能够接收的到。注意，这种方式的注册是常驻型的，也就是说当应用关闭后，如果有广播信息传来，MyBroadcastReceiverUtil 也会被系统调用而自动运行。

② 动态注册方式是在 Activity 或 Service 里面调用函数来注册，有以下几个关键参数：一个是 receiver，另一个是 IntentFilter，其中里面是要接收的 action。

```java
public class BroadcastReceiverDemo extends Activity {
    private BroadcastReceiver receiver;
    ……
    receiver = new CallReceiver();
    //动态注册广播地址
    registerReceiver(receiver,new IntentFilter("android.intent.action.EDIT")); }
    ……
    //解除注册广播地址
    unregisterReceiver(receiver);
    ……
```

推荐使用静态注册方式，由系统来管理 receiver，动态注册方式不是常驻型的，而且代码隐藏在代码中，比较难发现，在退出程序前要记得调用 unregisterReceiver()方法进行注销。

3）利用 sendBroadcast 发送广播。

可以根据以上任意一种方法完成注册，当注册完成之后，定义好的广播接收者就可以正常工作了，先构建 Intent 对象，然后调用 sendBroadcast(Intent)方法将广播发出。

```java
Intent it = new Intent("android.intent.action.EDIT ");   //指定 Action
intent.putExtra("msg", "hello receiver.");               //附加数据
sendBroadcast(it);                                       //进行广播
```

BroadcastReceiver 的生命周期只有十秒左右。

一个 BroadcastReceiver 对象只有在被调用 onReceive()方法才有效，当从该函数返回后，该对象就无效了，结束生命周期。因此，在所调用的 onReceive()方法里，不能有过于耗时的操作，对于耗时的操作，可用 Service 来完成，因为当得到其他异步操作所返回的结果时，BroadcastReceiver 可能已经无效了。

6.7.2 BroadcastReceiver 组件操作实例

实例 6-9：BroadcastReceiver 组件操作实例

建立一个 Android 项目，项目的名称为 exam6_9，包名称为"org.hnist.demo"，利用 BroadcastReceiver 组件发送广播。

1)新建布局管理文件 activity_main.xml,代码如下:

```xml
<?xml version="1.0" encoding="utf-8"?>
<LinearLayout
    xmlns:android="http://schemas.android.com/apk/res/android"
    android:layout_width="fill_parent"
    android:layout_height="fill_parent">
    <Button
        android:id="@+id/mybut"
        android:layout_width="wrap_content"
        android:layout_height="wrap_content"
        android:text="发送广播消息"/>
</LinearLayout>
```

2)新建 MyBroadcastReceiver.java 文件继承 BroadcastReceiver 类,并覆写 onReceive()方法,代码如下:

```java
package org.hnist.demo;
import android.content.BroadcastReceiver;
import android.content.Context;
import android.content.Intent;
import android.widget.Toast;
public class MyBroadcastReceiver extends BroadcastReceiver {//继承 BroadcastReceiver
    public void onReceive(Context context, Intent intent) {  //覆写 onReceive()方法
        String msg=intent.getStringExtra("msg");              //获得广播信息内容
        Toast.makeText(context, msg, Toast.LENGTH_LONG).show();}}//显示信息
```

3)新建 Activity 文件 MainActivity.java,代码如下:

```java
package org.hnist.demo;
import android.app.Activity;
import android.content.Intent;                               //导入 content.Intent 类
import android.os.Bundle;
import android.view.View;
import android.view.View.OnClickListener;
import android.widget.Button;
public class MainActivity extends Activity {
private Button mybut ;
    public void onCreate(Bundle savedInstanceState) {
      super.onCreate(savedInstanceState);
      super.setContentView(R.layout.activity_main);
      this.mybut = (Button) super.findViewById(R.id.mybut) ;
      this.mybut.setOnClickListener(new SendMsg());}         //设置监听事件
      private class SendMsg implements OnClickListener {    //定义事件
        public void onClick(View v) {
          Intent it = new Intent(Intent.ACTION_EDIT);        //启动 Action
          it.putExtra("msg", "明天上午 9:30 在 16307 开会");
          MainActivity.this.sendBroadcast(it); }}}          //进行广播
```

4)在 AndroidManifest.xml 文件之中注册广播组件,在<application>与</ application >之间写入如下代码:

```xml
<receiver
    android:name="MyBroadcastReceiver"
```

```
            android:enabled="true">
        <intent-filter>
            <action android:name="android.intent.action.EDIT" />
        </intent-filter>
</receiver>
```

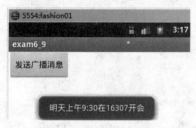

图 6.33　发送广播消息

保存所有文件，运行该项目，单击按钮，结果如图 6.33 所示。

上面的例子只是一个接收者来接收广播，如果有多个接收者都注册了相同的广播地址，又会是什么情况呢，能同时接收到同一条广播吗，相互之间会不会有干扰呢？这就涉及普通广播和有序广播的概念了。

普通广播（Normal Broadcast）对于多个接收者来说是完全异步的，通常每个接收者都无需等待即可以接收到广播，接收者相互之间不会有影响。对于这种广播，接收者无法终止广播，即无法阻止其他接收者的接收动作。

有序广播（Ordered Broadcast）比较特殊，它每次只发送到优先级较高的接收者那里，然后由优先级高的接受者再传播到优先级低的接收者那里，优先级高的接收者有能力终止这个广播。

6.7.3　通过 BroadCast 启动 Service

由前面的学习知道，可以通过 Activity 程序启动 Service，其实 Service 也可以通过 BroadCast 来启动，如果一个 Service 要通过 Broadcast 启动很简单，只需要在 Broadcast 中调用 startService()方法即可。

实例 6-10：BroadcastReceiver 组件操作实例

建立一个项目，利用 Broadcast 来启动 Service，Service 的主要功能就是播放一首歌，如何播放歌曲将在下一个章节介绍。

鼠标右键单击实例 6-9，选择 Copy，然后右击左边空白处，选择 "Paste"，输入项目名称 exam6_10，如图 6.34 所示。在上例的基础上，利用 BroadcastReceiver 组件启动 Service。

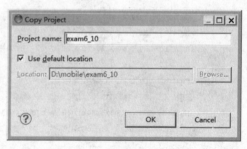

图 6.34　利用复制方法建立项目

1）建立一个服务 MyService.java，播放一首歌 abird.mp3，首先要将这首歌复制到 res\raw 文件夹下。

```
package org.hnist.demo;
import java.io.IOException;
import android.app.Service;
import android.content.Intent;
import android.media.MediaPlayer;
import android.os.IBinder;
public class MyService extends Service       //定义MyService.java继承Service类
{    private MediaPlayer player;             //定义MediaPlayer 变量player
public IBinder onBind(Intent intent)         //定义onBind方法
{    return null;}
public void onCreate()                       //覆写onCreate方法
{    super.onCreate();
    player = MediaPlayer.create(MyService.this, R.raw.abird);  //获得音频资源
```

```
    try {
      MyService.this.player.prepare();      //进入预备状态
    } catch (IllegalStateException e) {
      e.printStackTrace();
    } catch (IOException e) {
      e.printStackTrace();}
       MyService.this.player.start();       //播放文件
    player.setLooping(true);}               //反复播放
public void onDestroy()                     //覆写 onDestroy 方法
{   super.onDestroy();   }
public void onStart(Intent intent, int startId)   //覆写 onStart 方法
{   super.onStart(intent, startId); }}
```

2）修改 MyBroadcastReceiver.java 文件，斜体是增加的部分，代码如下：

```
package org.hnist.demo;
import android.content.BroadcastReceiver;
import android.content.Context;
import android.content.Intent;
import android.widget.Toast;
public class MyBroadcastReceiver extends BroadcastReceiver {//继承BroadcastReceiver
    public void onReceive(Context context, Intent intent) { //覆写 onReceive 方法
       Intent serviceIntent = new Intent(context, MyService.class);//定义 Intent
          context.startService(serviceIntent);           //启动指定服务
       String msg=intent.getStringExtra("msg");
       Toast.makeText(context, msg, Toast.LENGTH_LONG).show();/}}/ 显示信息
```

3）修改 AndroidManifest.xml 文件，在<application>与</application>之间增加如下代码：

```
<service android:enabled="true" android:name=".MyService" />
```

其他文件不做修改，保存所有文件，运行该项目，单击按钮，出现如图 6.33 所示结果的同时还会播放歌曲。

Broadcast 不仅可以用来启动 Service，还可以用来启动 Activity 程序，只需要在 Broadcast 中调用 startActivity ()方法即可。

经常会有开机就要启动某个服务或某个应用程序的需要，要实现这样的功能，可以使用系统"启动完成"这条广播，接收到这条广播后就可以启动自己的服务或自己的应用程序了，下面就来看具体的实现。

在实例 6-10 的基础上做适当修改，实现开机运行服务和指定应用程序，详细代码见实例 exam6_11，按照第 1 章的方法将项目编译成 d:\exam6_11.apk 文件，按照如下的方法将 apk 文件安装到模拟器中。

1）在 Eclipse 中启动模拟器；
2）运行 cmd，输入：adb install D:/exam6_11.apk，出现如图 6.35 所示的提示，表示安装成功。

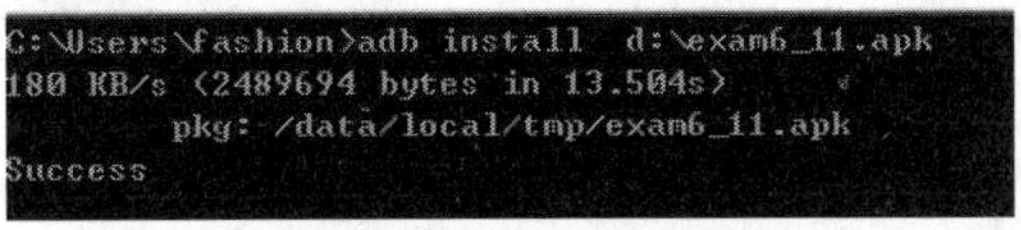

图 6.35　安装 apk 文件到模拟机

关机，重新开机就会自动启动指定的服务和指定的应用程序。

本章小结

本章详细介绍了 Intent 组件、Android 中的消息处理机制以及 Android 的基本组件 Activity、Service（服务）、BroadcastReceiver（广播接收者）的使用，（Content Provider 将在第 8 章介绍）。掌握这些知识后就可以实现不同组件之间的切换，数据的传递。

习题

1. Android 四大基本组件是什么？简述各自的大致功能。
2. 如何才能将信息从 SD 卡读出，给出实现的步骤和关键代码。
3. 如何才能将信息写入 SD 卡保存，给出实现的步骤和关键代码。
4. 查找资料，了解通话记录的 ContentProvider。
5. 参照实例 8-13 实现获取通信记录，写出关键代码。
6. 编程实现实例 8-14 显示的信息在一个 ListView 组件中，写出关键代码。
7. 如果要编程实现实例 8-15 中当歌曲播放时显示播放音量控制条和进度控制条，写出关键代码。

第 7 章 Android 多媒体技术

学习目标：
- 掌握基本图形绘制的方法和图形绘制常用类；
- 了解使用 Bitmap 和 Matrix 类对图片进行处理；
- 掌握 Android 中两种常见动画的使用；
- 了解动画操作组件的运用；
- 掌握使用 MediaPlayer 播放音频和视频文件；
- 掌握使用 Camera 捕获 SurfaceView 采集视频数据；
- 掌握 Android 中 MediaRecorder 常见用法。

多媒体信息包括：文本、声音、图形、图像、视频、动画等，多媒体技术就是指处理这些信息的程序和过程。Android 对这些多媒体信息的处理提供了支持。

7.1 Android 中图形的绘制

7.1.1 图形绘制基础

动态图形绘制的基本思路是，创建一个 View 类或者 SurfaceView 类，重写 onDraw()方法，使用 Canvas 对象在界面上绘制不同图形，使用 invalidate()方法刷新界面。动态绘制图形的常见类有 Canvas、Paint、Color、Path 等。

1）Canvas 类：Canvas 类是画布类，它的层次关系为：

```
java.lang.Object
    android.graphics.Canvas
```

要在 Android 的 Java 程序中使用 Canvas 类必须在程序中使用下面的语句。

```
import android.graphics.Canvas;;        //导入 graphics.Canvas 类
```

Canvas 类提供了各种图形的绘制方法，如矩形、圆、椭圆、点、线、文字等，具体方法见表 7-1。

表 7-1 Canvas 类常见方法

方法名称	方法描述
drawText(String text,float x,float y,Paint paint)	画文本
drawPoint(float x,float y,Paint paint)	画点
drawLine(float startX,float startY,float soptX,float stopY,Paint paint)	画线
drawCircle(float cx,float cy,flaot radius,Paint paint)	画圆
drawOval(rectF oval,paint paint)	画椭圆
drawRect(rectF rect,paint paint)	画矩形
drawRoundRect(rectF rect,float rx,float ry,paint paint)	圆角矩形
clipRect(float left,float top,float right,float bottom)	裁剪矩形
clipRegion(Region region)	裁剪区域

2）Paint 类：Paint 类是画笔类，它的层次关系为：

```
java.lang.Object
    android.graphics.Paint
```

要在 Android 的 Java 程序中使用 Paint 类必须在程序中使用下面的语句。

```
import android.graphics.Paint;                    //导入 graphics.Paint 类
```

Paint 类用来描述图形的颜色和风格，如线宽、颜色、字体等信息。Paint 类常用方法如表 7-2 所示。

表 7-2 Paint 类常见方法

方法名称	方法描述
Paint()	构造方法
setColor(int color)	设置颜色
setStrokeWidth(float width)	设置线宽
setTextAlign(Paint.Align align)	设置文字对齐
setTextSize(float textSize)	设置文字尺寸
setShader(Shader shader)	设置渐变
setAlpha(int a)	设置 Alpha 值
reset()	复位 Paint 默认设置

3）Color 类：Color 类中定义了一些颜色常量和创建颜色的方法，它的层次关系为：

```
java.lang.Object
    android.graphics.Color
```

要在 Android 的 Java 程序中使用 Paint 类必须在程序中使用下面的语句。

```
import android.graphics.Color;                    //导入 graphics.Color 类
```

Color 类颜色定义使用 RGB 定义，常见颜色的描述如表 7-3 所示。

表 7-3 Color 类常见颜色描述

颜色属性名称	描述
BLACK	黑色
BLUE	蓝色
CYAN	青色
DKGRAY	深灰色
GRAY	灰色
GREEN	绿色
LIGRAY	浅灰色
MAGENTA	紫色
RED	红色
TRANSSARENT	透明
WHITE	白色
YELLOW	黄色

4）Path 类：点与点之间的连线，它的层次关系为：

```
java.lang.Object
    android.graphics.Path
```

要在 Android 的 Java 程序中使用 Path 类必须在程序中使用下面的语句。

```
import android.graphics.Path;                //导入graphics.Path类
```

Path 类一般用来从某个点移动到另一个点连线，例如画梯形需要有点和连线，常使用的方法如表 7-4 所示。

表 7-4　Path 类常见方法

方法名称	方法描述
lineTo(float x,float y)	从最后点到指定点画线
moveTo(float x,float y)	移动到某一点
reset()	复位

5）Shader 类及其子类：

用来渲染图像使用的类，子类有 BitmapShader、ComposeShader、LinearGradient、RadialGradient、SweepGradient。

7.1.2　图形绘制实例

在一般的图形绘制中，创建一个 View 类，在类中覆写 onDraw()方法即可，然后在主程序中调用就可实现简单图形的绘制。

实例 7-1：绘制各种图形

新建一个项目，项目的命名为：exam7_1，包名称为：org.hnist.demo，绘制圆、正方形、长方形、椭圆、三角形、文字，并用不同的方式进行填充输出。

1）创建一个 MyView .java 继承 View 类。

```
package org.hnist.demo;
import android.app.Activity;
import android.view.View;
import android.content.Context;
import android.graphics.Color;
import android.graphics.Paint;
import android.graphics.Rect;
import android.graphics.RectF;
import android.graphics.Path;
import android.graphics.Shader;
import android.graphics.Canvas;
import android.util.AttributeSet;
import android.graphics.LinearGradient;
import android.graphics.SweepGradient;
public class MyView extends View {                //创建一个View类
public MyView(Context context){
   super(context);}
protected void onDraw(Canvas canvas) {            //覆写onDraw()方法
super.onDraw(canvas);
canvas.drawColor(Color.WHITE);                    //设置画布颜色
Paint paint = new Paint();
paint.setAntiAlias(true);                         //去除锯齿
```

```java
paint.setColor(Color.RED);                          //设置线条颜色
paint.setStyle(Paint.Style.STROKE);                 //设置样式
paint.setStrokeWidth(3);                            //设置画笔粗细
canvas.drawCircle(40, 40, 30, paint);               //画圆
canvas.drawRect(10, 90, 70, 150, paint);            //画矩形
canvas.drawRect(10, 170, 70, 200, paint);
RectF re = new RectF(10,220,70,250);                //声明矩形区域
canvas.drawOval(re, paint);                         //画椭圆
//实例化Path,画一个三角形
Path path = new Path();
path.moveTo(10,330);
path.lineTo(70, 330);
path.lineTo(40, 270);
path.close();
canvas.drawPath(path, paint);
paint.setStyle(Paint.Style.FILL);                   //设置填充
paint.setColor(Color.BLUE);                         //设置填充颜色
canvas.drawCircle(120, 40, 30, paint);              //画圆
canvas.drawRect(90, 90, 150, 150, paint);           //画矩形
canvas.drawRect(90, 170, 150, 200, paint);
RectF re2 = new RectF(90,220,150,250);              //声明矩形区域
canvas.drawOval(re2, paint);                        //画椭圆
//实例化Path,画一个三角形
Path path2 = new Path();
path2.moveTo(90,330);
path2.lineTo(150, 330);
path2.lineTo(120, 270);
path2.close();
canvas.drawPath(path2, paint);
//设置红、绿、蓝、黄四色填充
Shader lShader1 = new LinearGradient(0, 0, 100, 100, new int[]{
Color.RED,Color.GREEN,Color.BLUE,Color.YELLOW}, null, Shader.TileMode.REPEAT );
Shader lShader2 = new LinearGradient(0, 0, 100, 100,Color.RED,Color.GREEN,
                 Shader.TileMode.MIRROR);
//设置红、绿、蓝、黄四色填充
Shader sShader1 = new SweepGradient(0, 0,new int[] {
Color.RED,Color.GREEN,Color.BLUE,Color.YELLOW}, null);
Shader sShader2 = new SweepGradient(0, 0,Color.RED,Color.GREEN );
paint.setShader(lShader1);
canvas.drawCircle(200, 40, 30, paint);              //画圆
canvas.drawRect(170, 90, 230, 150, paint);          //画矩形
canvas.drawRect(170, 170, 230, 200, paint);
RectF re3 = new RectF(170,220,230,250);             //声明矩形区域
canvas.drawOval(re3, paint);                        //画椭圆
//实例化Path,画一个三角形
Path path4 = new Path();
path4.moveTo(170,330);
path4.lineTo(230, 330);
path4.lineTo(200, 270);
```

```
            path4.close();
            canvas.drawPath(path4, paint);
            paint.reset();
            paint.setColor(Color.BLACK);//
            paint.setTextSize(24);
            canvas.drawText("圆形", 240, 50, paint);           //绘制输出文字
            canvas.drawText("正方形", 240, 120, paint);
            canvas.drawText("长方形", 240, 190, paint);
            canvas.drawText("椭圆形", 240, 250, paint);
            canvas.drawText("三角形", 240, 320, paint);}}
```

2）创建一个主程序 MainActivity .java 调用 MyView。

```
        package org.hnist.demo;
        import android.os.Bundle;
        import android.app.Activity;
        public class MainActivity extends Activity {
            public void onCreate(Bundle savedInstanceState) {
                super.onCreate(savedInstanceState);
                setContentView(new MyView(this));     }}     //调用建立的类
```

保存文件，运行该项目，结果如图 7.1 所示。

7.2 Android 中图像的处理

Android 可以支持的图像格式有 png、jpg、gif 和 bmp 等。Android 手机中提供了用于操作图像资源的操作类，使用此类可以直接从资源文件之中进行图像资源的读取，并且对这些图像进行一些简单的修改。

7.2.1 图像的获取

访问图像一般有三种方式，一是使用在工程中保存的图像；二是使用 xml 定义 Drawable 属性；三是使用构造器来完成，但这种方法不经常被采纳。

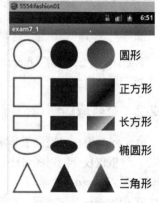

图 7.1 绘制各种图形

前面我们学习了 ImageView 组件，利用它可以方便地获取图像文件，将图像文件 test.jpg 复制到 res\ drawable 文件夹下，然后通过如下代码使用该图像。

1）在程序 MainActivity 中获取该图像。

```
        ImageView img = (ImageView)findViewById(R.id.img);
        img.setImageResource(R.drawable.test);
```

2）在布局管理文件 activity_main.xml 中获取该图像。

```
        <ImageView
            android:id="@+id/img"
            android:layout_width="fill-parent"
            android:layout_height="wrap_content"
            android:src="@drawable/test"/>
```

3）也可以利用 Bitmap 和 BitmapFactory 两个类来获取图像文件，这两个类的常用方法如表 7-5 和表 7-6 所示。

表 7-5　Bitmap 类的常用方法

方法	描述
public static Bitmap createBitmap (Bitmap src)	复制一个 Bitmap
public static Bitmap createBitmap(Bitmap source, int x, int y, int width, int height, Matrix m, boolean filter)	对一个 Bitmap 进行剪切
public final int getHeight()	取得图像的高
public final int getWidth()	取得图像的宽
public static Bitmap createScaledBitmap(Bitmap src, int dstWidth, int dstHeight, boolean filter)	创建一个指定大小的 Bitmap

表 7-6　BitmapFactory 类的常用方法

方法	描述
public static Bitmap decodeByteArray(byte[] data, int offset, int length)	根据指定的数据文件创建 Bitmap
public static Bitmap decodeFile(String pathName)	根据指定的路径创建 Bitmap
public static Bitmap decodeResource(Resources res, int id)	根据指定的资源创建 Bitmap
public static Bitmap decodeStream(InputStream is)	根据指定的 InputStream 创建 Bitmap

例如：

```
String path = "/sdcard/img.jpg";              //指定图像的路径
Bitmap bmp = BitmapFactory.decodeFile(path);  //根据指定路径创建图像
ImageView img = new ImageView(this);          //定义 ImageView 组件
img.setImageBitmap(bmp);                      //获得图像
this.setContentView(img); }}                  //显示图像
```

Bitmap 类还提供了 compress()接口支持压缩的图像，目前只支持 PNG、JPG 格式的压缩，读者可以自行查看 API 资料了解。

7.2.2　对获取的图像进行处理

Android 手机中提供了非常丰富的图像处理功能，例如：图像的缩放、平移、旋转、水印、倒影等等，这里仅仅介绍对图像进行平移、旋转、缩放、倾斜等变换的操作方法，其他的功能感兴趣的话可以参考 Android API 学习。

图像的缩放具体实现有如下几种方法。

1）将一个位图按照需求重画一遍，画后的位图就是我们需要的，与位图的显示几乎一样：drawBitmap(Bitmap bitmap, Rect src, Rect dst, Paint paint)。

2）在原有位图的基础上，缩放原位图，创建一个新的位图：CreateBitmap(Bitmap source, int x, int y, int width, int height, Matrix m, boolean filter)。

3）借助 Canvas 的 scale(float sx, float sy)，不过要注意此时整个画布进行了缩放。

4）借助 Matrix 类实现缩放，这个方法比较灵活。

在 Android 系统中 Matrix 类中有一个 3×3 的矩阵坐标，Matrix 类的方法如表 7-7 所示，通过这些方法可以实现图像的旋转、平移和缩放等操作。

表 7-7　Matrix 类常用方法

方法	描述
public void reset()	重设矩阵
public void setTranslate(float dx, float dy)	设置矩阵平移(dx, dy)
public void setScale(float sx, float sy)	设置矩阵缩放（sx,sy）

方法	描述
public void setScale(float sx, float sy, float px, float py)	设置矩阵以(px,py)坐标点为中心进行缩放（sx,sy)
public void setRotate(float degrees)	设置矩阵以原点为中心旋转 degrees
public void setRotate(float degrees, float px, float py)	设置矩阵以坐标点（px,py)为中心旋转
public void setSkew(float kx, float ky)	设置矩阵以原点为中心倾斜（kx,ky)
public void setSkew(float kx, float ky, float px, float py)	设置矩阵以坐标点（px,py)为中心倾斜

Matrix 的操作，总共分为 translate(平移)、rotate(旋转)、scale(缩放)和 skew(倾斜)四种，除了 translate，其他三种操作都可以指定中心点，每一种变换都提供了 set、post 和 pre 三种操作方式，以 Scale 为例，setScale ()方法、postScale ()方法和 preScale ()方法都可以实现缩放。post、pre 和 set 其实代表了 Matrix 中方法变换的次序，pre 是向前加入队列执行，post 从后面加入队列执行，set 方法一旦调用即会清空之前 matrix 中的所有变换，更详细的资料可以查看 Android API 文档。

例如：方框左边的执行顺序等价于右边方框，如图 7-2 所示。

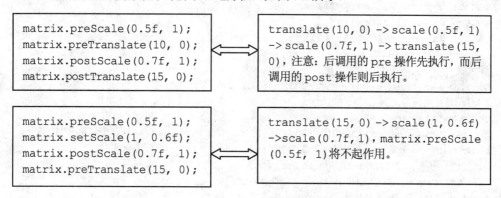

图 7-2 执行顺序

Canvas 里 scale, translate, rotate,方法都是 pre 方法，如果要进行更多的变换可以先从 canvas 获得 matrix，变换后再设置回 canvas。

Matrix 进行图像处理的基本步骤：

1）获取 Matrix 对象，可以新创建，也可以获取其他对象内封装的 Matrix。

例如：Matrix matrix = new Matrix();

2）调用 Matrix 的方法进行图像的平移、旋转、缩放、倾斜等操作。

例如：matrix.reset();//重置
　　　matrix.setTranslate()/setSkew()/setRotate()/setScale()//平移/倾斜/旋转/缩放

3）将程序对 Matrix 所做的变换应用到指定图像或组件。

例如：//获取图像（先将图像文件 love.jpg 复制到 res\ drawable 文件夹下）
　　　this.bitmap = BitmapFactory.decodeResource(super.getResources(),love);
　　　……
　　　protected void onDraw(Canvas canvas) {   //覆写 onDraw()方法
　　　　canvas.drawBitmap(this.bitmap, this.matrix, null);} //按照Matrix 要求画图

7.2.3 图像处理实例

实例 7-2：图像处理实例

新建一个项目，项目的命名为：exam7_2，包名称为：org.hnist.demo，显示一个指定大小的图片，然后显示对这个图片做了缩放、旋转、平移操作的图片。

1）创建一个 MyView.java 继承 View 类。

```java
package org.hnist.demo;
import android.content.Context;
import android.graphics.Bitmap;
import android.graphics.BitmapFactory;
import android.graphics.Canvas;
import android.graphics.Matrix;
import android.util.AttributeSet;
import android.graphics.drawable.BitmapDrawable;
import android.view.View;
import android.graphics.Paint;
public class MyView extends View {                    //继承 View
    private Bitmap bitmap = null ;                    //定义 bitmap
    private Matrix matrix = new Matrix();             //定义 matrix
    public MyView(Context context, AttributeSet attrs) {
        super(context, attrs);
        this.bitmap = BitmapFactory.decodeResource(super.getResources(),
                R.drawable.love);                     //取得项目中的 Bitmap
//创建一个指定大小的图片
        bitmap = Bitmap.createScaledBitmap(bitmap, 340, 230,true);
//对图片进行缩放、旋转、平移操作
        this.matrix.preScale(0.5f, 0.5f, 250, 150);   //缩小 1/2
        this.matrix.preRotate(60, 250, 150) ;         //在指定坐标翻转 60 度
        this.matrix.preTranslate(250, 400) ;}         //图像平移
    protected void onDraw(Canvas canvas) {            //覆写 onDraw()方法
        canvas.drawBitmap(this.bitmap, this.matrix, null); //画变换后的图
        canvas.drawBitmap(this.bitmap, 0,0, null); }} //画原图
```

2）创建一个布局管理文件 activity_main.xml，包含 MyView。

```xml
<?xml version="1.0" encoding="utf-8"?>
<LinearLayout
xmlns:android="http://schemas.android.com/apk/res/android"
android:orientation="vertical"
android:layout_width="fill_parent"
android:layout_height="fill_parent"
android:background="#000000">        背景颜色为黑色
< org.hnist.demo.MyView               引用建立的 MyView 类
    android:layout_width="wrap_content"
    android:layout_height="wrap_content" />
</LinearLayout>
```

3）创建一个主程序 MainActivity.java，调用布局管理文件 activity_main.xml。

```java
package org.hnist.demo;
import android.app.Activity;
import android.os.Bundle;
public class MainActivity extends Activity {
public void onCreate(Bundle savedInstanceState) {
    super.onCreate(savedInstanceState);
    super.setContentView(R.layout.activity_main); }}
```

保存所有文件，运行该项目，结果如图 7.2 所示。

7.3 Android 中的动画

在 Android 中，可以使用 Animation 组件对手机屏幕上的文字或者是图像等对象进行旋转、移动、淡入淡出等动画处理，Android 中的动画分为两类：

Tweened Animation（渐变动画）：该类的 Animation 可以完成控件的旋转、移动、伸缩、淡入淡出等特效；

Frame Animation（帧动画）：采用帧的方式进行动画效果的编排，将预先定义好的对象按照顺序播放。

7.3.1 Tween 动画

Tween 动画可以使视图组件移动、放大、缩小，以及产生透明的变化。例如在一个 ImageView 组件中，通过 Tween 动画可以使该视图实现缩放、旋转、渐变、平移等效果。Tween 动画相关的类在 android.view.animation 包中，如：

图 7.2 图像做旋转缩放平移处理

（1）Animation：抽象类，其他几个动画类继承自该类。
（2）ScaleAnimation 类：控制缩放变化动画类。
（3）AlphaAnimation：控制透明变化的动画类。
（4）RotateAnimation：控制旋转变化的动画类。
（5）TranslateAnimation：控制移动变化的动画类。
（6）AnimationSet：定义动画属性集合类。
（7）AnimationUtils：动画的工具类。
Animation 类常用的方法见表 7-8。

表 7-8 Animation 类常用的方法

方法	描述
Public Animation()	创建一个 Animation 对象
public void setDuration（long time）	设置动画持续时间（毫秒）
public void setFillAfter (boolean fillafter)	为 true 表示该动画在动画结束后被应用
public void setFillBefore (boolean fillbefore)	为 true 表示该动画在动画开始前被应用
public void setFillEnable (boolean fillenable)	setFillAfter、setFillBefore 方法是否有效
public void setRepeatCount（int repeatcount）	动画反复执行次数
public void setRepeatMode（int repeatmode）	动画反复执行模式
public void setStartOffset（long startoffset）	设置动画在多少毫秒后开始
public void setInterpolator(Interpolator i)	设置动画变化的速率
public boolean hasStarted()	判断动画是否已运行
public boolean hasEnded()	判断动画是否已结束
public long getDuration ()	返回动画持续时间
public void cancel()	取消动画
public void reset()	让动画回到初始状态
public void start()	动画开始执行
public void setAnimationListener(listener)	设置动画监听

其中还有一些常用常量，例如：RESTART：重新运行，INFINITE：反复运行，REVERSE：反转动画效果，等等。

Tween 动画一共有 4 种形式：Scale（缩放动画）、Alpha（渐变动画）、Translate（位置变化动画）、Rotate（旋转变化动画）。

缩放动画由 ScaleAnimation 类来实现，ScaleAnimation 类中常用的方法有：

Public   ScaleAnimation (context,Aattrs)，传入指定的属性创建 ScaleAnimation 对象。

Public   ScaleAnimation(float fromX,float toX,float fromY,float toY)，传入开始点和结束点坐标创建 ScaleAnimation 对象。其中：fromX 为动画开始时 X 坐标值，toX 为动画结束时 X 坐标值，fromY 为动画开始时 Y 坐标值，toY 为动画结束时 Y 坐标值。

渐变动画由 AlphaAnimation 类来实现，AlphaAnimation 类中常用的方法有：

Public   AlphaAnimation(context,Aattrs)，传入指定的属性创建 AlphaAnimation 对象。

Public   AlphaAnimation(float fromAlpha,float,toAlpha)，传入动画开始和结束值创建 AlphaAnimation 对象。其中：fromAlpha 为动画开始的透明度，toAlpha 为动画结束前的透明度（取值范围是 0.0～1.0）。

平移动画由 TranslateAnimation 类来实现，TranslateAnimation 类中常用的方法有：

Public   TranslateAnimation (context,Aattrs)，传入指定的属性创建 TranslateAnimation 对象。

Public TranslateAnimation(float fromXData,float toXData,float fromYData,float toYData)，传入移动的坐标创建 TranslateAnimation 对象。其中：fromXData 为动画开始平移前 X 坐标值，toXData 为动画开始平移后 X 坐标值，fromYData 为动画开始平移前 Y 坐标值，toYData 为动画开始平移后 Y 坐标值。

旋转动画由 RotateAnimation 类来实现，RotateAnimation 类中常用的方法有：

Public   RotateAnimation (context,Aattrs)，传入指定的属性创建 RotateAnimation 对象。

Public   RotateAnimation(float fromDegrees,float toDegrees)，传入指定旋转角度的范围创建 RotateAnimation 对象。其中：fromDegrees 为起始旋转角度，toDegrees 为结束旋转角度，正数表示顺时针旋转，负数表示逆时针旋转。

要创建一个 Tween 动画，一般有以下几个关键步骤：

1）定义一个 Animation 对象，例如：

```
Animation scaleAnimation = new ScaleAnimation(0.1f,1.0f,0.1f,1.0f);//初始化
```

2）设置动画运行时间，例如：

```
scaleAnimation.setDuration(500);         //设置动画时间
```

3）开始运行动画，例如：

```
startAnimation(scaleAnimation);    //开始动画
```

7.3.2　创建动画实例

实例 7-3：Tween 动画操作实例

新建一个项目，项目的命名为：exam7_3，包名称为：org.hnist.demo，通过一个图片缩放的例子来介绍动画的创建，其他的动画形式可以参考实现。

1）创建一个布局管理文件 activity_main.xml，代码如下：

```xml
<?xml version="1.0" encoding="utf-8"?>
<LinearLayout
    xmlns:android="http://schemas.android.com/apk/res/android"
```

```
        android:orientation="vertical"                    //所有组件垂直摆放
        android:layout_width="fill_parent"
        android:layout_height="fill_parent">
    <ImageView                                            //添加图片组件
        android:id="@+id/myimg"                           //设置组件 ID,程序中使用
        android:layout_width="wrap_content"
        android:layout_height="wrap_content"
        android:src="@drawable/love" />                   //图片为 drawable\love.jpg 文件
</LinearLayout>
```

2)建立一个 MainActivity.java 文件,代码如下:

```
package org.hnist.demo;
import android.app.Activity;
import android.os.Bundle;
import android.view.View;
import android.view.View.OnClickListener;
import android.view.animation.Animation;
import android.view.animation.ScaleAnimation;
import android.widget.ImageView;
public class MainActivity extends Activity {
private ImageView img = null;
@Override
public void onCreate(Bundle savedInstanceState) {
    super.onCreate(savedInstanceState);
    super.setContentView(R.layout. activity_main);
    this.img = (ImageView) super.findViewById(R.id.myimg);    //取得组件
    this.img.setOnClickListener(new OnClickListenerImpl());} //设置监听
private class OnClickListenerImpl implements OnClickListener {
public void onClick(View view) {
    Animation scaleAnimation = new ScaleAnimation(0.1f,1.0f,0.1f,1.0f);
                                        //初始化
    scaleAnimation.setDuration(3000) ;  //3 秒完成动画
    MainActivity.this.img.startAnimation(scaleAnimation) ; }}}//启动动画
```

保存所有文件,运行该项目,在手机上显示一个图片,单击这个图片,这个图片会展示从小逐渐变大,然后停在手机屏幕,如图 7.3 所示。

图 7.3　Tween 动画操作演示

只要修改 MainActivity.java 中的最后三条语句，可以实现其他动画效果，读者可以参照下面的语句自行完成，注意 import 中也要做适当的修改。

设置 AlphaAnimation

```
Animation alphaAnimation = new AlphaAnimation(0.1f, 1.0f);     //初始化
alphaAnimation.setDuration(3000);                              //设置动画时间为 3 秒
MainActivity.this.img.startAnimation(alphaAnimation) ; }}}     //启动动画
```

设置 RotateAnimation

```
Animation rotateAnimation = new RotateAnimation(0f, 360f);     //初始化
rotateAnimation.setDuration(1000);                             //设置动画时间
MainActivity.this.img.startAnimation(rotateAnimation) ; }}}    //启动动画
```

设置 TranslateAnimation

```
Animation tranAnimation=new TranslateAnimation(0.1f,10.0f,0.1f,10.0f);
                                                               //初始化
tranAnimation.setDuration(1000);                               //设置动画时间
MainActivity.this.img.startAnimation(tranAnimation) ; }}}      //启动动画
```

复杂动画创建

前面介绍了单个动画效果的实现，可不可以让上面设置的多个动画效果在一个对象上都生效呢？利用 AnimationSet 就可以轻松实现，AnimationSet 类常用方法如表 7-9 所示。

表 7-9 AnimationSet 类的常用方法

方法名称	描述
public AnimationSet(boolean shareInterpolator)	如果设置为 true,则表示使用 AnimationSet 所提供的 Interpolator（速率），如果为 false，则使用各个动画效果自己的 Interpolator
public void addAnimation(Animation a)	增加一个 Animation 组件
public List<Animation> getAnimations()	取得所有的 Animation 组件
public long getDuration()	取得动画的持续时间
public long getStartTime()	取得动画的开始时间
public void reset()	重置动画
public void setDuration(long durationMillis)	设置动画的持续时间
public void setStartTime(long startTimeMillis)	设置动画的开始时间

要创建多个 Tween 动画效果，一般有以下几个关键步骤。

1）建立多个动画对象，例如：

```
//初始化 Translate 动画
Animation tran=new TranslateAnimation(0.1f,10.0f,0.1f,10.0f);
Animation alpha=new AlphaAnimation(0.1f, 1.0f);     //初始化 Alpha 动画
Animation rotate=new RotateAnimation(0f, 360f);     //初始化 Rotate 动画
Animation scale = new ScaleAnimation(0.1f,1.0f,0.1f,1.0f); //初始 Scale 动画
```

2）建立动画集，例如：

```
AnimationSet set = new AnimationSet(true);          //定义一个动画集
set.addAnimation(tranAnimation);                    //增加动画 tran
set.addAnimation(alphaAnimation);                   //增加动画 alpha
……
```

3）设置动画时间，作用到每个动画，例如：

```
set.setDuration(1000);                        //设置动画时间
this.startAnimation(set);                     //启动动画
```

实例 7-4：复杂动画的创建实例

新建一个项目，项目的命名为：exam7_4，包名称为：org.hnist.demo，通过含有多个 Tween 动画效果的例子来介绍复杂动画的创建。

1）创建布局管理文件 activity_main.xml，代码与实例 7-3 中的 activity_main.xml 相同。

2）建立一个 MainActivity.java 文件，代码如下：

```
package org.hnist.demo;
import android.app.Activity;
import android.os.Bundle;
import android.view.View;
import android.view.View.OnClickListener;
import android.view.animation.AlphaAnimation;
import android.view.animation.Animation;
import android.view.animation.AnimationSet;
import android.view.animation.RotateAnimation;
import android.view.animation.ScaleAnimation;
import android.view.animation.TranslateAnimation;
import android.widget.ImageView;
public class MainActivity extends Activity {
private ImageView img = null;
@Override
public void onCreate(Bundle savedInstanceState) {
    super.onCreate(savedInstanceState);
    super.setContentView(R.layout. activity_main);
    this.img = (ImageView) super.findViewById(R.id.myimg);    //取得组件
    this.img.setOnClickListener(new OnClickListenerImpl());}  //设置监听
 private class OnClickListenerImpl implements OnClickListener {
    public void onClick(View view) {
    AnimationSet set = new AnimationSet(true);          //定义一个动画集
    Animation tran=new TranslateAnimation(0.1f,10.0f,0.1f,10.0f);
                                                        //初始化 Translate 动画
    Animation alpha=new AlphaAnimation(0.1f, 1.0f);     //初始化 Alpha 动画
    Animation rotate=new RotateAnimation(0f, 360f);     //初始化 Rotate 动画
    Animation scale = new  ScaleAnimation(0.1f,1.0f,0.1f,1.0f);
                                                        //初始化 Scale 动画
        scale.setRepeatCount(2) ;                       //动画重复2次
        set.addAnimation(tran) ;                        //增加动画 tran
        set.addAnimation(scale) ;                       //增加动画 scale
        set.addAnimation(alpha) ;                       //增加动画 alpha
        set.addAnimation(rotate) ;                      //增加动画 rotate
        set.setDuration(3000) ;                         //动画持续时间为3秒
        MainActivity.this.img.startAnimation(set) ;  }}} //启动动画
```

3）保存所有文件，运行该项目，在手机上显示一个图片，单击这个图片，这个图片会逐个展示动画效果，然后停在手机屏幕，如图 7.3 所示。

7.3.3 通过 XML 文件来创建动画

上面在 Activity 中通过 Java 代码，实现了 4 中不同的 Tween 动画，其实在 Android 中完全可以通过 XML 文件来实现动画，这样做更加简洁、清晰，也更利于重用，使得程序与配置相分离，建议采用这种形式。

<alpha>节点的常用属性有：fromAlpha 为动画开始的透明度，toAlpha 为动画结束前的透明度（取值范围是 0.0～1.0）。

<scale>节点的常用属性有：fromXScale 为动画开始时 X 坐标值，toXScale 为动画结束时 X 坐标值，fromYScale 为动画开始时 Y 坐标值，toYScale 为动画结束时 Y 坐标值。

表 7-10 可定义的动画效果元素

可配置的元素	描述
<set>	为根节点，定义全部的动画元素
<alpha>	定义渐变动画效果
<scale>	定义缩放动画效果
<translate>	定义平移动画效果
<rotate>	定义旋转动画效果

<translate>节点的常用属性有：fromXData 为动画开始平移前 X 坐标值，toXData 为动画开始平移后 X 坐标值，fromYData 为动画开始平移前 Y 坐标值，toYData 为动画开始平移后 Y 坐标值。

<rotate>节点的常用属性有：fromDegrees 为起始旋转角度，toDegrees 为结束旋转角度，正数表示顺时针旋转，负数表示逆时针旋转。

可以配置的公共属性见表 7-11。

表 7-11 可以配置的公共属性

可配置的属性	数据类型	描述
android:duration	long	定义动画的持续时间，以毫秒为单位
android:fillAfter	boolean	设置为 true 表示该动画转化在动画结束后被应用
android:fillBefore	boolean	当设置为 true，该动画转化在动画开始前被应用
android:interpolator	String	动画插入器，例如：decelerate_interpolator（减速动画）
android:repeatCount	int	动画重复执行的次数
android:repeatMode	String	动画重复的模式（restart、reverse）
android:startOffset	long	动画之间的间隔
android:zAdjustment	int	动画的 Z Order 配置：0（保持 Z Order 不变）、1（保持在最上层）、-1（保持在最下层）
android:interpolator	String	指定动画的执行速率

Interpolator 对象的配置情况见表 7-12。

表 7-12 Interpolator 对象的配置情况

可配置的属性	描述
@android:anim/accelerate_decelerate_interpolator	先加速再减速
@android:anim/accelerate_interpolator	加速
@android:anim/anticipate_interpolator	先回退一小步然后加速前进
@android:anim/anticipate_overshoot_interpolator	在上一基础上超出终点一点再到终点
@android:anim/bounce_interpolator	最后阶段弹球效果
@android:anim/cycle_interpolator	周期运动
@android:anim/decelerate_interpolator	减速
@android:anim/linear_interpolator	匀速
@android:anim/overshoot_interpolator	快速到达终点并超出一点最后到终点

在 Android 中所有定义好的动画配置 XML 文件都要求保存在 res/anim 文件夹之中，然后使用 android.view.animation.AnimationUtils 类中的 loadAnimation 方法读取这些动画配置文件。

实例 7-5：通过 XML 文件来创建动画实例

新建一个项目，项目的命名为：exam7_5，包名称为：org.hnist.demo，通过设置 XML 文件来实现实例 7-4 的效果。

1）在项目建立一个文件夹 anim 位于 res 下，在 res\anim\下建立一个 scale.xml 文件，做如下配置：

```xml
<?xml version="1.0" encoding="utf-8"?>
<set xmlns:android="http://schemas.android.com/apk/res/android">
< scale                              定义缩放动画
    android:fromXScale="0.1"         组件从 X 轴缩小 0.1 倍开始
    android:toXScale="1.0"           组件到 X 轴满屏显示结束
    android:fromYScale="0.1"         组件从 Y 轴缩小 0.1 倍开始
    android:toYScale="1.0"           组件到 Y 轴满屏显示结束
    android:duration="3000" />       动画持续的时间为 3 秒
</set>
```

2）定义布局管理器文件 activity_main.xml，其配置与实例 7-3activity_main.xml 相同。
3）定义 Activity 程序 MainActivity.java，读取 scale.xml 文件，代码如下：

```java
package org.hnist.demo;
import android.app.Activity;
import android.os.Bundle;
import android.view.View;
import android.view.View.OnClickListener;
import android.view.animation.Animation;
import android.view.animation.AnimationUtils;
import android.view.animation.ScaleAnimation;
import android.widget.ImageView;
public class MainActivity extends Activity {
private ImageView img = null;
public void onCreate(Bundle savedInstanceState) {
    super.onCreate(savedInstanceState);
    super.setContentView(R.layout. activity_main);
    this.img = (ImageView) super.findViewById(R.id.myimg);    //取得组件
    this.img.setOnClickListener(new OnClickListenerImpl());} //设置监听
private class OnClickListenerImpl implements OnClickListener {
    public void onClick(View view) {
        Animation anim = AnimationUtils.loadAnimation(MainActivity.this,
                R.anim.scale);    //读取动画配置文件 scale.xml
        MainActivity.this.img.startAnimation(anim) ; }}}        //启动动画
```

保存所有文件，运行结果与实例 7-3 一致。

可以设置多个不同的动画配置 XML 文件，然后通过语句：

Animation anim = AnimationUtils.loadAnimation(MainActivity.this,R.anim.all);来调用所有的动画配置文件，实现多个动画的效果。

7.3.4 Frame 动画

Frame 动画是采用帧的方式进行动画效果的编排，所有的动画会按照事先定义好的顺序执行，这样的顺序播放图像来产生一种动画效果类似于电影。例如要实现一个人走路的动画效果，可以通过 3 张图像重复不停播放来实现：第一张是两脚着地；第二张是右脚抬起，左脚着地；第三张是左脚抬起，右脚着地。

如果要想使用这种动画则需要利用 android.graphics.drawable.AnimationDrawable 类来实现。该类中有两个重要的方法：start()和 stop()，分别用来开始和停止动画。

动画一般通过 xml 配置文件来进行配置。在 res/anim 目录中定义 xml 文件，该文件的根元素是 <animation-list>，子元素是<item>，子元素可以有多个。

对于 Frame 动画也可以在 XML 文件之中进行动画的配置，同样需要将配置文件保存在 res/anim 文件夹之中，但是此配置文件的根节点为："<animation-list>"，里面包含多个"<item>"元素，用于定义每一帧动画，这两个元素可以配置的属性如表 7-13 所示。

表 7-13　animation-list 常用配置的属性

属性	描述
android:drawable	每一帧动画的资源
android:duration	动画的持续时间
android:oneshot	是否只显示一次，true 为只显示一次，false 为重复显示
android:visible	定义 drawable 是否初始可见

要创建一个 Frame 动画，一般有以下几个关键步骤。

1）定义一个动画配置文件，例如：frame.xml

```xml
<animation-list                                 //定义动画集合
xmlns:android="http://schemas.android.com/apk/res/android"
android:oneshot="true">                         //默认为显示一次
<item                                           //定义动画帧
    android:drawable="@drawable/tiger1"         //引入的图片文件
    android:duration="200" />                   //动画持续时间为 0.2 秒
  ......                                        //可以有多个<item>项，将多个图片文件引入到动画
</animation-list>
```

2）定义布局管理文件，例如定义两个按钮，通过单击按钮启动和停止动画。

3）定义 Activity 程序，操作帧动画。

```java
private class OnClickListenerstart implements OnClickListener {
    @Override
    public void onClick(View view) {
      //设置动画资源
    MainActivity.this.img.setBackgroundResource(R.anim. framedemo);
      //取得 Drawable
    MainActivity.this.draw= (AnimationDrawable)MainActivity.this.
                    img.getBackground();
    MainActivity.this.draw.setOneShot(false);          //动画执行次数
    MainActivity.this.draw.start();}}                  //开始动画
private class OnClickListenerstop implements OnClickListener {
    @Override
    public void onClick(View view) {
        MainActivity.this.draw.stop();}} }             //停止动画
```

实例 7-6：Frame 动画实例

新建一个项目，项目的命名为：exam7_6，包名称为：org.hnist.demo，建立帧动画，实现如下功能：单击"开始动画"按钮，一只老虎奔跑，单击"停止动画"按钮，老虎立刻停止不动。

1）定义一个动画配置文件，例如：frame.xml，放置在 res\anim 文件夹下。

```xml
<animation-list                                  定义动画集合
    xmlns:android="http://schemas.android.com/apk/res/android"
    android:oneshot="true">                      默认为显示一次
<item                                            定义动画帧
    android:drawable="@drawable/tiger1"          引入的图片文件
    android:duration="200" />                    动画持续时间为0.2秒
<item
    android:drawable="@drawable/tiger2"
    android:duration="200" />
<item
    android:drawable="@drawable/tiger3"
    android:duration="200" />
<item
    android:drawable="@drawable/tiger4"
    android:duration="200" />
</animation-list>
</animation-list>
```

2) 定义布局管理文件,例如定义两个按钮,通过单击按钮启动和停止动画。

```xml
<?xml version="1.0" encoding="utf-8"?>
<LinearLayout
    android:id="@+id/layout"
    xmlns:android="http://schemas.android.com/apk/res/android"
    android:orientation="vertical"
    android:layout_width="fill_parent"
    android:layout_height="fill_parent">
<ImageView
    android:id="@+id/img"
    android:layout_width="wrap_content"
    android:layout_height="wrap_content"/>
<Button
    android:id="@+id/start"
    android:layout_width="wrap_content"
    android:layout_height="wrap_content"
    android:text="开始动画"/>
<Button
    android:id="@+id/stop"
    android:layout_width="wrap_content"
    android:layout_height="wrap_content"
    android:text="停止动画"/>
</LinearLayout>
```

3) 定义 Activity 程序 MainActivity.java 操作帧动画,代码如下:

```java
package org.hnist.demo;
import android.app.Activity;
import android.graphics.drawable.AnimationDrawable;
import android.os.Bundle;
import android.view.View;
import android.view.View.OnClickListener;
import android.widget.Button;
import android.widget.ImageView;
```

```java
public class MainActivity extends Activity {
    private ImageView img = null;
    private Button start = null;
    private Button stop = null;
    private AnimationDrawable draw = null;                    //定义动画操作
    @Override
    public void onCreate(Bundle savedInstanceState) {
        super.onCreate(savedInstanceState);
        super.setContentView(R.layout.main);
        this.img = (ImageView) super.findViewById(R.id.img);
        this.start = (Button) super.findViewById(R.id.start);
        this.stop = (Button) super.findViewById(R.id.stop);
        this.start.setOnClickListener(new OnClickListenerstart()) ;
                                                              //设置开始按钮监听
        this.stop.setOnClickListener(new OnClickListenerstop()) ;}
                                                              //设置停止按钮监听
    private class OnClickListenerstart implements OnClickListener {
        public void onClick(View view) {
                                                              //设置动画资源
            MainActivity.this.img.setBackgroundResource(R.anim.frame);
                                                              //取得背景的Drawable
            MainActivity.this.draw=(AnimationDrawable)MainActivity.this.
                          img.getBackground();
            MainActivity.this.draw.setOneShot(false);         //动画执行次数
            MainActivity.this.draw.start(); }}                //开始动画
    private class OnClickListenerstop implements OnClickListener {
        public void onClick(View view) {
            MainActivity.this.draw.stop();   }} }             //停止动画
```

保存所有文件，运行该项目，单击"开始动画"按钮，帧动画开始，单击"停止动画"按钮，帧动画停止，如图 7.4 所示。

7.3.5 动画监听器：AnimationListener

在对动画的操作过程之中，还可以对动画的一些操作状态进行监听，例如：动画是否启动、动画是否正重复执行、动画是否结束，在 Android 中专门提供了一个用于这些动作监听的接口：android.view.animation.Animation.AnimationListener，在此接口中定义了三个监听动画的操作方法。

动画开始时触发：public abstract void onAnimationStart (Animation animation)。

动画重复时触发：public abstract void onAnimationRepeat (Animation animation)。

图 7.4 Tween 动画操作演示

动画结束时触发：public abstract void onAnimationEnd(Animation animation)。

实例 7-7：动画监听器实例

新建一个项目，项目的命名为：exam7_7，包名称为：org.hnist.demo，先建立一个缩放动画，动画会重复执行 2 次，设置动画监听器，当缩放动画启动后再增加一个渐变动画。

1)定义布局管理文件 activity_main.xml,代码如下:

```xml
<?xml version="1.0" encoding="utf-8"?>
<LinearLayout
android:id="@+id/layout"
xmlns:android="http://schemas.android.com/apk/res/android"
android:orientation="vertical"
android:layout_width="fill_parent"
android:layout_height="fill_parent">
<ImageView
    android:id="@+id/img"
    android:layout_width="fill_parent"
    android:layout_height="wrap_content"
    android:src="@drawable/love" />
</LinearLayout>
```

2)建立 Activity 文件 MaiinActivity.java,代码如下:

```java
package org.hnist.demo;
import android.app.Activity;
import android.os.Bundle;
import android.view.animation.Animation;
import android.view.animation.Animation.AnimationListener;
import android.view.animation.AnimationSet;
import android.view.animation.RotateAnimation;
import android.view.animation.ScaleAnimation;
import android.widget.ImageView;
public class MainActivity extends Activity {
private ImageView img = null;
public void onCreate(Bundle savedInstanceState) {
    super.onCreate(savedInstanceState);
    super.setContentView(R.layout.activity_main);
    this.img = (ImageView) super.findViewById(R.id.img);
    AnimationSet set = new AnimationSet(true);         //定义一个动画集
    Animation scale = new  ScaleAnimation(0.1f,1.0f,0.1f,1.0f);
                                                        //初始化Scale动画
    scale.setRepeatCount(2) ;                           //动画重复2次
    set.addAnimation(scale) ;                           //增加动画scale
    set.setDuration(3000) ;                             //动画持续时间为3秒
    MainActivity.this.img.startAnimation(set) ;         //启动动画
    set.setAnimationListener(new AnimationListenerImpl()) ;  //设置动画监听
    this.img.startAnimation(set) ;  }                   //启动动画
private class AnimationListenerImpl implements AnimationListener {
    public void onAnimationEnd(Animation animation) {  }     //动画结束时触发
    public void onAnimationRepeat(Animation animation) {}    //动画重复执行时触发
    public void onAnimationStart(Animation animation) {      //动画开始时触发
        if(animation instanceof AnimationSet) {              //判断类型
            AnimationSet set = (AnimationSet) animation ;
            Animation rotate=new RotateAnimation(0f, 360f);  //初始化Rotate动画
            rotate.setDuration(3000) ;                       //3秒完成动画
            set.addAnimation(rotate) ;  }  }  }}             //增加动画
```

保存所有文件，运行该项目，发现在动画开始时，会增加一个旋转的效果，而重复执行动画时，没有旋转效果，是因为设置了动画开始时触发旋转动画的效果。

7.3.6 动画操作组件：LayoutAnimationController

LayoutAnimationController 表示的是在 Layout 组件上使用动画的操作效果，例如：可以在使用 ListView 增加一些渐变、缩放、旋转、平移动画效果，LayoutAnimationController 可以通过配置文件轻松实现，当然也可以利用程序代码完成，其常用配置的属性及常量和方法见表 7-14 和表 7-15。

表 7-14　LayoutAnimationController 常用配置的属性

属性	对应的方法	描述
android:animation	setAnimation(Animation)	要引入的动画配置文件
android:animationOrder	setOrder(int)	动画执行顺序，noraml 顺序执行，reverse 逆序执行，random 随机执行
android:delay		多个动画时间间隔
android:interpolator	setInterpolator(Context,int)	配置动画的执行速率

表 7-15　LayoutAnimationController 常用常量及方法

常量及方法	描述
public static final int ORDER_NORMAL	动画采用顺序效果完成
public static final int ORDER_RANDOM	动画采用随机顺序效果完成
public static final int ORDER_REVERSE	动画采用逆序效果完成
public void setDelay(float delay)	设置动画间隔
public void setAnimation(Animation animation)	设置要使用的动画效果
public void setAnimation(Context context, int resourceID)	设置要使用的动画效果的配置文件
public void setOrder(int order)	设置动画的执行顺序
public void start()	开始动画

可以利用动画操作组件实现指定组件中的每条信息都带有动画效果，详细代码见实例 exam7_8。

7.4　Android 中的媒体播放

智能手机中播放音乐或者是视频已经司空见惯了，在 Android 操作系统之中使用 android.media.MediaPlayer 类就可以用来播放音频、视频和流媒体，可以利用该类提供的方法建立自己的音频和视频播放器，其常用方法如表 7-16 所示。

表 7-16　MediaPlayer 类常用的方法

方法名称	描述
public static MediaPlayer create(Context context, Uri uri)	通过 Uri 创建一个 MediaPlayer 对象
public static MediaPlayer create(Context context, int resid)	通过资源 ID 创建一个 MediaPlayer 对象
public static MediaPlayer create(Context context, Uri uri, SurfaceHolder holder)	通过 Uri 创建一个 MediaPlayer 对象，并显示该视频
public int　getCurrentPosition()	返回 Int，得到当前播放位置
public int　getDuration()	返回 Int，得到文件的时间
public int　getVideoHeight()	返回 Int，得到视频的高度
public int　getVideoWidth()	返回 Int，得到视频的宽度
public boolean　isLooping()	返回 boolean，是否循环播放
public boolean　isPlaying()	返回 boolean，是否正在播放

续表

方法名称	描述
public void pause()	暂停播放
public void prepare()	准备同步播放，在播放前调用
public void prepareAsync()	准备异步播放，在播放前调用
public void release()	释放 MediaPlayer 对象所占资源
public void reset()	重置 MediaPlayer 对象
public void seekTo(int msec)	指定播放的位置（以毫秒为单位的时间）
public void setAudioStreamType(int streamtype)	指定流媒体的类型
public void setDataSource(String path)	设置多媒体数据来源【根据 路径】
public void setDataSource(FileDescriptor fd, long offset, long length)	设置多媒体数据来源【根据 文件系统】
public void setDataSource(FileDescriptor fd)	设置多媒体数据来源【根据 文件系统】
public void setDataSource(Context context, Uri uri)	设置多媒体数据来源【根据 Uri】
public void setDisplay(SurfaceHolder sh)	设置视频显示多媒体信息
public void setLooping(boolean looping)	设置是否循环播放
public void setOnBufferingUpdateListener (MediaPlayer.OnBufferingUpdateListener listener)	监听事件，网络流媒体的缓冲更新时触发
public void setOnCompletionListener (MediaPlayer. OnCompletionListener listener)	监听事件，网络流媒体播放结束触发
public void setOnErrorListener (MediaPlayer.OnErrorListener listener)	监听事件，设置出现错误时触发
public void setOnVideoSizeChangedListener (MediaPlayer.OnVideoSizeChangedListener listener)	监听事件，视频尺寸改变后触发
public void setScreenOnWhilePlaying(boolean screenOn)	设置是否使用 SurfaceHolder 显示
public void setVolume(float leftVolume, float rightVolume)	设置音量
public void start()	开始播放
public void stop()	停止播放

在介绍利用 MediaPlayer 类播放音频和视频之前，要介绍 MediaPlayer 操作的生命周期，先来看看 MediaPlayer 的几个状态。

1）Idle 状态：当使用关键字 new 实例化一个 MediaPlayer 对象或者是调用了类中的 reset()方法会进入到此状态，通过 create()方法创建的 MediaPlayer 对象不是处于 Idle 状态，如果成功调用了重载的 create()方法，那么这个对象已经是 Prepare 状态了。

2）End 状态：当调用 release()方法之后将进入到此状态，此时会释放掉所有占用的硬件和软件资源，并且不会再进入到其他的任何一种状态了，建议一旦一个 MediaPlayer 对象不再被使用，应调用 release()方法来释放资源。

3）Initialized 状态：当 MediaPlayer 对象设置好了要播放的媒体文件（setDataSource()）之后进入到此状态。

4）Prepared 状态：进入到预播放状态（prepare()、prepareAsync()），进入到此状态则表示目前的媒体文件没有任何的问题，可以使用 OnPreparedListener 监听此状态。

5）Started 状态：正在进行媒体播放（start()），此时可以使用 seekTo()方法指定媒体播放的位置。

6）Paused 状态：在 Started 状态下使用 Paused 状态可以暂停 MediaPlayer 的播放，暂停之后可以通过 start()方法将其变回到 Started 状态，继续播放。

7）Stop 状态：在 Started 和 Paused 状态下都可以通过 stop()方法停止 MediaPlayer 的播放，在 Stop 状态下要想重新进行播放，则可以使用 prepare()和 prepareAsync()方法进入到就绪状态。

8）PlaybackCompleted 状态：当媒体播放完毕之后会进入到此状态，用户可以使用 OnCompletionListener 监听此状态，此时可以使用 start()方法重新播放，也可以使用 stop()方法停止播放，或者使用 seekTo()方法来重新定位播放位置。

9）Error 状态：当用户播放操作之中出现了某些错误（文件格式不正确、播放文件过大等）则进入到此状态，用户可以使用 OnErrorListener 来监听此状态，如果 MediaPlayer 进入到了此状态后可以使用 reset()方法重新变回 Idle 状态。

MediaPlayer 操作的生命周期如图 7.5 所示。

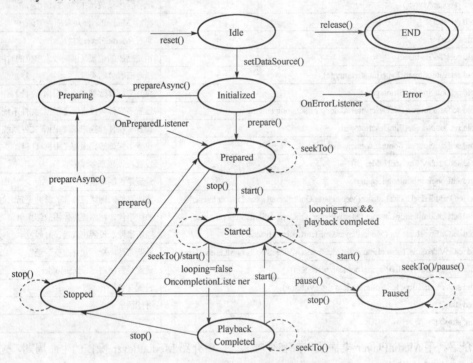

图 7.5 MediaPlayer 操作的生命周期图

7.4.1 Android 中音频播放

MediaPlayer 类支持以下几种不同的音频来源有：res 中的音频、文件系统中的音频、网络资源中的音频。支持的音频格式有：WAV、AAC、MP3、WMA、AMR、OGG、MIDI 等。要在 Android 中实现音频播放的功能一般有以下几个步骤：

1）构建 MediaPlayer 对象，例如：

```
MediaPlayer mp = new MediaPlayer();
```

2）设置数据源，例如：

```
mp.setDataSource("/mnt/sdcard/test.mp3" );                      //从 SD 卡取得音频文件
MediaPlayer mp = MediaPlayer.create(this, R.raw.test);          //从资源中取得音频文件
Uri uri = Uri.parse("http://www.xxxx.com/xxx.mp3");             //定义 uri
player = MediaPlayer.create(this,uri);                          //从网络取得音频文件
```

3）准备播放，例如：

```
mp.prepared();              //准备播放
player.prepared();          //网络中的音频准备播放
```

4）开始播放，例如：

```
mp.start();                    //开始播放
player.start();                //网络中的音频开始播放
```

如果要使用 MediaPlayer 来播放基于互联网内容的流,那么应用程序就必须申请互联网访问的权限。

```
<uses-permission android:name="android.permission.INTERNET" />
```

5)MediaPlayer 音量管理:

```
//获得音量控制器
AudioManager AM = (AudioManager) getSystemService(Context.AUDIO_SERVICE);
//获得最大音量
int maxVolume = AM.getStreamMaxVolume(AudioManager.STREAM_MUSIC);
//获得当前音量
int currentVolume = AM.getStreamVolume(AudioManager.STREAM_MUSIC);
```

6)控制音量大小,可以通过触摸会触发 dispatchTouchEvent(Activity 中的方法)来控制,例如:

```
public boolean dispatchKeyEvent(KeyEvent event) {
    int action = event.getAction();              //获得是按下还是弹起操作
    int keyCode = event.getKeyCode();            //获得按下键盘对应的按键值
    switch (keyCode) {
    case KeyEvent.KEYCODE_VOLUME_UP:
        if (action == KeyEvent.ACTION_UP) {
            if (currentVolume < maxVolume) {
            currentVolume = currentVolume + 1;
    AM.setStreamVolume(AudioManager.STREAM_MUSIC,currentVolume, 0); }
            else {
    AM.setStreamVolume(AudioManager.STREAM_MUSIC,currentVolume, 0);}}
            return false;
    case KeyEvent.KEYCODE_VOLUME_DOWN:
        if (action == KeyEvent.ACTION_UP) {
            if (currentVolume > 0) {
                currentVolume = currentVolume - 1;
    AM.setStreamVolume(AudioManager.STREAM_MUSIC,currentVolume, 0); }
            else {
    AM.setStreamVolume(AudioManager.STREAM_MUSIC,currentVolume, 0);}}
            return false;
    default:
            return super.dispatchKeyEvent(event);}}
```

7)进度控制,可以设置一个 SeekBar 组件,通过拖曳来实现在指定位置进行音频播放,例如:

```
//设置拖动条长度为媒体长度
MainActivity.this.seekbar.setMax(MainActivity.this.myMediaPlayer.getDuration());
UpdateSeekBar update = new UpdateSeekBar() ;              //更新拖动条
update.execute(1000) ;                                    //休眠1秒
MainActivity.this.seekbar.setOnSeekBarChangeListener(new
        OnSeekBarChangeListenerImpl());                   //拖动条改变音乐播放位置
    if (MainActivity.this.myMediaPlayer != null) {
        MainActivity.this.myMediaPlayer.stop();  }        //停止播放
        try { MainActivity.this.myMediaPlayer.prepare(); //进入预备状态
            MainActivity.this.myMediaPlayer.start();}}}   //播放文件
    //更新拖动条
private class UpdateSeekBar extends AsyncTask<Integer, Integer, String> {
```

```
            protected void onPostExecute(String result) {      }
                            //任务执行完后执行
            protected void onProgressUpdate(Integer... progress) {
                            //更新之后的数值
        MainActivity.this.seekbar.setProgress(progress[0]) ; }   //更新拖动条
            protected String doInBackground(Integer... params) { //处理后台任务
                while (MainActivity.this.playFlag) {             //进度条累加
                    try {
                        Thread.sleep(params[0]); }               //延缓执行
                    catch (InterruptedException e) {
                        e.printStackTrace();        }
                    this.publishProgress(MainActivity.this.myMediaPlayer
                            .getCurrentPosition()); }            //修改拖动条
                return null;}        }                           //返回执行结果
        //设置 SeekBar 改变是监听,改变播放位置,实现播放进度的改变
        private class OnSeekBarChangeListenerImpl implements OnSeekBarChangeListener
        {
            public void onProgressChanged(SeekBar seekBar,int progress,boolean
                                    fromUser) {   ……}
            public void onStartTrackingTouch(SeekBar seekBar) { ……}  //进度条开始拖曳
            public void onStopTrackingTouch(SeekBar seekBar) {       //进度条停止拖曳
            //定义播放位置
            MainActivity.this.myMediaPlayer.seekTo(seekBar.getProgress());}}
```

实例 7-9:音频播放实例

新建一个项目,项目的命名为:exam7_9,包名称为:org.hnist.demo,编写一个 MP3 播放器,可以播放程序资源中提供的 mp3 格式的音乐,可以使用开始、暂停、停止按钮来控制音乐的播放。

打开 Eclipse 后,将 abird.mp3 文件复制到 res\raw 文件夹下,将三个图片文件:play.jpg、pause.jpg、stop.jpg 复制到 res\drawable 文件夹下。

1)定义布局管理文件 activity_main.xml,放置一个文本显示框,三个图片按钮:开始、暂停、停止,一个进度控制条。

```
<?xml version="1.0" encoding="utf-8"?>
<LinearLayout
xmlns:android="http://schemas.android.com/apk/res/android"
android:orientation="vertical"
android:layout_width="fill_parent"
android:layout_height="fill_parent">
<TextView
    android:id="@+id/info"
    android:layout_width="fill_parent"
    android:layout_height="wrap_content"
    android:text="等待播放……" />
<LinearLayout
    android:orientation="horizontal"
    android:layout_width="wrap_content"
    android:layout_height="wrap_content">
    <ImageButton
        android:id="@+id/play"
        android:layout_width="wrap_content"
```

```xml
            android:layout_height="wrap_content"
            android:src="@drawable/play" />
    <ImageButton
            android:id="@+id/pause"
            android:layout_width="wrap_content"
            android:layout_height="wrap_content"
            android:src="@drawable/pause" />
    <ImageButton
            android:id="@+id/stop"
            android:layout_width="wrap_content"
            android:layout_height="wrap_content"
            android:src="@drawable/stop" />
</LinearLayout>
<SeekBar
    android:id="@+id/seekbar"
    android:layout_width="fill_parent"
    android:layout_height="wrap_content" />
</LinearLayout>
```

2) 定义 Activity 文件 MainActivity.java，控制音频的播放。

```java
package org.hnist.demo;
import android.app.Activity;
import android.media.MediaPlayer;
import android.media.MediaPlayer.OnCompletionListener;
import android.os.AsyncTask;
import android.os.Bundle;
import android.view.View;
import android.view.View.OnClickListener;
import android.widget.ImageButton;
import android.widget.SeekBar;
import android.widget.SeekBar.OnSeekBarChangeListener;
import android.widget.TextView;
public class MainActivity extends Activity {
private ImageButton play = null;
private ImageButton pause = null;
private ImageButton stop = null;
private TextView info = null;
private MediaPlayer myMediaPlayer = null;           //定义 MediaPlayer
private boolean pauseFlag = false;                  //暂停播放标记
private boolean playFlag = true ;                   //是否播放的标记
private SeekBar seekbar = null;                     //定义拖动条
@Override
public void onCreate(Bundle savedInstanceState) {
    super.onCreate(savedInstanceState);
    super.setContentView(R.layout.activity_main);
    this.info = (TextView) super.findViewById(R.id.info);
    this.play = (ImageButton) super.findViewById(R.id.play);
    this.pause = (ImageButton) super.findViewById(R.id.pause);
    this.stop = (ImageButton) super.findViewById(R.id.stop);
    this.seekbar = (SeekBar) super.findViewById(R.id.seekbar);
                                            //取得组件
```

```java
        this.play.setOnClickListener(new PlayOnClickListenerImpl()) ;
                                                                    //设置播放事件
        this.pause.setOnClickListener(new PauseOnClickListenerImpl());
                                                                    //设置暂停单击事件
        this.stop.setOnClickListener(new StopOnClickListenerImpl());}
                                                                    //设置结束事件
    private class PlayOnClickListenerImpl implements OnClickListener {
        @Override
        public void onClick(View view) {
            MainActivity.this.myMediaPlayer = MediaPlayer.create(
              MainActivity.this, R.raw.abird);//找到指定的资源
MainActivity.this.myMediaPlayer
                .setOnCompletionListener(new OnCompletionListener() {
                    @Override
                    public void onCompletion(MediaPlayer media) {
                        MainActivity.this.playFlag = false ; //播放完毕
                        media.release();     }});            //释放所有状态
//设置拖动条长度为媒体长度
MainActivity.this.seekbar.setMax(MainActivity.this.myMediaPlayer.getDuration());
            UpdateSeekBar update = new UpdateSeekBar() ; //更新拖动条
            update.execute(1000) ;                       //休眠1秒
            MainActivity.this.seekbar.setOnSeekBarChangeListener(new
                OnSeekBarChangeListenerImpl());          //拖动条改变音乐播放位置
        if (MainActivity.this.myMediaPlayer != null) {
            MainActivity.this.myMediaPlayer.stop(); }    //停止播放
        try { MainActivity.this.myMediaPlayer.prepare(); //进入预备状态
            MainActivity.this.myMediaPlayer.start();     //播放文件
            MainActivity.this.info.setText("正在播放音频文件...");  }
                                                         //设置显示的文字
        catch (Exception e) {                            //异常处理
            MainActivity.this.info.setText("文件播放出现异常," + e);//设置文字
        }    }    }
    private class UpdateSeekBar extends AsyncTask<Integer, Integer, String> {
        @Override
        protected void onPostExecute(String result) {     }//任务执行完后执行
        protected void onProgressUpdate(Integer... progress) { //更新之后的数值
MainActivity.this.seekbar.setProgress(progress[0]) ; } //更新拖动条
        protected String doInBackground(Integer... params) {  //处理后台任务
            while (MainActivity.this.playFlag) {          //进度条累加
                try {
                    Thread.sleep(params[0]); }            //延缓执行
                catch (InterruptedException e) {
                    e.printStackTrace();    }
                this.publishProgress(MainActivity.this.myMediaPlayer
                    .getCurrentPosition()); }             //修改拖动条
            return null;}    }                            //返回执行结果
    private class OnSeekBarChangeListenerImpl implements OnSeekBarChangeListener {
        public void onProgressChanged(SeekBar seekBar,
            int progress, boolean fromUser) {        }
        public void onStartTrackingTouch(SeekBar seekBar) {       }
        public void onStopTrackingTouch(SeekBar seekBar) {    //进度条停止拖曳
```

```
                    MainActivity.this.myMediaPlayer.seekTo(seekBar
                            .getProgress());} }                              //定义播放位置
        private class PauseOnClickListenerImpl implements OnClickListener {
            public void onClick(View view) {
                if (MainActivity.this.myMediaPlayer != null) {
                    if (MainActivity.this.pauseFlag) {     //true 表示由暂停变为播放
                        MainActivity.this.myMediaPlayer.start();       //播放文件
                        MainActivity.this.pauseFlag = false;           //修改标记位
                    } else {                              //false 表示由播放变为暂停
                        MainActivity.this.myMediaPlayer.pause();       //暂停播放
                        MainActivity.this.pauseFlag = true; }}}}       //修改标记位
        private class StopOnClickListenerImpl implements OnClickListener {
            public void onClick(View view) {
                if (MainActivity.this.myMediaPlayer != null) {
                    MainActivity.this.myMediaPlayer.stop();            //停止播放
                    MainActivity.this.info.setText("停止播放音频文件..."); }}}}
```

保存所有文件,运行该项目,如图 7.6 所示,各个按钮都能实现相应的功能,拖动滚动条能达到预期效果。

使用 MediaPlayer 类可以播放音频,但是该类占用资源较多,这对于游戏应用可能不是很适合,这个时候一般采用 SoundPool 类(android.media.SoundPool),它主要播放一些较短的声音片段,可以从程序的资源或文件系统加载。

SoundPool 最大只能支持 1MB 大小的音频文件,但是效率相对来说比 MediaPlayer 要高。

图 7.6 简单的音频播放器

SoundPool 播放音频的基本步骤:

1) 创建一个 SoundPool;

```
new SoundPool(int maxStreams, int streamType, int srcQuality);
```

其中:maxStreams:同时播放的流的最大数量;streamType:流的类型,一般为 STREAM_MUSIC(具体在 AudioManager 类中列出);srcQuality:采样率转化质量,当前无效果,使用 0 作为默认值。例如:

```
SoundPool soundPool = new SoundPool(3, AudioManager.STREAM_MUSIC, 0);
```

创建了一个最多支持 3 个流同时播放的,类型标记为音乐的 SoundPool。

2) 从资源或者文件载入音频流;

```
int  load(Context context, int resId, int priority)        //从 res 资源载入
int  load(FileDescriptor fd, long offset, long length, int priority)
                                              //从 FileDescriptor 对象载入
int  load(AssetFileDescriptor afd, int priority)          //从 Asset 对象载入
int  load(String path, int priority)                //从完整文件路径名载入
```

priority 参数为优先级。

3) 播放声音。

```
play(int soundID, float leftVolume, float rightVolume, int priority, int loop,
    float rate),
```

其中 leftVolume 和 rightVolume 表示左右音量,priority 表示优先级,loop 表示循环次数,rate 表示速率,如://速率最低 0.5,最高为 2,1 代表正常速度。

例如：

```
sp.play(soundId, 1, 1, 0, 0, 1);
```

停止播放可以使用 pause(int streamID) 方法，这里的 streamID 和 soundID 均在构造 SoundPool 类的第一个参数中指明了总数量，而 id 从 0 开始。

7.4.2 Android 中视频播放

MediaPlayer 除了可以对音频播放之外，也可以对视频进行播放，它支持以下几种不同的视频来源：res 中的视频、文件系统中的视频、网络资源中的视频，支持的视频格式有：MP4、H.263 (3GP)、H.264 (AVC)等。

要播放视频只依靠 MediaPlayer 是不够的，还需要依靠 android.view.SurfaceView 组件完成，SurfaceView 是 View 的子类，里面含有一个专门用于绘制的 Surface 对象，使用它对于像视频、3D 图形等需要快速更新的地方有很大的帮助。

SurfaceView 类使用 SurfaceView()方法创建，用得最多的一个操作就是 getHolder()方法，这个方法可以获取 SurfaceHolder 接口，此外需要重写的方法有：

1）在 surface 的大小发生改变时激发。

```
public void surfaceChanged(SurfaceHolder holder,int format,int width,int height){ }
```

2）在创建时激发。

```
Public void surfaceCreated(SurfaceHolder holder){ }
```

3）在销毁时激发。

```
Public void surfaceDestroyed(SurfaceHolder holder) { }
```

这里一个重要的概念 SurfaceHolder，它可以说是 surface 的控制器，用来操纵 surface，处理它的 Canvas 上画的效果和动画，控制表面、大小、像素等，它的常用方法见表 7-17。

表 7-17 SurfaceHolder 类常用的方法

方法名称	描述
public abstract void addCallback (SurfaceHolder. Callback callback)	给 SurfaceView 当前的持有者一个回调对象
public abstract Canvas lockCanvas()	锁定画布，返回的画布对象 Canvas，在其上面画图等操作了
public abstract Canvas lockCanvas(Rect dirty)	锁定画布的某个区域进行画图
public abstract void unlockCanvasAndPost(Canvas canvas)	结束锁定画图，并提交改变
public abstract void setFixedSize(int width,int height)	设置一个 Video 大小区域
public abstract void setType(int type)	设置 SurfaceView 的类型

实例 7-10：视频播放实例 1

新建一个项目，项目的命名为：exam7_10，包名称为：org.hnist.demo，使用 MediaPlayer 和 SurfaceView 建立一个简单的视频播放器，播放 SD 卡中的媒体文件。

1）在模拟机里怎样才能把文件放到 SD 卡中呢？

① 打开 DDMS 页面。

② 打开 File Explorer 页，如果没有，可执行 Window → Show View → File Explorer 命令打开。

③ 一般就在 mnt → sdcard 中。

④ 在 sdcard 中，单击你要将文件放到的目的文件夹。

⑤ 单击文件夹后，在 File Explorer 页的右边有两个图标，如图 7.7 所示，一个是 pull，将 SD 卡里面的文件保存到 PC，一个是 push，是将外面（如硬盘）的文件放到 SD 卡中。

图 7.7 复制视频文件到 SD 卡

2）建立一个布局管理文件 Activity_main.xml，代码如下：

```
<?xml version="1.0" encoding="utf-8"?>
<LinearLayout xmlns:android="http://schemas.android.com/apk/res/android"
android:orientation="vertical"
android:layout_width="fill_parent"
android:layout_height="fill_parent">
<LinearLayout
    xmlns:android="http://schemas.android.com/apk/res/android"
    android:orientation="horizontal"
    android:layout_width="wrap_content"
    android:layout_height="wrap_content">
    <ImageButton
        android:id="@+id/play"
        android:layout_width="wrap_content"
        android:layout_height="wrap_content"
        android:src="@drawable/play" />
    <ImageButton
        android:id="@+id/pause"
        android:layout_width="wrap_content"
        android:layout_height="wrap_content"
        android:src="@drawable/pause" />
    <ImageButton
        android:id="@+id/stop"
        android:layout_width="wrap_content"
        android:layout_height="wrap_content"
        android:src="@drawable/stop" />
</LinearLayout>
<SurfaceView
    android:id="@+id/surfaceView"
    android:layout_width="fill_parent"
    android:layout_height="fill_parent" />
</LinearLayout>
```

3）建立一个 Activity 文件 MainActivity.java，代码如下：

```
package org.hnist.demo;
import android.app.Activity;
import android.media.AudioManager;
import android.media.MediaPlayer;
import android.os.Bundle;
import android.view.SurfaceHolder;
import android.view.SurfaceView;
```

```java
import android.view.View;
import android.view.View.OnClickListener;
import android.widget.ImageButton;
public class MainActivity extends Activity {
private ImageButton play = null;
private ImageButton pause = null;
private ImageButton stop = null;
private MediaPlayer media = null;
private MediaPlayer myMediaPlayer = null;
private SurfaceView surfaceView = null;
private SurfaceHolder surfaceHolder = null;
public void onCreate(Bundle savedInstanceState) {
    super.onCreate(savedInstanceState);
    super.setContentView(R.layout.activity_main);
    this.play = (ImageButton) super.findViewById(R.id.play);
    this.pause = (ImageButton) super.findViewById(R.id.pause);
    this.stop = (ImageButton) super.findViewById(R.id.stop);
    this.surfaceView = (SurfaceView) super.findViewById(R.id.surfaceView);
    this.surfaceHolder = this.surfaceView.getHolder(); //取得SurfaceHolder
    //设置SurfaceView的类型
    this.surfaceHolder.setType(SurfaceHolder.SURFACE_TYPE_PUSH_BUFFERS);
    this.media = new MediaPlayer();                   //创建MediaPlayer对象
    try {
        this.media.setDataSource("/sdcard/test.3gp"); //设置播放文件的路径
    } catch (Exception e) {
        e.printStackTrace();}
    this.play.setOnClickListener(new PlayOnClickListenerImpl());
                                                //播放单击事件
    this.pause.setOnClickListener(new PauseOnClickListenerImpl());
                                                //暂停单击事件
    this.stop.setOnClickListener(new StopOnClickListenerImpl()); }
                                                //停止单击事件
private class PlayOnClickListenerImpl implements OnClickListener {
    public void onClick(View arg0) {
     //设置音频类型
    MainActivity.this.media    .setAudioStreamType(AudioManager.STREAM_MUSIC);
    //设置显示的区域
    MainActivity.this.media.setDisplay(MainActivity.this.surfaceHolder);
        try {
            MainActivity.this.media.prepare();      //预备状态
            MainActivity.this.media.start();        //播放视频
        } catch (Exception e) {
            e.printStackTrace();}}}
private class PauseOnClickListenerImpl implements OnClickListener {
    public void onClick(View arg0) {
        MainActivity.this.media.pause();}}
private class StopOnClickListenerImpl implements OnClickListener {
    public void onClick(View arg0) {
MainActivity.this.media.stop();}}} //停止播放
```

保存所有文件,运行该项目,如图7.8所示,各个按钮都能实现相应的功能。

图 7.8 简单的视频播放器

除了可以用 MediaPlayer 来实现视频播放外,还可通过 android.widget.VideoView 类播放视频文件,它的常用方法见表 7-18,与 VideoView 一起使用的还有 MediaController 类,它的作用是提供一个友好的控制界面通过该控件来控制视频的播放。

表 7-18 VideoView 类常用的方法

方法名称	描述
public VideoView (Context context)	创建一个默认属性的 VideoView 实例
Public VideoView (Context context, AttributeSet attrs)	创建一个带有 attrs 属性的 VideoView 实例
Public VideoView (Context context, AttributeSet attrs, int defStyle)	创建一个带有 attrs 属性,并且指定其默认样式的 VideoView 实例
public boolean canPause()	判断是否能够暂停播放视频
public void pause()	播放暂停
public boolean canSeekBackward()	判断是否能够倒退
public boolean canSeekForward()	判断是否能够快进
public int getBufferPercentage()	获得缓冲区的百分比
public int getCurrentPosition()	获得当前的位置
public int getDuration()	获得所播放视频的总时间
public boolean isPlaying()	判断是否正在播放视频
public boolean onKeyDown (int keyCode, KeyEvent event)	如果处理了事件,返回"真"。如果允许下一个事件接收器处理该事件,可以返回"假"
public boolean onTouchEvent (MotionEvent ev)	该方法来处理触屏事件
public void resume()	恢复挂起的播放器
public void seekTo(int msec)	设置播放位置
public void setMediaController (MediaController controller)	设置媒体控制器
public void setOnCompletionListener (MediaPlayer. OnCompletionListener l)	在媒体文件播放完毕时触发
public void setOnErrorListener (MediaPlayer.OnErrorListener l)	出现错误时触发
public void setOnPreparedListener (MediaPlayer.OnPreparedListener l)	在媒体文件加载完毕时触发,可以播放时指定调用的回调函数
public void setVideoPath(String path)	设置视频文件的路径名
public void setVideoURI(Uri uri)	设置视频文件的统一资源标识符
public void start()	开始播放视频文件
public void stopPlayback()	停止回放视频文件
public void suspend()	挂起视频文件的播放

实例 7-11：视频播放实例 2

当然类似地也可以使用 VideoView 和 MediaController 建立一个简单的视频播放器，播放 SD 卡中的指定的媒体文件，详细代码见 exam7_11，利用 VideoView 类播放视频的速度可能不是很理想。

7.5 Android 中的照相机

前面介绍了使用 SurfaceView 组件可以播放视频文件，其实还可以利用 SurfaceView 实现拍照的功能，SurfaceView 的操作核心就是在于 android.view.SurfaceHolder 对象的操作，如果要想实现拍照的功能，首先用户必须手工实现 android.view.SurfaceHolder.Callback 这个操作接口，在此接口中定义了高速图像浏览时的各个操作，具体方法见表 7-19。

表 7-19 SurfaceHolder.Callback 接口中定义的方法

方法	描述
public abstract void surfaceChanged(SurfaceHolder holder, int format, int width, int height)	当预览界面的格式和大小发生改变时会触发此操作
public abstract void surfaceCreated(SurfaceHolder holder)	当预览界面被创建时会触发此操作
public abstract void surfaceDestroyed(SurfaceHolder holder)	当预览界面关闭时会触发此操作

除了拍照的预览界面之外还有一个调用摄像头的操作类 android.hardware.Camera，此类主要负责完成拍照图片的参数设置及保存，常用方法见表 7-20。

表 7-20 Camera 类的常用操作方法

方法	描述
public final void autoFocus (Camera. AutoFocusCallback cb)	自动对焦
public final void cancelAutoFocus()	取消自动对焦
public static int getNumberOfCameras()	得到摄像头的个数
public Camera.Parameters getParameters()	得到摄像头的各个参数
public final void lock()	锁定设备
public static Camera open(int cameraId)	打开指定的摄像头，以获得 Camera 对象
public static Camera open()	打开默认的摄像头
public final void reconnect()	重新连接摄像头
public final void release()	释放摄像头资源
public void setParameters(Camera.Parameters params)	设置摄像头的若干参数
public final void startPreview()	开始预览
public final void stopPreview()	停止预览
public final void takePicture(Camera. ShutterCallback shutter, Camera.PictureCallback raw, Camera.PictureCallback jpeg)	捕获图像
public final void unlock()	设备解除锁定
public final void setZoomChangeListener (Camera.OnZoomChangeListener listener)	显示区域发生改变时触发
public final void setDisplayOrientation(int degrees)	设置摄像头角度

Camera 类中定义了若干内部接口，这些接口的作用如表 7-21 所示。

表 7-21　Camera 类中定义的内部接口

接口名称	描述
android.hardware.Camera.AutoFocusCallback	自动对焦的回调操作
android.hardware.Camera.ErrorCallback	错误出现时的回调操作
android.hardware.Camera.OnZoomChangeListener	显示区域改变时的回调操作
android.hardware.Camera.PictureCallback	图片生成时的回调操作
android.hardware.Camera.PreviewCallback	预览时的回调操作
android.hardware.Camera.ShutterCallback	按下快门后的回调操作

通过 Camera 控制摄像头拍照的步骤如下：

1）调用 Camera 的 open()方法打开相机；

```
private Camera cam = null;                           //定义拍照组件
MyCameraDemo.this.cam = Camera.open(0);              //取得摄像头，打开相机
```

2）调用 Camera 的 setParameters()方法获取拍照参数。该方法返回一个 Camera.Parameters 对象；

```
Parameters param = MyCameraDemo.this.cam.getParameters();  //取得照相机参数
```

3）调用 Camera.Paramers 对象方法设置拍照参数；

```
param.setPreviewSize(display.getWidth(), display.getHeight());//设置预览大小
param.setPreviewFrameRate4);                         //每秒显示 4 帧的数据
param.setPictureFormat(PixelFormat.JPEG);            //设置图片格式
param.set("jpeg-quality", 85);                       //设置图片质量，最高为 100
parameters.setPictureSize(screenWidth, screenHeight);//设置照片的大小
```

4）调用 Camera 的 setParameters，并将 Camera.Paramers 作为参数传入，这样即可对相机的拍照参数进行控制；

```
MyCameraDemo.this.cam.setParameters(param);          //设置参数
```

5）调用 Camera 的 startPreview()方法开始预览取景，在预览取景之前需要调用 Camera 的 setPreViewDisplay(SurfaceHolder holder)方法设置使用哪个 SurfaceView 显示取景图片；

```
MyCameraDemo.this.cam.setPreviewDisplay(MyCameraDemo.this.holder);
MyCameraDemo.this.cam.startPreview();                //开始预览
MyCameraDemo.this.cam.autoFocus(afcb);  }            //自动对焦
```

6）调用 Camera 的 takePicture()方法进行拍照；

```
MyCameraDemo.this.cam.takePicture(sc, pc, jpgcall);  //获取图片
```

7）结束程序时，调用 Camera 的 StopPriview()结束取景预览，并调用 release()方法释放资源；

```
MyCameraDemo.this.cam.stopPreview();                 //结束取景预览
MyCameraDemo.this.cam.release();                     //释放摄像头
```

8）配置 AndroidManifest.xml 文件。

```
<uses-feature android:name="android.hardware.camera" />    //调用摄像头功能
<uses-feature android:name="android.hardware.camera.autofocus" />
                                                           //调用自动对焦
<uses-permission android:name="android.permission.CAMERA" />
                                                           //设置允许拍照
<uses-permission android:name="android.permission.MOUNT_UNMOUNT_FILESYSTEMS" />
```

```
                                                //SD 卡创建与删除文件权限
<uses-permission android:name="android.permission.WRITE_EXTERNAL_STORAGE" />
                                                //配写 SD 卡权限
```

注意:模拟器上不能实现照相功能,要在实体机上才能真正实现照相功能,所以下面的代码要在真正的手机上才能达到预期效果。

实例 7-12:Android 中照相机实例

新建一个项目,项目的命名为:exam7_12,包名称为:org.hnist.demo,使用手机摄像头进行拍照,并将照片保存。

1)布局文件 activity_main.xml,代码如下:

```xml
<?xml version="1.0" encoding="utf-8"?>
<LinearLayout
xmlns:android="http://schemas.android.com/apk/res/android"
android:orientation="vertical"
android:layout_width="fill_parent"
android:layout_height="fill_parent">
<Button
    android:id="@+id/but"
    android:layout_width="fill_parent"
    android:layout_height="wrap_content"
    android:text="开始拍照" />
<SurfaceView
    android:id="@+id/surface"
    android:layout_width="fill_parent"
    android:layout_height="match_parent" />
</LinearLayout>
```

2)建立一个 Activity 文件 MainActivity.java,代码如下:

```java
package org.hnist.demo;
package org.hnist.demo;
import java.io.BufferedOutputStream;
import java.io.File;
import java.io.FileOutputStream;
import java.io.IOException;
import android.app.Activity;
import android.content.Context;
import android.graphics.Bitmap;
import android.graphics.BitmapFactory;
import android.graphics.PixelFormat;
import android.hardware.Camera;
import android.hardware.Camera.AutoFocusCallback;
import android.hardware.Camera.Parameters;
import android.hardware.Camera.PictureCallback;
import android.hardware.Camera.ShutterCallback;
import android.os.Bundle;
import android.os.Environment;
import android.view.Display;
import android.view.SurfaceHolder;
import android.view.SurfaceView;
```

```java
import android.view.View;
import android.view.View.OnClickListener;
import android.view.Window;
import android.view.WindowManager;
import android.widget.Button;
import android.widget.Toast;
public class MainActivity extends Activity {
    private SurfaceHolder holder = null;              //定义SurfaceHolder
    private SurfaceView surface = null;               //定义SurfaceView
    private Camera camera01 = null;                   //拍照组件
    private Button startbtn = null;                   //按钮组件
    private boolean previewRunning = true;            //预览结束的标记
    public void onCreate(Bundle savedInstanceState) {
        super.onCreate(savedInstanceState);
        super.getWindow().setFlags(WindowManager.LayoutParams.FLAG_FULLSCREEN,
            WindowManager.LayoutParams.FLAG_FULLSCREEN);  //全屏显示
        super.getWindow().addFlags(
            WindowManager.LayoutParams.FLAG_KEEP_SCREEN_ON);    //高亮显示
        super.setContentView(R.layout.activity_main);      //布局管理器
        this.startbtn = (Button) super.findViewById(R.id.but);  //取得组件
        this.surface = (SurfaceView) findViewById(R.id.surface);  //取得组件
        this.holder = surface.getHolder();                 //设置Holder
        this.holder.addCallback(new MySurfaceViewCallback());   //加入回调
        this.holder.setType(SurfaceHolder.SURFACE_TYPE_PUSH_BUFFERS);//设置缓冲
        this.holder.setFixedSize(500, 350);                //设置分辨率
        this.startbtn.setOnClickListener(new StartCameraimpl()); }//单击事件
//接口SurfaceHolder.Callback被用来接收摄像头预览界面变化的信息
    private class MySurfaceViewCallback implements SurfaceHolder.Callback {
//当预览界面的格式和大小发生改变时,该方法被调用
    public void surfaceChanged(SurfaceHolder holder, int format, int width,
                    int height) { }
//初次实例化预览界面被创建时,该方法被调用
    public void surfaceCreated(SurfaceHolder holder) {
        MainActivity.this.camera01 = Camera.open(0);        //取得摄像头
        WindowManager manager = (WindowManager)MainActivity.this.getSystemService
            (Context.WINDOW_SERVICE);                       //取得窗口服务
        Display display = manager.getDefaultDisplay();      //取得Display对象
        Parameters param = MainActivity.this.camera01.getParameters();
                                                            //取得相机参数
        param.setPreviewSize(display.getWidth(), display.getHeight());
                                                            //设置预览大小
        param.setPreviewFrameRate(5);                       //每秒显示5帧的数据
        param.setPictureFormat(PixelFormat.JPEG);           //设置图片格式
        param.set("jpeg-quality", 80);                      //设置图片质量
        MainActivity.this.camera01.setParameters(param);    //设置参数
        try {                                               //通过SurfaceView显示
            MainActivity.this.camera01.setPreviewDisplay
                (MainActivity.this.holder);
        } catch (IOException e) {e.printStackTrace();}
        MainActivity.this.camera01.startPreview();          //开始预览
        MainActivity.this.previewRunning = true;}           //修改预览标记
```

```java
            public void surfaceDestroyed(SurfaceHolder holder) {
                                            //当预览界面被关闭时，该方法被调用
                if (MainActivity.this.camera01 != null) {
                    if (MainActivity.this.previewRunning) {         //如果正在预览
                        MainActivity.this.camera01.stopPreview();   //停止预览
                        MainActivity.this.previewRunning = false;}  //修改标记
                    //摄像头只能被一个Activity程序使用，所以要释放摄像头
                    MainActivity.this.camera01.release();} }}       //释放摄像头
        private PictureCallback jpgcall = new PictureCallback() {
            public void onPictureTaken(byte[] data, Camera camera) {
                try {//定义BitMap
                    Bitmap bmp = BitmapFactory.decodeByteArray(data, 0, data.length);
                    String fileName = Environment.getExternalStorageDirectory()
                        .toString()+ File.separator + "Cameraphoto"+
                        File.separator  +"exam7_12_"+System.currentTimeMillis()+
                        ".jpg";                     定义输出文件名称
                    File file = new File(fileName);         //定义File对象
                    if (!file.getParentFile().exists()) {   //父文件夹不存在
                        file.getParentFile().mkdirs();}     //创建父文件夹
                    BufferedOutputStream bos = new BufferedOutputStream(new
                        eOutputStream(file));               //节缓存流
                    bmp.compress(Bitmap.CompressFormat.JPEG, 80, bos); //图片压缩
                    bos.flush();                            //清空缓冲
                    bos.close();                            //关闭
                    Toast.makeText(MainActivity.this,
                        "照片保存在" + fileName + "文件中", Toast.LENGTH_SHORT)
                        .show();                            //显示Toast
                    MainActivity.this.camera01.stopPreview();   //停止预览
                    MainActivity.this.camera01.startPreview();  //开始预览
                } catch (Exception e) { }}};
        private class StartCameraimpl implements OnClickListener {
            public void onClick(View v) {
                if (MainActivity.this.camera01 != null) {   //如果存在Camera对象
        MainActivity.this.camera01.autoFocus(new AutoFocusCallbackImpl());   }}}
                                                            //自动对焦
        private class AutoFocusCallbackImpl implements AutoFocusCallback {
            public void onAutoFocus(boolean success, Camera camera01) {
                if (success) {                              //如果对焦成功
                    MainActivity.this.camera01.takePicture(sc, pc, jpgcall);
                                                            //获取图片
                    MainActivity.this.camera01.stopPreview();    }}}  //停止预览
        private ShutterCallback sc = new ShutterCallback() {
            public void onShutter() {}  };
        private PictureCallback pc = new PictureCallback() {
            public void onPictureTaken(byte[] arg0, Camera arg1) {}  };}
```

3）配置AndroidManifest.xml文件，在AndroidManifest.xml文件后面添加下述语句：

```xml
<uses-feature android:name="android.hardware.camera" />    //调用摄像头功能
<uses-feature android:name="android.hardware.camera.autofocus" />
                                                           //调用自动对焦
<uses-permission android:name="android.permission.CAMERA" /> //设置允许拍照
```

```
<uses-permission android:name="android.permission.MOUNT_UNMOUNT_FILESYSTEMS" />
                                                       //SD 卡创建与删除文件权限
<uses-permission android:name="android.permission.WRITE_EXTERNAL_STORAGE" />
                                                       //配写 SD 卡权限
```

保存所有文件,运行该项目,如图 7.9 所示,单击"开始拍照"按钮,可以拍摄照片,将照片保存在指定地方,如图 7.10 所示。注意在模拟机上不能正常拍照,要在真实手机上才可以拍照。

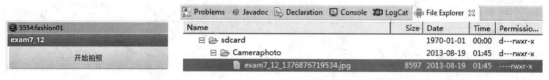

图 7.9　简单照相机　　　　　　　　　图 7.10　照片保存位置

7.6　Android 中的媒体录制

在 Android 中提供了负责进行媒体录制的 android.media.MediaRecorder 类,此类可以实现音频和视频文件的录制操作。MediaRecorder 作为状态机运行。需要设置不同的参数,比如源设备和格式,设置后,可执行任何时间长的录制,直到用户停止。MediaRecorder 类的常用方法见表 7-22。

表 7-22　MediaRecorder 类的常用方法

方法	描述
public MediaRecorder()	定义一个默认的 MediaRecorder 对象
public static final int getAudioSourceMax()	得到最大音量
public void prepare()	进入到就绪操作状态
public void release()	释放资源
public void reset()	重置设备
public void setAudioEncoder(int audio_encoder)	设置音频编码
public void setAudioSource(int audio_source)	设置音频源
public void setCamera(Camera c)	设置 Camera 对象
public void setOutputFile(String path)	设置输出文件
public void setPreviewDisplay(Surface sv)	设置预览显示
public void setVideoEncoder(int video_encoder)	设置视频编码
public void setVideoFrameRate(int rate)	设置视频帧率
public void setVideoSize(int width, int height)	设置视频分辨率
public void setVideoSource(int video_source)	设置视频源
public void start()	开始录制
public void stop()	停止录制
public void release()	释放该对象资源

在介绍利用 MediaRecorder 类录制音频和视频之前,要介绍 MediaRecorder 操作的生命周期,如图 7.12 所示,先来看看 MediaRecorder 的几个状态。

Initial 状态:当用户通过 MediaRecorder 类的构造方法实例化 MediaRecorder 类对象时将处于初始化状态,即便此时没有任何的操作 MediaRecorder 也会占用着系统资源,而所有的状态都可以通过 reset() 方法返回到此状态;

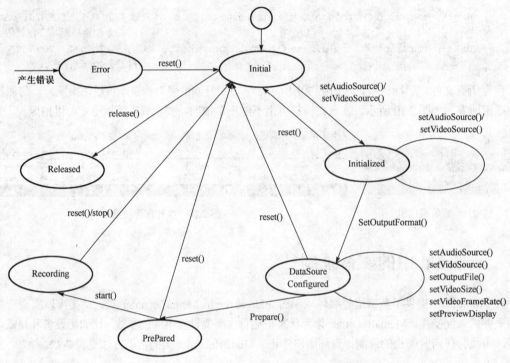

图 7.12 MediaRecorder 类的生命周期示意图

Initialized 状态：当用户使用 setAudioSource()或者是 setVideoSource()方法之后将进入到音频录制或者是视频录制，并可以指定一些音频或视频的文件属性，设置完成后将进入到 DataSourceConfigured 状态；

Prepared 状态：当用户使用 MediaRecorder 类中的 prepare()方法时将进入到就绪状态，录制前的状态已经准备就绪；

Recording 状态：当用户使用 MediaRecorder 类中的 start()方法时将进入到录制状态，并且一直持续到录音或录像操作完毕；

7.6.1 Android 中的录音

一般情况下，要实现音频的录制有如下步骤：

1）初始化 MediaRecorder 对象。

```
MediaRecorder mrecorder = new MediaRecorder();
```

2）设置声音来源，利用 MediaRecorder.AudioSource 设置声音来源。一般设置的是手机上的麦克风，MediaRecorder.AudioSource.MIC，例如：

```
mrecorder.setAudioSource(MediaRecorder.AudioSource.MIC);
```

3）设置输出文件格式，该语句必须在 setAudioSource 之后，在 prepare 之前，利用 OutputFormat 定义音频输出的格式，主要包含 MPEG_4、THREE_GPP、RAW_AMR……等，默认格式为 amr。例如：

```
mrecorder.setOutputFormat(MediaRecorder.OutputFormat.THREE_GPP);
```

或

```
mrecorder.setOutputFormat(MediaRecorder.OutputFormat.DEFAULT GPP);
```

4）设置音频编码，利用 AudioEncoder 定义两种编码，一般有两种编码：AudioEncoder.DEFAULT 和 AudioEncoder.AMR_NB。

第 7 章　Android 多媒体技术

例如：
```
mrecorder.setAudioEncoder(MediaRecorder.AudioEncoder.AMR_NB);
```

5）设置音频文件保存路径。例如：
```
mrecorder.setOutputFile(PATH_NAME);
```

6）准备录制，例如：
```
mrecorder.prepare();
```

7）开始录制，例如：
```
mrecorder.start();
mrecorder.stop();           //停止
mrecorder.reset();          //重置
mrecorder.release();        //释放资源
```

8）在 AndroidManifest.xml 文件中添加使用记录音频的权限和往 SD 卡写的权限。
```
<uses-permission android:name="android.permission.RECORD_AUDIO" />
<uses-permission android:name="android.permission.WRITE_
    EXTERNAL_STORAGE"></uses-permission>
```

实例 7-13：Android 中音频录制实例

新建一个项目，项目的命名为：exam7_13，包名称为：org.hnist.demo，使用 MediaRecorder 录制音频，并将音频文件保存在 SD 卡。

1）布局文件 activity_main.xml，代码如下：
```xml
<?xml version="1.0" encoding="utf-8"?>
<LinearLayout
  xmlns:android="http://schemas.android.com/apk/res/android"
  android:orientation="vertical"
  android:layout_width="fill_parent"
  android:layout_height="fill_parent" >
<ImageButton
   android:id="@+id/btnStart"
     android:layout_width="wrap_content"
     android:layout_height="wrap_content"
     android:src="@drawable/record"/>
<ImageButton
     android:id="@+id/btnStop"
     android:layout_width="wrap_content"
     android:layout_height="wrap_content"
     android:src="@drawable/stop"/>
<TextView
     android:id="@+id/msg"
     android:layout_width="wrap_content"
     android:layout_height="wrap_content" />
</LinearLayout>
```

2）建立一个 Activity 文件 MainActivity.java，代码如下：
```java
package org.hnist.demo;
import java.io.File;
import android.app.Activity;
import android.media.MediaRecorder;
import android.os.Bundle;
```

```java
import android.os.Environment;
import android.view.View;
import android.view.View.OnClickListener;
import android.widget.ImageButton;
import android.widget.TextView;
import android.widget.Toast;
public class MainActivity extends Activity implements OnClickListener {
    private TextView msg;
    private ImageButton btnStart;
    private ImageButton btnStop;
    private MediaRecorder mediaRecorder;
    private File soundFile;
    private String strTempFile = "exam07_13_";
    private File myRecAudioDir;
    public void onCreate(Bundle savedInstanceState) {
        super.onCreate(savedInstanceState);
        setContentView(R.layout.activity_main);
        msg = (TextView) this.findViewById(R.id.msg);
        btnStart = (ImageButton) this.findViewById(R.id.btnStart);
        btnStop = (ImageButton) this.findViewById(R.id.btnStop);
        btnStart.setOnClickListener(this);
        btnStop.setOnClickListener(this); }
    public void onClick(View v) {
        switch (v.getId()) {
        case R.id.btnStart:
            msg.setText("---开始录音---");
            //先检测是否含有 SDCard
            if (!Environment.getExternalStorageState().equals (Environment.
                MEDIA_MOUNTED)) {
                Toast.makeText(MainActivity.this, "SD卡不存在,请插入SD卡",
                        Toast.LENGTH_LONG).show();
                return; }
            try { mediaRecorder = new MediaRecorder();
                myRecAudioDir = Environment.getExternalStorageDirectory();
                soundFile = File.createTempFile(strTempFile, ".amr",
                    myRecAudioDir);
mediaRecorder.setAudioSource(MediaRecorder.AudioSource.MIC);
                            //设置录音的来源为麦克风
mediaRecorder.setOutputFormat(MediaRecorder.OutputFormat.DEFAULT);
                            //设置录制的声音输出格式
mediaRecorder.setAudioEncoder(MediaRecorder.AudioEncoder.DEFAULT);
                            //设置声音的编码格式
mediaRecorder.setOutputFile(soundFile.getAbsolutePath());
                            //设置录音的输出文件路径
mediaRecorder.prepare();        //做预期准备
mediaRecorder.start();          //开始录音
} catch (Exception e) { }
    break;
case R.id.btnStop:
    msg.setText("---停止录音---");
    if (soundFile != null ) {
        mediaRecorder.stop();               //停止录制
        mediaRecorder.release();            //释放资源
```

第 7 章 Android 多媒体技术

```
            mediaRecorder = null; }
    break;
default:
    break; } }
protected void onDestroy() {
        if (soundFile != null && soundFile.exists()) {
            mediaRecorder.stop();              //停止录制
            mediaRecorder.release();           //释放资源
            mediaRecorder = null; }
        super.onDestroy();   }  }
```

3）配置 AndroidManifest.xml 文件，在 AndroidManifest.xml 文件后面添加下述语句：

```
<uses-permission android:name="android.permission.RECORD_AUDIO" />
<uses-permission android:name="android.permission.WRITE_EXTERNAL_
                        STORAGE"></uses-permission>
```

保存所有文件，运行该项目，如图 7.13 所示，单击"开始录音"按钮，开始录音，单击"停止录音"按钮，停止录音，在 File Explorer 中有了录音文件，如图 7.14 所示。

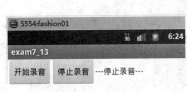

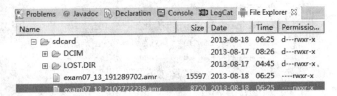

图 7.13 音频录制   图 7.14 在 SD 卡建立的录音文件

注意这里没有实现对录音文件的播放，可以这样来实现，建立一个 ListView 用来放置你的录音文件，然后对这个 ListView 中的选项设置相应的操作事件，详细代码见 exam7_13_1，限于篇幅这里不给出具体代码。

7.6.2 Android 中的录像

Android 系统提供的录像和录音大部分类似，最大的区别就是录像要定义一个 SurfaceView 组件，用来捕获摄像头采集的视频数据。

实例 7-14：Android 中视频录制实例

新建一个项目，项目的命名为：exam7_14，包名称为：org.hnist.demo，使用 MediaRecorder 和 SurfaceView 录制视频，并将视频文件保存。

1）布局文件 activity_main.xml，代码如下：

```
<?xml version="1.0" encoding="utf-8"?>
<LinearLayout
xmlns:android="http://schemas.android.com/apk/res/android"
android:orientation="vertical"
android:layout_width="fill_parent"
android:layout_height="fill_parent">
<SurfaceView android:id="@+id/videoView"
android:visibility="visible"
android:layout_width="320px"
android:layout_height="240px">
```

```xml
</SurfaceView>
<RelativeLayout
android:layout_width="fill_parent"
android:layout_height="wrap_content">
<Button
android:layout_width="wrap_content"
android:layout_height="wrap_content"
android:text="开始录像"
android:id="@+id/start"/>
<Button
android:layout_width="wrap_content"
android:layout_height="wrap_content"
android:layout_toRightOf="@id/start"
android:text="停止录像"
android:id="@+id/stop"/>
</RelativeLayout>
<TextView
    android:id="@+id/msg"
    android:layout_width="100dp"
    android:layout_height="wrap_content"
    android:textSize="20dp"
    android:text="" />
</LinearLayout>
```

2）建立一个 Activity 文件 MainActivity.java，代码如下：

```java
package org.hnist.demo;
import java.io.File;
import java.io.IOException;
import android.app.Activity;
import android.media.MediaRecorder;
import android.os.Bundle;
import android.os.Environment;
import android.view.SurfaceHolder;
import android.view.SurfaceView;
import android.view.View;
import android.view.View.OnClickListener;
import android.widget.Button;
import android.widget.TextView;
public class MainActivity extends Activity {
  private File myRecAudioFile;
  private SurfaceView mSurfaceView;
  private SurfaceHolder mSurfaceHolder;
  private TextView msg ;
  private Button btnStart;
  private Button btnStop;
  private File dir;
  private MediaRecorder recorder;
  public void onCreate(Bundle savedInstanceState) {
    super.onCreate(savedInstanceState);
```

```java
        setContentView(R.layout.activity_main);
        mSurfaceView = (SurfaceView) findViewById(R.id.videoView);
        mSurfaceHolder = mSurfaceView.getHolder();
        mSurfaceHolder.setType(SurfaceHolder.SURFACE_TYPE_PUSH_BUFFERS);
        btnStart=(Button)findViewById(R.id.start);
        btnStop=(Button)findViewById(R.id.stop);
        msg=(TextView) findViewById(R.id.msg);
        File defaultDir = Environment.getExternalStorageDirectory();
        //创建文件夹存放视频
        String path = defaultDir.getAbsolutePath()+File.separator+"MyVideo"
                    +File.separator;
        dir = new File(path);
        if(!dir.exists()){
          dir.mkdir(); }
     recorder = new MediaRecorder();
     btnStart.setOnClickListener(new OnClickListener() {
       public void onClick(View v) {
         msg.setText("开始录制");
         recorder(); } });
     btnStop.setOnClickListener(new OnClickListener() {
       public void onClick(View v) {
         msg.setText("停止录制");
         recorder.stop();
         recorder.reset();
         recorder.release();
         recorder=null; } }); }
public void recorder() {
try {
myRecAudioFile = File.createTempFile("video", ".3gp",dir);//创建临时文件
recorder.setPreviewDisplay(mSurfaceHolder.getSurface());//预览
recorder.setVideoSource(MediaRecorder.VideoSource.CAMERA);//视频源
recorder.setAudioSource(MediaRecorder.AudioSource.MIC);  //录音源为麦克风
recorder.setOutputFormat(MediaRecorder.OutputFormat.THREE_GPP);//格式为3gp
recorder.setVideoSize(800, 480);//视频尺寸
recorder.setVideoFrameRate(15);//视频帧频率
recorder.setVideoEncoder(MediaRecorder.VideoEncoder.H263);//视频编码
recorder.setAudioEncoder(MediaRecorder.AudioEncoder.AMR_NB);//音频编码
recorder.setMaxDuration(10000);//最大期限
recorder.setOutputFile(myRecAudioFile.getAbsolutePath());//保存路径
recorder.prepare();
recorder.start();
} catch (IOException e) {
e.printStackTrace(); } }
```

3) 配置 AndroidManifest.xml 文件,在 AndroidManifest.xml 文件后面添加下述语句:

```xml
<uses-permission android:name="android.permission.CAMERA"/>
<uses-permission android:name="android.permission.RECORD_AUDIO"/>
<uses-permission android:name="android.permission.WRITE_EXTERNAL_STORAGE"/>
```

保存所有文件，运行该项目，如图 7.15 所示，单击"开始录像"按钮，开始录制，单击"停止录像"按钮，停止录制，在 File Explorer 中有了录像文件，如图 7.16 所示，注意这里的录像大小为 0，因为是在模拟机上运行的缘故，这里需要真正的摄像头设备。

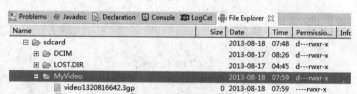

图 7.15　视频录制　　　　　　　　图 7.16　在 SD 卡建立的视频文件

注意这里没有实现对视频文件的播放，实例 exam7_14_1，给出了一个实例，限于篇幅这里不给出具体代码。

Android 中的还有许多多媒体方面的功能，例如：语音功能、手势控制屏幕、多点触控屏幕功能等等，感兴趣的读者可以查阅 Android API 学习了解。

本章小结

本章着重介绍了几种常见基本图形的绘制，常见的两种动画处理、使用 MediaPlayer 播放音频和视频文件、利用摄像头进行拍照、进行媒体的录制等技术，简要介绍了动画操作组件、图片的简单处理技术，要注意的是拍照和摄像要借助硬件设备，在模拟机上不能正常运行。

习题

1. 要绘制一个蓝色的圆，需要用到哪些类？给出关键代码。
2. 利用 XML 文件，实现实例 7-4 的动画效果。
3. 播放网络上的音频的文件，用什么语句指定网络上的音频文件？
4. 利用摄像头拍照时要怎样配置 AndroidManifest.xml 文件？
5. 参照音频文件录制的步骤，写出录制视频文件的步骤。
6. 使用 MediaRecorder 录制音频，利用什么方法可以设置录音的输出文件路径？
7. 视频文件录制完毕，如果想要播放录制的视频文件，应该如何操作？写出代码。

第 8 章 Android 数据存储技术

学习目标：
- 掌握使用 SharedPreferences 存储数据
- 掌握对文件的保存和读取操作
- 了解 SQLite 数据库的基本作用
- 掌握在 Activity 文件中操作 SQLite 数据库
- 掌握 ContentProvider 的使用

一个应用程序，经常需要与用户进行交互，需要保存用户的设置和用户数据，这些都离不开数据的存储。Android 系统提供了以下五种主要的数据存储方式。

1. 使用 SharedPreferences 存储数据；
2. 文件存储数据；
3. SQLite 数据库存储数据；
4. 使用 ContentProvider 存储数据；
5. 网络存储数据。

8.1 使用 SharedPreferences 存储数据

对于软件配置参数的保存，Windows 系统通常会采用 ini 文件保存，Java 程序采用 properties 属性文件或 xml 文件保存。类似的，Android 平台为我们提供了一个 SharedPreferences 接口用于保存参数设置等较为简单的数据，例如：字符串、整型、布尔型等。使用 SharedPreferences 进行保存的数据，采用 key=value 键值对的形式进行保存一些简单的配置信息，信息以 xml 文件的形式存储在/data/data/<包名>/shared_prefs 目录下。

SharedPreferences 类似 Windows 系统上的 ini 配置文件，但是它分为多种权限，可以全局共享访问，以 xml 方式来保存，这样占用的内存资源比较少。

要在 Android 程序中使用 SharedPreferences 组件必须在程序中使用下面的语句。

```
import android.content.SharedPreferences; //导入 content.SharedPreferences 类
```

SharedPreferences 接口的常用方法见表 8-1。

表 8-1 SharedPreferences 接口的常用方法

方法	描述
public abstract SharedPreferences.Editor edit()	使其处于可编辑状态
public abstract boolean contains(String key)	判断某一个 key 是否存在
public abstract Map<String, ?> getAll()	取出全部的数据
public abstract boolean getBoolean(String key, boolean defValue)	取出 boolean 型数据，并指定默认值
public abstract float getFloat(String key, float defValue)	取出 float 型数据，并指定默认值
public abstract int getInt(String key, int defValue)	取出 int 型数据，并指定默认值
public abstract long getLong(String key, long defValue)	取出 long 型数据，并指定默认值
public abstract String getString(String key, String defValue)	取出 String 型数据，并指定默认值

SharedPreferences 对象本身只能获取数据而不支持存储和修改，存储修改是通过 android.content.SharedPreferences.Editor 接口来实现。SharedPreferences.Editor 接口的常用方法见表 8-2。

表 8-2 SharedPreferences.Editor 接口的常用方法

方法	描述
public abstract SharedPreferences.Editor clear()	清除所有的数据
public abstract boolean commit()	提交更新的数据
public abstract SharedPreferences.Editor putBoolean(String key, boolean value)	保存一个 boolean 型数据
public abstract SharedPreferences.Editor putFloat(String key, float value)	保存一个 float 型数据
public abstract SharedPreferences.Editor putInt(String key, int value)	保存一个 int 型数据
public abstract SharedPreferences.Editor putLong(String key, long value)	保存一个 long 型数据
public abstract SharedPreferences.Editor putString(String key, String value)	保存一个 String 型数据
public abstract SharedPreferences.Editor remove(String key)	删除指定 key 的数据

由于 SharedPreferences 和 SharedPreferences.Editor 两个都是接口，所以要想取得 SharedPreferences 接口的实例化对象，还需要 Activity 类中的几个常量和方法的支持。取得 SharedPreferences 接口的几个常量和方法见表 8-3。

表 8-3 取得 SharedPreferences 接口的几个常量和方法

常量及方法	描述
public static final int MODE_PRIVATE	常量，创建的文件只能被一个应用程序调用，或者被具有相同 id 的应用程序访问
public static final int MODE_WORLD_READABLE	常量，允许其他应用程序读取文件
public static final int MODE_WORLD_WRITEABLE	常量，允许其他应用程序修改文件
public SharedPreferences getSharedPreferences (String name, int mode)	指定保存操作的文件名称，同时指定操作的模式，可以是 MODE_PRIVATE、MODE_WORLD_READABLE、MODE_WORLD_WRITEABLE

其中，String name 用于指定文件名称，不能包含路径分隔符 "/"，如果文件不存在，Android 会自动创建它。

int mode 用于指定操作模式，有如下几种模式：

1）Context.MODE_PRIVATE=0：为默认操作模式，代表该文件是私有数据，只能被应用本身访问，在该模式下，写入的内容会覆盖原文件的内容；

2）Context.MODE_WORLD_READABLE =1：表示当前文件可以被其他应用读取；

3）Context.MODE_WORLD_WRITEABLE =2：表示当前文件可以被其他应用写入。

8.1.1 使用 SharedPreferences 存储数据

实现 SharedPreferences 存储的步骤如下：

1）根据 Context 获取 SharedPreferences 对象，例如：

```
SharedPreferences sh= super.getSharedPreferences("my",Activity.MODE_PRIVATE);
```

2）利用 edit()方法获取 Editor 对象，例如：

```
SharedPreferences.Editor edit = sh.edit();     //获取 Editor 对象
```

3）通过 Editor 对象存储 key-value 键值对数据，例如：

```
edit.putString("college", "hnist") ;           //保存字符串
edit.putInt("students", 20000);                //保存整型
```

4）通过 commit()方法提交数据，例如：

```
                edit.commit() ;                                    //提交更新
```

实例 8-1：使用 SharedPreferences 保存数据实例

新建一个项目，项目的名称为 **exam8_1**，包名称为 "**org.hnist.demo**"，使用 SharedPreferences 存储数据到指定文件。

1）修改 Activity 文件 MainActivity.java，代码如下：

```
package org.hnist.demo;
import android.app.Activity;
import android.content.SharedPreferences;
import android.os.Bundle;
public class SaveData extends Activity {
    private static final String FILENAME = "hnist";        //定义文件名称
    public void onCreate(Bundle savedInstanceState) {
        super.onCreate(savedInstanceState);
        setContentView(R.layout.activity_main);
//指定操作的文件名称
SharedPreferences sh=super.getSharedPreferences(FILENAME,Context.
            MODE_WORLD_READABLE);
        SharedPreferences.Editor edit = sh.edit();         //编辑文件
        edit.putString("college", "hnist") ;               //保存字符串
        edit.putInt("students", 20000);                    //保存整型
        edit.commit() ; }}                                 //提交更新
```

2）查看保存的数据。

选择菜单栏的 "Window→Open Perspective→DDMS" 选项，打开 DDMS 视图，如图 8.1 所示。在 Eclipse 的下方会弹出如图 8.2 所示的栏目，如果没有 File Explorer 选项，则需要添加该项内容。

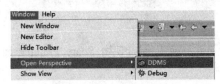

图 8.1　打开 DDMS 视图　　　　　　　　图 8.2　打开 DDMS 后

选择菜单栏的 "Window→Show View→Others" 选项，弹出如图 8.3 所示的界面。

选择其中的 "File Explorer" 选项，单击 "OK" 按钮，这时在 Eclipse 的下方会弹出 File Explorer 选项如图 8.4 所示。

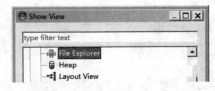

图 8.3　添加 File Explorer 选项　　　　　图 8.4　添加 File Explorer 选项后

选择 "File Explorer"，打开后找到 "/data/data/org.hnist.demo/shared_prefs/"，如图 8.5 所示。

选择 DDMS 工具栏上的 "Pull a file from the device"，如图 8.6 所示，导出指定文件，用记事本打开这个文件，如图 8.7 所示。

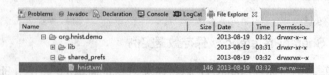

图 8.5 保存的文件

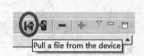

图 8.6 导出文件　　　　　　　图 8.7 记事本打开文件的内容

8.1.2 使用 SharedPreferences 读取数据

实现 SharedPreferences 读取数据的步骤如下：
1）定义 TextView 对象，用于显示读取的数据，例如：

```
private TextView college = null ;           //定义文本显示组件
private TextView students = null ;          //定义文本显示组件
```

2）根据 Context 获取 SharedPreferences 对象，例如：

```
SharedPreferences sh = super.getSharedPreferences("my",Activity.MODE_PRIVATE);
```

3）通过 getString()方法获得字符数据，getInt()获得数字数据，例如：

```
this.college.setText("学校: " + sh.getString("college ", "没有学校信息。"));
this.students.setText("学生数: " + sh.getInt("students ", 0)); }}
```

实例 8-2：使用 SharedPreferences 读取数据实例

新建一个项目，项目的命名为：exam8_2，包名称为：org.hnist.demo，使用 SharedPreferences 读取指定文件。

1）修改布局管理文件 activity_main.xml，代码如下：

```xml
<?xml version="1.0" encoding="utf-8"?>
<LinearLayout
xmlns:android="http://schemas.android.com/apk/res/android"
android:orientation="vertical"
android:layout_width="fill_parent"
android:layout_height="fill_parent">
<TextView
    android:id="@+id/ college "
    android:layout_width="fill_parent"
    android:layout_height="wrap_content"/>
<TextView
    android:id="@+id/ students "
    android:layout_width="fill_parent"
    android:layout_height="wrap_content"/>
</LinearLayout>
```

2）修改 Activity 文件 MainActivity.java，代码如下：

```
package org.hnist.demo;
import android.app.Activity;
import android.content.SharedPreferences;
import android.os.Bundle;
import android.widget.TextView;
public class LoadData extends Activity {
private static final String FILENAME = "hnist";
private TextView college = null ;
private TextView students = null ;
@Override
public void onCreate(Bundle savedInstanceState) {
    super.onCreate(savedInstanceState);  )
    setContentView(R.layout.main);
    this. college = (TextView) super.findViewById(R.id.college) ;
this. students = (TextView) super.findViewById(R.id.students) ;
    //指定操作的文件名称
SharedPreferences sh = super.getSharedPreferences
                (FILENAME,Activity.MODE_PRIVATE);
this. college.setText("学校: " + sh.getString("college ", "没有学校信息。"));
this. students.setText("学生数: " + sh.getInt("students ", 0)); }}
```

保存文件，运行该项目，结果如图 8.8 所示。

SharedPreferences 对象实现数据的存储和读取的数据比较简单，只能是 boolean，int，float，long 和 String 五种简单的数据类型，如果想存储更多类型的数据，则可以使用文件的存储操作。

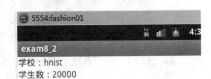

图 8.8　读取指定文件数据

8.2　使用文件存储数据

Android 系统基于 Java 语言，在 Java 语言中提供了一套完整的输入/输出流操作体系，与文件有关的 FileInputStream、FileOutputStream 等，通过这些类可以方便地访问磁盘上的文件，Android 也支持这种方式来访问手机上的文件。Activity 类对文件操作的常用方法见表 8-4。

表 8-4　Activity 类对文件操作的常用方法

方法	描述
public FileInputStream openFileInput(String name)	设置要打开的文件输入流
public FileOutputStream openFileOutput(String name, int mode)	设置要打开文件的输出流，指定操作的模式
public Resources getResources()	返回 Resources 对象

其中，String name 用于指定文件名称，不能包含路径分隔符"/"，如果文件不存在，Android 会自动创建它。int mode 用于指定操作模式，有四种模式。

Context.MODE_PRIVATE=0：为默认操作模式，代表该文件是私有数据，只能被应用本身访问，在该模式下，写入的内容会覆盖原文件的内容；

Context.MODE_APPEND=32768：模式会检查文件是否存在，存在就往文件追加内容，否则就创建新文件再写入内容；

Context.MODE_WORLD_READABLE =1：表示当前文件可以被其他应用读取；

Context.MODE_WORLD_WRITEABLE =2：表示当前文件可以被其他应用写入。

如果希望文件被其他应用读和写,可以传入:Context.MODE_WORLD_READABLE + Context.MODE_WORLD_WRITEABLE,或者直接传入数值 3 也可以。这 4 种模式除了 Context.MODE_APPEND 外,其他都会覆盖掉原文件的内容。

应用程序的数据文件默认保存在/data/data/<包名称>/files 目录下,文件的后缀名随意。

Android 手机中的文件有两个存储位置:内置存储空间和外部 SD 卡,相应的存储方式稍有不同,在后面的章节会有提到。

8.2.1 手机内存中的文件存储和读取

创建及写文件的步骤如下:

1)调用 OpenFileOutput()方法,传入文件的名称和操作的模式,该方法将返回一个文件输出流,例如:

```
FileOutputStream output = null ;              //定义文件输出流对象
output = super.openFileOutput("my.txt", Activity.MODE_PRIVATE);
                                              //设置输出的文件名称,及文件创建模式
```

2)调用 write()或 println()方法,向该文件输出流写入数据,例如:

```
PrintStream out = new PrintStream(output) ;   //封装打印流
out.println("学院:信息学院,");                 //输出数据
out.println("专业:信息工程,") ;               //输出数据
out.println("班级:信工 12-2BF。") ;            //输出数据
```

3)调用 Close()方法,关闭文件输出流,例如:

```
out.close() ;                                 //关闭输出流
```

实例 8-3:创建文件实例

新建一个项目,项目的命名为:exam8_3,包名称为:org.hnist.demo,利用 Activity 类来创建文件。

1)修改 Activity 文件 MainActivity.java,代码如下:

```
package org.hnist.demo;
import java.io.FileNotFoundException;
import java.io.FileOutputStream;
import java.io.PrintStream;
import android.app.Activity;
import android.os.Bundle;
public class MainActivity extends Activity {
    @Override
    public void onCreate(Bundle savedInstanceState) {
        super.onCreate(savedInstanceState);
        super.setContentView(R.layout.activity_main);
        FileOutputStream output = null ;              //定义文件输出流
        try {    //设置输出的文件名称,及文件创建模式
            output = super.openFileOutput("my.txt",Activity.MODE_PRIVATE);
        } catch (FileNotFoundException e) {
            e.printStackTrace();            }
        PrintStream out = new PrintStream(output) ;   //打印流包装
        out.println("学院:信息学院,");                 //输出数据
        out.println("专业:信息工程,") ;               //输出数据
        out.println("班级:信工 12-2BF。") ;            //输出数据
        out.close() ;      }}                         //关闭输出流
```

2）直接打开 DDMS 视图，刚才建立的文件保存在"/data/data/org.hnist.filedemo/files/"文件夹之中，如图 8.9 所示，导出这个文件，其内容如图 8.10 所示。

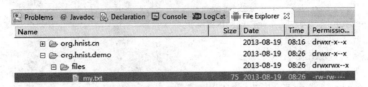

图 8.9 存储文件的位置

图 8.10 记事本打开文件的内容

读取文件的步骤：

1）定义 TextView 对象，用于显示读取的文件内容，例如：

```
private TextView fileread = null ;
```

2）调用 OpenFileInput()方法，传入需要读取数据的文件名，该方法将会返回一个文件输入流对象，例如：

```
FileInputStream input = null;                //定义文件输入流对象
input = super.openFileInput("my.txt");       //获得指定的文件"my.txt"
```

3）调用 read()方法或 scannext()方法读取文件的内容，例如：

```
Scanner scan = new Scanner(input) ;              //定义 Scanner
while(scan.hasNext()){                           //循环读取
this. fileread.append(scan.next() + "\n") ; }//设置文本
```

4）调用 Close()方法，关闭文件输入流。

```
scan.close() ;                                   //关闭输入流
```

实例 8-4：读取文件实例

新建一个项目，项目的命名为：exam8_4，包名称为：org.hnist.demo，利用 Activity 类来读取指定文件内容。

1）修改布局管理文件 activity_main.xml，代码如下：

```
<?xml version="1.0" encoding="utf-8"?>
<LinearLayout
xmlns:android="http://schemas.android.com/apk/res/android"
android:orientation="vertical"
android:layout_width="fill_parent"
android:layout_height="fill_parent">
<TextView
    android:id="@+id/msg"
    android:layout_width="fill_parent"
    android:layout_height="wrap_content" />
</LinearLayout>
```

2）修改 Activity 文件 MainActivity.java，代码如下：

```java
package org.hnist.demo;
import java.io.FileInputStream;
import java.io.FileNotFoundException;
import java.util.Scanner;
import android.app.Activity;
import android.os.Bundle;
import android.widget.TextView;
public class MainActivity extends Activity {
private TextView msg = null ;                           //文本组件
@Override
public void onCreate(Bundle savedInstanceState) {
    super.onCreate(savedInstanceState);
    super.setContentView(R.layout.activity_main);       //调用布局文件
    this.msg = (TextView) super.findViewById(R.id.msg) ;
       FileInputStream input = null;                    //文件输入流
    try {    //找到指定文件的输入流对象
        input = super.openFileInput("my.txt");
    } catch (FileNotFoundException e) {
       e.printStackTrace();}
       Scanner scan = new Scanner(input) ;              //定义 Scanner
       while(scan.hasNext()){                           //循环读取
       this.msg.append(scan.next() + "\n") ;    }       //设置文本
       scan.close() ;    }}                             //关闭输入流
```

保存文件，运行该项目，结果如图 8.11 所示。

8.2.2 SD 卡中的文件存储和读取

手机内存相对较小，其空间的大小会影响到手机的运行速度，不建议将一些大数据保存到手机内存中，通常建议将这些资源存放在外存设备上，最常见的就是 SD 卡。

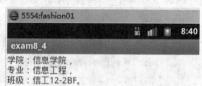

图 8.11 读取指定文件

在模拟器中使用 SD 卡，需要先创建一张 SD 卡（只是镜像文件）。SD 卡可以在 Eclipse 创建模拟器时随同创建。在 Android 中并没有提供单独的 SD 卡文件操作类，直接使用 Java 中的文件操作即可，关键是如何确定文件的位置。

在访问 SD 卡的文件之前，需要验证 SD 卡已被正确安装，如果没有安装 SD 卡可能会出现错误。因此要先判断 SD 卡是否存在，可以通过 android.os.Environment 类取得目录的信息来实现，具体常量及方法见表 8-5。一般情况下，文件保存在 SD 卡的 mnt\sdcard 文件夹下，当然也可以存放在指定文件夹中。

表 8-5　Environment 定义的常量及方法

常量及方法	描述
public static final String MEDIA_MOUNTED	SD 卡允许进行读/写访问
public static final String MEDIA_CHECKING	SD 卡处于检查状态
public static final String MEDIA_MOUNTED_READ_ONLY	SD 卡处于只读状态
public static final String MEDIA_REMOVED	SD 卡不存在
public static final String MEDIA_UNMOUNTED	没有找到 SD 卡
public static File getDataDirectory()	取得 Data 目录

第 8 章 Android 数据存储技术

续表

常量及方法	描述
public static File getDownloadCacheDirectory()	取得下载的缓存目录
public static File getExternalStorageDirectory()	取得扩展的存储目录
public static String getExternalStorageState()	取得 SD 卡的状态
public static File getRootDirectory()	取得 Root 目录
public static boolean isExternalStorageRemovable()	判断扩展的存储目录是否被删除

另外，File 还有下面一些常用的操作，例如：

```
String Name = File.getName();              //获得文件或文件夹的名称：
String parentPath = File.getParent();      //获得文件或文件夹的父目录
String path = File.getAbsoultePath();      //绝对路经
String path = File.getPath();              //相对路经
File.createNewFile();                      //建立文件
File.mkDir();                              //建立文件夹
File.isDirectory();                        //判断是文件或文件夹
File[] files = File.listFiles();           //列出文件夹下的所有文件和文件夹名
File.renameTo(dest);                       //修改文件夹和文件名
File.delete();                             //删除文件夹或文件
```

读、写 SD 卡上的文件步骤如下：

1）调用 Environment 的 getExternalStorageState()方法判断手机上是否插入了 SD 卡，并且应用程序具有读写 SD 卡的权限。Environment.getExternalStorageState()方法用于获取 SD 卡的状态，如果手机装有 SD 卡，并且可以进行读写，那么方法返回的状态等于 Environment.MEDIA_MOUNTED；例如：

```
if(Environment.getExternalStorageState().equals( Environment.MEDIA_MOUNTED))
```

2）调用 Environment 的 getExternalStorageDirectory()方法来获取外部存储器，也就是 SD 卡的目录（这里的目录是 mnt\sdcard\hnist，文件是 hnist.txt）；例如：

```
File file = new File(Environment.getExternalStorageDirectory().toString()
        + File.separator + hnist + File.separator + hnist.txt) ;
        //定义File 类对象
```

3）使用 FileInputStream、FileOutputStream、FileReader、FileWriter 等类读、写 SD 卡里的文件；例如：

```
PrintStream out = null ;                   //打印流对象用于输出
out = new PrintStream(new FileOutputStream(file, true));//追加文件
out.println("湖南理工学院信息学院信工 12-2BF");
```

4）调用 Close()方法，关闭文件输入流。

```
out.close() ;                              //关闭输入流
```

5）为了读、写 SD 卡上的数据，必须在应用程序的清单文件（AndroidManifest.xml）中添加读、写 SD 卡的权限。

```
<!-- 在 SD 卡中创建与删除文件权限 -->
<uses-permission android:name="android.permission.MOUNT_UNMOUNT_FILESYSTEMS"/>
<!-- 往 SD 卡写入数据权限 -->
<uses-permission android:name="android.permission.WRITE_EXTERNAL_STORAGE"/>
```

实例 8-5：往 SD 上写入文件

新建一个项目，项目的命名为：exam8_5，包名称为：org.hnist.demo，在 SD 卡上建立文件并写入信息。

1）修改 Activity 文件 MainActivity.java，代码如下：

```java
package org.hnist.demo;
import java.io.File;
import java.io.FileOutputStream;
import java.io.PrintStream;
import android.app.Activity;
import android.os.Bundle;
import android.os.Environment;
import android.widget.Toast;
public class MainActivity extends Activity {
@Override
public void onCreate(Bundle savedInstanceState) {
    super.onCreate(savedInstanceState);
    super.setContentView(R.layout.activity_main);
    if(Environment.getExternalStorageState().equals(
        Environment.MEDIA_MOUNTED)){         //如果 sdcard 存在
        File file = new File(Environment
            .getExternalStorageDirectory().toString()
            + File.separator
            + "hnist" + File.separator + "hnist.txt") ;//定义File类对象
        if (! file.getParentFile().exists()) {        //父文件夹不存在
            file.getParentFile().mkdirs() ; }         //创建文件夹
        PrintStream out = null ;                      //打印流对象用于输出
        try {
            out = new PrintStream(new FileOutputStream(file, true));
                                                      //追加文件
           out.println("湖南理工学院,信息学院,信工 12-2BF");
        } catch (Exception e) { e.printStackTrace();}
finally {
            if (out != null) {
                out.close() ;    }} }                  //关闭打印流
    else {   //SDCard 不存在,提示用户
        Toast.makeText(this, " SD 卡不存在！", Toast.LENGTH_LONG).show();}}}
```

2）修改 AndroidManifest.xml 文件，添加读、写 SD 卡的权限。

```xml
<uses-permission android:name="android.permission.MOUNT_UNMOUNT_FILESYSTEMS"/>
<uses-permission android:name="android.permission.WRITE_EXTERNAL_STORAGE"/>
```

保存文件，运行，发现在 mnt\sdcard\下建立了 hnist 文件夹和 hnist.txt 文件，如图 8.12 所示，再运行两次，将文件导出，打开文件发现是按照追加方式建立文件的，如图 8.13 所示。

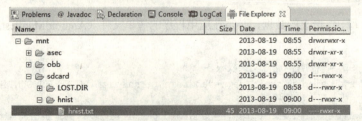

图 8.12　往 SD 卡写文件

第 8 章 Android 数据存储技术

图 8.13 读取指定文件

实例 8-6：从 SD 卡上读取文件

新建一个项目，项目的命名为：exam8_6，包名称为：org.hnist.demo，从 SD 卡上读取指定文件 hnist.txt 的内容。

1）修改布局管理文件 activity_main.xml，代码如下：

```xml
<?xml version="1.0" encoding="utf-8"?>
<LinearLayout
xmlns:android="http://schemas.android.com/apk/res/android"
android:orientation="vertical"
android:layout_width="fill_parent"
android:layout_height="fill_parent">
<TextView
    android:id="@+id/mytxt"
    android:layout_width="fill_parent"
    android:layout_height="wrap_content" />
</LinearLayout>
```

2）修改 Activity 文件 MainActivity.java，代码如下：

```java
package org.hnist.demo;
import java.io.File;
import java.io.FileInputStream;
import java.util.Scanner;
import android.app.Activity;
import android.os.Bundle;
import android.os.Environment;
import android.widget.TextView;
import android.widget.Toast;
public class FileOperate extends Activity {
private TextView msg = null ;
@Override
public void onCreate(Bundle savedInstanceState) {
    super.onCreate(savedInstanceState);
    super.setContentView(R.layout. activity_main);
    this.msg = (TextView) super.findViewById(R.id.mytxt) ;
    if(Environment.getExternalStorageState().equals(
        Environment.MEDIA_MOUNTED)){         //如果 sdcard 存在
        File file = new File(Environment
            .getExternalStorageDirectory().toString()
            + File.separator
            + "hnist"+ File.separator + "hnist.txt") ; //定义 File 类对象
        if (! file.getParentFile().exists()) {   //如果父文件夹不存在
            file.getParentFile().mkdirs() ; }    //创建文件夹
        Scanner scan = null ;                    //扫描输入
        try {
```

```
                    scan = new Scanner(new FileInputStream(file)) ;//实例化 Scanner
                    while(scan.hasNext()){                          //循环读取
                        this.msg.append(scan.next() + "\n") ; }     //设置文本
                } catch (Exception e) { e.printStackTrace();}
                finally {
                    if (scan != null) {
                        scan.close() ;           }}}               //关闭打印流
        else {    //SDCard 不存在,提示用户
             Toast.makeText(this, " SD 卡不存在! ", Toast.LENGTH_LONG).show();}}}
```

运行结果如图 8.14 所示。

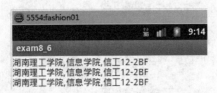

图 8.14 读取指定文件到手机

8.2.3 读取资源文件

在 Android 中,还可以对资源文件进行读取(注意不能写入),这些资源文件的 ID 都会自动通过 R.java 这个类生成,如果要对这些文件读取,使用 android.content.res.Resources 类即可完成,常见的资源文件有两种,使用两种不同的方式进行打开使用。

Raw 中的资源文件使用 getResources().openRawResource(int id)读取,asset 中的资源文件使用 getResources(). getAssets().open(fileName)读取。

实例 8-7:读取 asset 中的资源文件

新建一个项目,项目的命名为:exam8_7,包名称为:org.hnist.demo, 从 resource 的 asset 中读取文件数据。

1)建立一个文件,例如:my.txt,复制到项目的 asset 文件夹中。
2)修改布局管理文件 activity_main.xml,代码与实例 8-6 的 activity_main.xml 一致。
3)修改 Activity 文件 MainActivity.java,代码如下:

```
package org.hnist.demo;
import java.io.IOException;
import java.io.InputStream;
import android.app.Activity;
import android.content.res.Resources;
import android.os.Bundle;
import android.widget.TextView;
import org.apache.http.util.EncodingUtils;
public class RawResourceDemo extends Activity {
private TextView msg = null;
@Override
public void onCreate(Bundle savedInstanceState) {
    super.onCreate(savedInstanceState);
    super.setContentView(R.layout.activity_main);
```

```
            this.msg = (TextView) super.findViewById(R.id.mytxt);
            Resources res = super.getResources();           //定义操作资源
              String resa = "";
            try{//得到资源中的 Raw 数据流
                InputStream in = getResources().getAssets().open("my.txt");
                                                            //读取资源文件
                int length = in.available();                //得到文件的大小
                byte [] buffer = new byte[length];
                in.read(buffer);                            //读取数据
                resa = EncodingUtils.getString(buffer, "UTF-8");  //选择合适的编码
                in.close();                                 //关闭
            }catch(Exception e){
               e.printStackTrace();    }
            this.msg.setText(resa); }}                      //设置显示的文字
```

保存所有文件，运行该项目，结果如图 8.15 所示。

注意：选择合适的编码很重要，因为可能出现乱码，读者可以把"UTF-8"改为"BIG-5"试试。这里也可以用 scan.next()方法来获取文件数据，可参考前面的代码自行完成。

图 8.15　读取 asset 中的文件

实例 8-8：读取 raw 中的资源文件

新建一个项目，项目的命名为：exam8_8，包名称为：org.hnist.demo，从 resource 的 raw 中读取文件数据。

1）建立一个文件，例如：my.txt，复制到项目的 res\raw 文件夹中。
2）修改布局管理文件 activity_main.xml，代码与实例 8-6 的 activity_main.xml 一致。
3）修改 Activity 文件 MainActivity.java，代码与实例 8-7 的 MainActivity.java 类似，只需要将语句：
InputStream in = getResources().getAssets().open("my.txt");

修改为：InputStream in = res.openRawResource(R.raw.my);

其他代码不变，运行结果与图 8.15 一致。

8.3　使用数据库存储数据

对于大量的数据处理，前面介绍的方法就显得力不从心了，在 Android 平台上，集成了一个嵌入式关系型数据库 SQLite，SQLite 支持 SQL 语言，可以方便地实现数据增加、修改、删除、查询等操作，支持常见的五种数据类型。

SQLite 数据库是 D.Richard Hipp 用 C 语言编写的开源嵌入式数据库，支持的数据库大小为 2TB，具有如下特征：

1. 轻量级

SQLite 和 C/S 模式的数据库软件不同，它不存在数据库的客户端和服务器。使用 SQLite 一般只需要带上它的一个动态库，就可以享受它的全部功能。而且动态库的文件很小。

2. 独立性

SQLite 数据库的核心引擎本身不依赖第三方软件，使用它也不需要"安装"，所以在使用的时候能够省去不少麻烦。

3. 隔离性

SQLite 数据库中的所有信息（比如表、视图、触发器）都包含在一个文件内，方便管理和维护。

4. 跨平台

SQLite 数据库支持大部分操作系统，除了我们在电脑上使用的操作系统之外，很多手机操作系统同样可以运行，比如 Android、Windows Mobile、Symbian、Palm 等。

5. 多语言接口

SQLite 数据库支持很多语言编程接口，比如 C/C++、Java、Python、dotNet、Ruby、Perl 等。

6. 安全性

SQLite 数据库通过数据库级上的独占性和共享锁来实现独立事务处理，多个进程可以在同一时间从同一数据库读取数据，但只有一个可以写入数据。在某个进程或线程向数据库执行写操作之前，必须获得独占锁定。

Android 集成了 SQLite 数据库，所以每个 Android 应用程序都可以使用 SQLite 数据库，在 SDK 中的 samples/NotePad 下可以找到关于如何使用数据库的例子。

在 Android 系统中，如果要进行 SQLite 数据库的操作，主要使用 Android 系统提供的下面几个类或接口：

（1）android.database.sqlite.SQLiteDatabase，完成数据的增、删、修改、查询操作；

（2）android.database.sqlite.SQLiteOpenHelper，完成数据库的创建及更新操作；

（3）android.database.Cursor，保存所有的查询结果；

（4）android.database.ContentValues，对传递的数据进行封装。

下面简要介绍这几个类。

1．android.database.sqlite.SQLiteDatabase 类简介

在 Android 系统中，通过 android.database.sqlite.SQLiteDatabase 类可以执行 SQL 语句，以完成对数据表的增加、修改、删除、查询等操作，在此类之中定义了基本的数据库执行 SQL 语句的操作方法以及一些操作的模式常量。SQLiteDatabase 类定义的常用操作方法见表 8-6。

表 8-6 SQLiteDatabase 类定义的常用操作方法

常量或方法	描述
public static final int OPEN_READONLY	常量，以只读方式打开数据库
public static final int OPEN_READWRITE	常量，以读/写方式打开数据库
public static final int CREATE_IF_NECESSARY	常量，如果指定的数据库文件不存在，则创建新的数据库文件
public static final int NO_LOCALIZED_COLLATORS	常量，打开数据库时，不对数据进行基于本地化语言的排序
public void beginTransaction()	开始事务
public void endTransaction()	结束事务，提交或者是回滚数据
public void close()	关闭数据库
public void execSQL(String sql)	执行 SQL 语句
public void execSQL(String sql, Object[] bindArgs)	执行 SQL 语句，同时绑定参数
public static SQLiteDatabase openDatabase(String path, SQLiteDatabase.CursorFactory factory, int flags)	以指定的模式打开指定路径下的数据库文件
public static SQLiteDatabase openOrCreateDatabase(File file, SQLiteDatabase.CursorFactory factory)	打开或者是创建一个指定路径下的数据库

续表

常量或方法	描述
public static SQLiteDatabase openOrCreateDatabase(String path, SQLiteDatabase.CursorFactory factory)	打开或者是创建一个指定路径下的数据库
public long insert(String table, String nullColumnHack, ContentValues values)	插入数据，table 为表名称，nullColumnHack 表示传入的 valuesnull，则列被设为 null，values 表示所有要插入的数据
public long insertOrThrow(String table, String nullColumnHack, ContentValues values)	插入数据，但会抛出 SQLException 异常
public int update(String table, ContentValues values, String whereClause, String[] whereArgs)	修改数据，table 为表名称，values 为更新数据，whereCluause 指明 WHERE 子句，whereArgs 为 WHERE 子句参数，用于替换 "？"
public int delete(String table, String whereClause, String[] whereArgs)	删除数据，table 为表名称，whereClause 指明 WHERE 子句，whereArgs 为参数，用于替换 "？"
public boolean isOpen()	判断数据库是否是已打开
public void setVersion(int version)	设置数据库的版本
public Cursor query (boolean distinct, String table, String[] columns, String selection, String[] selectionArgs, String groupBy, String having, String orderBy, String limit)	执行数据表查询操作，其中的参数有 distinct（是否去掉重复行）、table（表名称）、columns（列名称）、selection（WHERE 子句）、selectionArgs（WHERE 条件）、groupBy（分组）、having（分组过滤）、orderBy（排序）、limit（LIMIT 子句）
public Cursor query(String table, String[] columns, String selection, String[] selectionArgs, String groupBy, String having, String orderBy)	执行数据表查询操作
public Cursor rawQuery(String sql, String[] selectionArgs)	执行指定的 SQL 查询语句

2. android.database.sqlite.SQLiteOpenHelper 类简介

SQLiteDatabase 类本身只是一个数据库的操作类，但是如果要想进行数据库的操作，还需要一个 android.database.sqlite.SQLiteOpenHelper 数据库操作辅助类才可以进行，SQLiteOpenHelper 类是一个抽象类，它的常用方法如表 8-7 所示。使用的时候需要定义其子类，并且在子类中要覆写相应的抽象方法。

表 8-7 SQLiteOpenHelper 类定义的方法

方法	描述
public SQLiteOpenHelper(Context context, String name, SQLiteDatabase.CursorFactory factory, int version)	通过此构造方法指明要操作的数据库名称以及数据库的版本编号
public synchronized void close()	关闭数据库
public synchronized SQLiteDatabase getReadableDatabase()	以只读的方式创建或者打开数据库
public synchronized SQLiteDatabase getWritableDatabase()	以修改的方式创建或者打开数据库
public abstract void onCreate(SQLiteDatabase db)	创建数据表
public void onOpen(SQLiteDatabase db)	打开数据表
public abstract void onUpgrade(SQLiteDatabase db, int oldVersion, int newVersion)	更新数据表

3. android.database.Cursor 接口简介

数据库的操作除了上述操作之外，还有一项数据的查询操作，当 Android 程序需要进行数据查询操作的时候需要保存全部的查询结果，而保存查询结果就可以使用 android.database.Cursor 接口完成，Cursor 接口的常用方法如表 8-8 所示。

表 8-8 Cursor 接口的常用方法

方法	描述
public abstract void close()	关闭查询
public abstract int getCount()	返回查询的数据量

续表

方法	描述
public abstract int getColumnCount()	返回查询结果之中列的总数
public abstract String[] getColumnNames()	得到查询结果之中全部列的名称
public abstract String getColumnName(int columnIndex)	得到指定索引位置列的名称
public abstract boolean isAfterLast()	判断结果集指针是否在最后一行数据之后
public abstract boolean isBeforeFirst()	判断结果集指针是否在第一行记录之前
public abstract boolean isClosed()	判断结果集是否已关闭
public abstract boolean isFirst()	判断结果集指针是否指在第一行
public abstract boolean isLast()	判断结果集指针是否指在最后一行
public abstract boolean moveToFirst()	将结果集指针移到第一行
public abstract boolean moveToLast()	将结果集指针移动到最后一行
public abstract boolean moveToNext()	将结果集指针向下移动一行
public abstract boolean moveToPrevious()	将结果集指针向前移动一行
public abstract boolean requery()	更新数据后刷新结果集中的内容
public abstract int getXxx(int columnIndex)	根据指定列的索引取得指定的数据

4．android.database.ContentValues 类简介

在 SQLiteDatabase 类之中专门提供了增加、删除、修改、查询等方法，在使用这些方法进行数据库操作的时候，所有的数据必须使用 android.database.ContentValues 类进行封装，ContentValues 类常用的方法如表 8-9 所示。

表 8-9　ContentValues 类常用的方法

方法	描述
public ContentValues()	创建 ContentValues 类实例
public void clear()	清空全部的数据
public void put(String key，包装类 value)	设置指定字段（key）数据
public Integer getAs 包装类(String key)	根据 key 取得数据，例如：getInteger()
public int size()	返回保存数据的个数

8.3.1　创建数据库及表

1．创建数据库

在 Android 应用程序中使用 SQLite，必须自己创建数据库，然后创建表、索引，填充数据。Android 提供了 SQLiteOpenHelper 帮助创建数据库，只要继承 SQLiteOpenHelper 类，就可以创建数据库了，在 SQLiteOpenHelper 的子类中，需要实现下面几个方法：

1）构造函数，调用父类 SQLiteOpenHelper 的构造函数，例如：

```
DatabaseHelper(Context context, String name, CursorFactory cursorFactory, int
            version) { super(context, name, cursorFactory, version); }
```

这个方法需要四个参数：上下文环境（例如，一个 Activity），数据库名字，一个可选的游标工厂（通常是 Null），一个正在使用的数据库模型版本的整数。

2）onCreate()方法，生成相应的数据表，它需要一个 SQLiteDatabase 对象作为参数，根据需要对这个对象填充表和初始化数据，例如：

```
public void onCreate(SQLiteDatabase db) {…} //创建数据库后，对数据库的操作
```

3）onUpgrage()方法，当数据库版本要升级时会调用此方法，例如：

```
public void onUpgrade(SQLiteDatabase db, int oldVersion, int newVersion) { …}
```

它需要三个参数，一个 SQLiteDatabase 对象，一个旧的版本号和一个新的版本号，一般可以在此方法中将数据表删除。

4）onopen()方法，当数据库打开时会调用这个方法，例如：

```
public void onOpen(SQLiteDatabase db) {super.onOpen(db); }}
            //每次成功打开数据库后首先被执行
```

2．创建数据表

调用 getReadableDatabase()或 getWriteableDatabase()方法，可得到 SQLiteDatabase 实例，具体调用哪个方法，取决于是否需要改变数据库的内容，例如：

```
db=(new DatabaseHelper(getContext())).getWritableDatabase();
return (db == null) ? false : true;
```

上面这段代码会返回一个 SQLiteDatabase 类的实例，使用这个对象就可以查询或者修改数据库。当完成了对数据库的操作（例如 Activity 已经关闭），需要调用 SQLiteDatabase 的 Close()方法来释放掉数据库连接。

创建表和索引

为了创建表和索引，需要调用 SQLiteDatabase 的 execSQL()方法来执行相应的语句，例如：创建一个名为 mytable 的表，表有一个列名为_id，并且是主键，这列的值会自动增长，另外还有两列：title(字符)和 value(浮点数)，代码如下：

```
db.execSQL("CREATE TABLE mytable(_id INTEGER PRIMARY KEY AUTOINCREMENT, title
            TEXT, value REAL);");
```

通常情况下，SQLite 会自动为主键列创建索引。

实例 8-9：创建数据库及表实例

新建一个项目，项目的命名为：exam8_9，包名称为：org.hnist.demo，创建一个数据库及表。

1）建立一个类继承 SQLiteOpenHelper 类，右键单击该项目的包，选择 New→Class，输入类的名称，例如：DataBaseHelp.java。在其中定义数据库的名称、表名称以及表中各字段的名称、类型、长度等，具体代码如下：

```
package org.hnist.demo;
import android.content.Context;
import android.database.sqlite.SQLiteDatabase;
import android.database.sqlite.SQLiteOpenHelper;
//定义 DataBaseHelp 继承 SQLiteOpenHelper 类
public class DataBaseHelp extends SQLiteOpenHelper {
private static final String DATABASENAME = "hnist.db" ;    //定义数据库名称
private static final int DATABASEVERSION = 1 ;             //定义数据库版本
private static final String TABLENAME = "personinfo" ;     //定义数据表名称
public DataBaseHelp(Context context) {                     //定义构造
   super(context, DATABASENAME, null, DATABASEVERSION); }  //调用父类构造
public void onCreate(SQLiteDatabase db) {                  //创建数据表
    String sql = "CREATE TABLE " + TABLENAME + " (" +
```

```
            "id         INTEGER       PRIMARY KEY ," +        //设置为自动增长列
            "name       VARCHAR(50)   NOT NULL ," +
            "sex        VARCHAR(10)   NOT NULL ," +
            "DateofBorth   DATE       NOT NULL ," +
            "email      VARCHAR(50)   NOT NULL)";             //定义SQL语句
    db.execSQL(sql) ;    }                                    //执行SQL语句
    public void onUpgrade(SQLiteDatabase db, int oldVersion, int newVersion) {
        String sql = "DROP TABLE IF EXISTS " + TABLENAME ;    //SQL语句
        db.execSQL(sql);                                      //执行SQL语句
        this.onCreate(db); }}                                 //创建表
```

2）修改 Activity 文件 MainActivity.java，代码如下：

```
package org.hnist.demo;
import android.app.Activity;
import android.database.sqlite.SQLiteOpenHelper;
import android.os.Bundle;
public class MainActivity extends Activity {
    public void onCreate(Bundle savedInstanceState) {
        super.onCreate(savedInstanceState);
        super.setContentView(R.layout.activity_main);
        SQLiteOpenHelper helper = new DataBaseHelp(this) ;  //定义数据库辅助类
        helper.getWritableDatabase() ; }}                   //以修改方式打开数据库
```

3）保存所有文件，运行该项目，会创建数据库 hnist.db，并建立含有指定字段的数据表 personinfo，这些信息存储在 data/data/org.hnist.demo/databases/文件夹下，通过 DDMS 视图中的 File Explorer 选项可以查看，如图 8.16 所示。

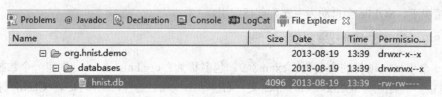

图 8.16 建立的数据库文件

4）将 hnist.db 导出，可以使用 adb shell 命令来查看，也可以利用 SQLite Manager 工具来查看。下载安装 SQLite Manager 后运行，选择"File→Open"，找到导出的 hnist.db 文件，选择其中的 Table 选项，可以看到表中的字段信息，如图 8.17 所示。

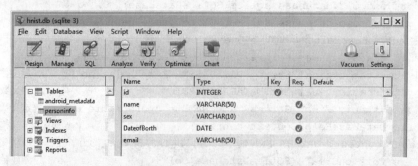

图 8.17 新建数据表的结构

8.3.2 操作数据库

数据库以及表建立好了，表中还没有记录，怎样往其中添加记录、对记录做一些简单的操作呢？这些可以借助 SQL 语句来实现。

插入语句：insert into 表名(字段列表) values(值列表)。如：

```
insert into personinfo(name, sex) values("张小君","男")
```

更新语句：update 表名 set 字段名=值 where 条件子句。如：

```
update personinfo set name='张小君' where id=10
```

删除语句：delete from 表名 where 条件子句。如：

```
delete from personinfo where id=10
```

查询语句：select * from 表名 where 条件子句 group by 分组字句 having ... order by 排序子句。如：

```
select * from personinfo                //获取表中的所有记录
select * from Account limit 5 offset 2   //获取5条记录，跳过前面2条记录
```

使用 SQLiteDatabase 进行数据库操作的步骤如下：

1）获取 SQLiteDatabase 对象，它代表了与数据库的连接；

```
MainActivity.this.helper.getWritableDatabase(); //取得可写的数据库
```

2）定义要执行的 SQL 语句；例如：

```
String sql = "INSERT INTO " + TABLENAME + " (name,sex,DateofBorth,email) VALUES ('"+ name + "','" + sex + "','" + DateofBorth + "','" + email + "')";
```

3）执行定义好的 SQL 语句；例如：

```
this.db.execSQL(sql);           //执行 SQL 语句
```

4）关闭 SQLiteDatabase，回收资源。例如：

```
this.db.close() ;
```

实例 8-10：操作数据库实例

新建一个项目，项目的命名为：exam8_10，包名称为：org.hnist.demo，往数据表中添加记录，修改、删除记录，对数据库进行简单操作。

1）建立布局管理文件 activity_main.xml，定义三个按钮，代码如下：

```xml
<?xml version="1.0" encoding="utf-8"?>
<LinearLayout
xmlns:android="http://schemas.android.com/apk/res/android"
android:orientation="vertical"
android:layout_width="fill_parent"
android:layout_height="fill_parent">
<Button
    android:id="@+id/insert"
    android:layout_width="fill_parent"
    android:layout_height="wrap_content"
    android:text="增加数据" />
<Button
    android:id="@+id/update"
    android:layout_width="fill_parent"
```

```xml
        android:layout_height="wrap_content"
        android:text="修改数据" />
    <Button
        android:id="@+id/delete"
        android:layout_width="fill_parent"
        android:layout_height="wrap_content"
        android:text="删除数据" />
</LinearLayout>
```

2）建立一个类 DataBaseHelp.java 继承 SQLiteOpenHelper 类，定义数据库的名称、表名称以及表的结构，具体代码与实例 8-9 的 DataBaseHelp.java 一致。

3）建立一个表的操作类 OperateTable.java，进行表中数据的添加、更新和删除操作，具体代码如下：

```java
package org.hnist.demo;
import android.database.sqlite.SQLiteDatabase;
public class OperateTable {
    private static final String TABLENAME = "personinfo" ;        //表名称
    private SQLiteDatabase db = null ;                            //SQLiteDatabase
    public OperateTable(SQLiteDatabase db) {                      //构造方法
        this.db = db ;    }
    public void insert(String name, String sex,String DateofBorth,String email) {
        String sql = "INSERT INTO " + TABLENAME + " (name,sex,DateofBorth,email)
            VALUES ('"+ name + "','" + sex + "','" + DateofBorth + "',
            '" + email + "')";                                    //SQL 语句
        this.db.execSQL(sql);                                     //执行 SQL 语句
        this.db.close() ;    }                                    //关闭数据库操作
    public void update(int id, String name, String sex,String DateofBorth,String
        email) {String sql = "UPDATE " + TABLENAME + " SET name='" + name+ "',sex=
        '" + sex + "',email='" + email   + "',DateofBorth='" + DateofBorth + "'
        WHERE id=" + id;                                          //SQL 语句
        this.db.execSQL(sql);                                     //执行 SQL 语句
        this.db.close() ;              }                          //关闭数据库操作
    public void delete(int id) {
    String sql = "DELETE FROM " + TABLENAME + " WHERE id=" + id;//SQL 语句
        this.db.execSQL(sql) ;                                    //执行 SQL 语句
        this.db.close() ;     }}                                  //关闭数据库操作
```

4）定义 Activity 文件 MainActivity.Java，代码如下：

```java
package org.hnist.demo;
import android.app.Activity;
import android.database.sqlite.SQLiteOpenHelper;
import android.os.Bundle;
import android.view.View;
import android.view.View.OnClickListener;
import android.widget.Button;
public class MainActivity extends Activity {
    private DataBaseHelp helper = null ;            //数据库操作
    private OperateTable mytable = null ;           //mytab 表操作类
    private Button insBut = null ;
```

```java
        private Button updBut = null ;
        private Button delBut = null ;
        private static int count = 0 ;                    //计数统计
          @Override
          public void onCreate(Bundle savedInstanceState) {
              super.onCreate(savedInstanceState);
              super.setContentView(R.layout.activity_main);
              this.insBut = (Button) super.findViewById(R.id.insert);
              this.delBut = (Button) super.findViewById(R.id.delete);
              this.updBut = (Button) super.findViewById(R.id.update);
              this.helper = new DataBaseHelp(this) ;        //定义数据库辅助类
              this.insBut.setOnClickListener(new InsOnClickListenerImpl()) ;
                                                            //设置监听
              this.delBut.setOnClickListener(new DelOnClickListenerImpl());
                                                            //设置监听
              this.updBut.setOnClickListener(new UpnClickListenerImpl()) ; }
                                                            //设置监听
        private class InsOnClickListenerImpl implements OnClickListener {
            public void onClick(View view) {
                MainActivity.this.mytable = new OperateTable (MainActivity.this.
                helper.getWritableDatabase());                //取得可写的数据库
                MainActivity.this.mytable.insert("zhang" + count ++ ,
                "男","1978-10-12","zxj@163.com") ; }}         //添加记录
        private class DelOnClickListenerImpl implements OnClickListener {
            public void onClick(View view) {
                MainActivity.this.mytable = new OperateTable (MainActivity.this.
                    helper.getWritableDatabase());            //取得可写的数据库
                MainActivity.this.mytable.delete(2) ;}}       //删除第二条记录数据
        private class UpnClickListenerImpl implements OnClickListener {
            public void onClick(View view) {
                MainActivity.this.mytable = new OperateTable (MainActivity.this.
                    helper.getWritableDatabase());            //取得可写的数据库
//更新第一条记录数据
MainActivity.this.mytable.update(1, "胡晓莲", "女","1981-06-27",
                                "hxl@163.com") ; }}}
```

保存所有文件，运行项目，弹出如图 8.18 所示界面，先单击三次 "添加记录" 按钮，然后分别单击另外两个按钮，最后导出 hnist.db，用 SQLite Manager 打开表，如图 8.19 所示。

图 8.18 新建数据表的结构

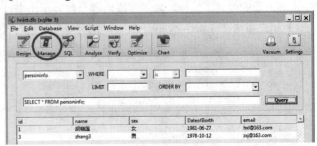

图 8.19 对表做了添加、修改和删除操作后的数据

8.3.3 数据查询操作

数据库的操作除了上述简单的操作之外,还有一个很重要的数据查询操作,当 Android 程序需要进行数据查询操作的时候需要保存全部的查询结果,而保存查询结果可以使用 android.database.Cursor 接口完成。

Cursort 接口没有提供方法直接控制指针移动,如果要想从前到后依次取得全部数据记录就要使用 isAfterLast()、moveToNext()、moveToFirst()三个方法,按照如下步骤进行:

1)使用 moveToFirst()方法将结果集的指针放在第一行数据;
2)使用 isAfterLast()方法判断是否还有数据,如果还有数据,则进行取出;
3)利用 moveToNext()将指针向下移动,并继续使用 isAfterLast()方法判断;

记录集指针控制示意图如图 8.20 所示。

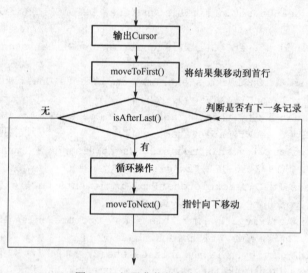

图 8.20 记录集指针控制示意图

用如下代码表示:

```
for (result.moveToFirst(); !result.isAfterLast(); result.moveToNext()){循环体;}
```

其中 result 表示查询到记录的集合。

实例 8-11:数据库查询操作实例

新建一个项目,项目的命名为:exam8_11,包名称为:org.hnist.demo,将 hnist.db 中的 personinfo 表中所有记录的姓名、性别、邮箱字段内容全部显示出来。

1)建立布局文件 activity_main.xml,定义一个按钮,代码如下:

```xml
<?xml version="1.0" encoding="utf-8"?>
<LinearLayout
xmlns:android="http://schemas.android.com/apk/res/android"
android:id="@+id/mylayout"
android:orientation="vertical"
android:layout_width="fill_parent"
android:layout_height="fill_parent">
<Button
```

```
        android:id="@+id/findall"
        android:layout_width="fill_parent"
        android:layout_height="wrap_content"
        android:text="查询全部记录" />
</LinearLayout>
```

2）建立一个类 DataBaseHelp.java 继承 SQLiteOpenHelper 类，定义数据库的名称、表名称以及表的结构，具体代码与实例 8-9 的 DataBaseHelp.java 一致。

3）建立一个表的查询操作类 SearchTable.java，定义查询操作。

```
package org.hnist.demo;
import java.util.ArrayList;
import java.util.List;
import android.database.Cursor;
import android.database.sqlite.SQLiteDatabase;
public class SearchTable {
private static final String TABLENAME = "personinfo" ;      //数据表名称
private SQLiteDatabase db = null ;                          //SQLiteDatabase
public SearchTable(SQLiteDatabase db) {                     //构造方法
    this.db = db ; }                                        //接收 SQLiteDatabase
public List<String> find() {                                //查询数据表
    List<String> all = new ArrayList<String>() ;            //定义 List 集合
    String sql = "SELECT name,sex,email FROM " + TABLENAME; //定义 SQL
        Cursor result = this.db.rawQuery(sql,null);         //不设置查询参数
    for (result.moveToFirst(); !result.isAfterLast(); result.moveToNext()) {
        all.add( result.getString(0) + " " + result.getString(1)
               + " "+ result.getString(2));        }        //设置集合数据
    this.db.close() ;                                       //关闭数据库连接
    return all ; }}
```

4）定义 Activity 文件 MainActivity.Java，代码如下：

```
package org.hnist.demo;
import android.app.Activity;
import android.database.sqlite.SQLiteOpenHelper;
import android.os.Bundle;
import android.view.View;
import android.view.View.OnClickListener;
import android.widget.ArrayAdapter;
import android.widget.Button;
import android.widget.LinearLayout;
import android.widget.ListView;
public class MainActivity extends Activity {
private SQLiteOpenHelper helper = null ;               //定义数据库辅助类组件
private Button findAll = null ;
private LinearLayout mylayout = null ;                 //定义布局管理器组件
    public void onCreate(Bundle savedInstanceState) {
        super.onCreate(savedInstanceState);
        super.setContentView(R.layout.activity_main);
        this.findAll = (Button) super.findViewById(R.id.findall) ;
        System.out.println("**" + super.findViewById(R.id.mylayout).getClass()) ;
        this.mylayout = (LinearLayout) super.findViewById(R.id.mylayout) ;
                                                       //取得组件
        this.helper = new DataBaseHelp(this) ;         //定义数据库辅助类
        this.findAll.setOnClickListener(new OnClickListenerfindall()) ; }
```

```
                                                   //设置监听
private class OnClickListenerfindall implements OnClickListener {
    public void onClick(View view) {
        MainActivity.this.helper = new DataBaseHelp(MainActivity.this) ;
        ListView listView = new ListView(MainActivity.this) ;
                                                   //定义 ListView
        listView.setAdapter(new ArrayAdapter<String>(MainActivity.this,
                                                   //将数据包装
        android.R.layout.simple_list_item_1,       //每行显示一条数据
        new SearchTable(MainActivity.this.helper.getReadableDatabase()).find()));
                                                   //显示数据
        MainActivity.this.mylayout.addView(listView) ; }}} //追加组件
```

保存所有文件,运行该项目,单击按钮,显示如图 8.21 所示的结果。

如果要查询含有性别为"女"的记录,应该如何操作呢?只需要修改查询操作类 SearchTable.java 中的代码:

```
String sql = "SELECT name,sex,email FROM " +
             TABLENAME;
```

修改为:

```
String sql = "SELECT name,sex,email FROM " +
             TABLENAME + " WHERE sex = '女'";
```

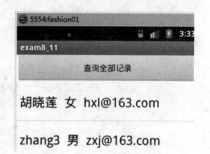

图 8.21　查询所有记录

读者可以在上面的基础上自行完成带有多个条件的查询操作。

8.4　使用 ContentProvider 存储数据

在 Android 当中,每一个应用程序的数据都是采用私有的形式进行操作的,一般不能被外部应用程序访问。为了使其他应用程序能够操作本程序的数据,可以通过 ContentProvider 提供数据操作的接口。可以将底层数据封装成 ContentProvider,有效地屏蔽底层操作的细节,并且使程序保持良好的扩展性和开放性。

例如:打电话程序和发短信程序都需要使用联系人的数据,因此必须将这些数据以 ContentProvider 存放。如果不需要在多个应用程序中共享数据,那么可以直接在应用程序中使用 SQLiteDatabase。

8.4.1　ContentProvider 基础

ContentProvider(内容提供者)它是一个类,这个类主要是对 Android 系统中进行共享的数据进行包装,并提供统一的访问接口供其他程序调用。这些被共享的数据,可以是系统提供的也可以是某个应用程序中的数据,ContentProvider 使用表的形式来组织数据。

一个程序可以通过实现一个 Content Provider 的抽象接口将自己的数据暴露出去。外界可以通过这些接口和程序里的数据打交道,可以读取程序的数据,也可以删除程序的数据,当然,中间也会涉及一些权限的问题。

1. ContentProvider 类

ContentProvider 类主要制定数据的操作标准,常用的操作方法如表 8-10 所示。

表 8-10 ContentProvider 类常用的方法

方法	描述
public abstract boolean onCreate()	当启动此组件的时候调用
public abstract int delete(Uri uri, String selection, String[] selectionArgs)	根据指定的 Uri 删除数据，并返回删除数据的行数
public final Context getContext()	返回 Context 对象
public abstract String getType(Uri uri)	根据指定 Uri，返回操作的 MIME 类型
public abstract Uri insert(Uri uri, ContentValues values)	根据指定的 Uri 进行增加数据的操作，并且返回增加后的 Uri，在此 Uri 中会附带有新数据的_id
public abstract Cursor query(Uri uri, String[] projection, String selection, String[] selectionArgs, String sortOrder)	根据指定的 Uri 执行查询操作，所有的查询结果通过 Cursor 对象返回
public abstract int update(Uri uri, ContentValues values, String selection, String[] selectionArgs)	根据指定的 Uri 进行数据的更新操作，并返回更新数据的行数

2. Uri 类

使用 ContentProvider 类进行数据操作，用 Uri 的形式进行数据的交换，Uri 是标识资源的逻辑位置，并不提供资源的具体位置，Uri 代表了要操作的数据，Uri 主要包含了两部分信息：①需要操作的 ContentProvider；②对 ContentProvider 中的什么数据进行操作。

一个 Uri 一般由以下几部分组成：

协议部分：ContentProvider（内容提供者）访问协议，已经由 Android 所规定为：content://；

主机名（或 Authority）：用于唯一标识这个 ContentProvider，外部调用者可以根据这个标识来找到它，一般都为程序的"包.类"名称，要采用小写字母的形式表示；

Path 部分：访问的路径，一般都为要操作的数据表的名称。

例如：

content://org.hnist.demo.personprovider/person 的含义就是访问 person 表中所有记录。

content://org.hnist.demo.personprovider/person/5 的含义就是访问 person 表中的 id 为 5 的记录。

content://org.hnist.demo.personprovider/person/5/name 的含义就是访问 person 表中的 id 为 5 的记录的 name 字段数据。

当然要操作的数据不一定来自数据库，也可以是文件、xml 或网络等其他存储方式，例如要操作 xml 文件中 person 节点下的 name 节点，可以构建这样的路径：/person/name。

前面介绍的 Uri 都是以字符串的形式出现的，在 Android 中要利用 Uri 类对这些字符串进行封装才能被访问，Uri 类常用的操作方法见表 8-11。

表 8-11 Uri 类常用的操作方法

方法	描述
public static String decode(String s)	对字符串进行编码
public static String encode(String s)	对编码后的字符串进行解码
public static Uri fromFile(File file)	从指定的文件之中读取 URI
public static Uri withAppendedPath(Uri baseUri, String pathSegment)	在已有地址之后添加数据
public static Uri parse(String uriString)	将给出的字符串地址变为 Uri 对象

如果要把一个字符串转换成 Uri，可以使用 Uri 类中的 parse()方法，如下：

```
Uri uri = Uri.parse("content://org.hnist.demo.personprovider/person/5")
```

3. 使用 ContentResolver 操作 ContentProvider 中的数据

当外部应用需要对 ContentProvider 中的数据进行添加、删除、修改和查询操作时，可以使用

ContentResolver 类来完成，其常用方法如表 8-12 所示，要获取 ContentResolver 对象，可以使用 Activity 提供的 public ContentResolver getContentResolver()方法。

表 8-12 ContentResolver 类常用方法

方法	描述
public final int delete(Uri url, String where, String[] selectionArgs)	调用指定 ContentProvider 对象中的 delete()方法
public final Uri insert(Uri url, ContentValues values)	调用指定 ContentProvider 对象中的 insert()方法
public final Cursor query(Uri uri, String[] projection, String selection, String[] selectionArgs, String sortOrder)	调用指定 ContentProvider 对象中的 query()方法
public final int update(Uri uri, ContentValues values, String where, String[] selectionArgs)	调用指定 ContentProvider 对象中的 update()方法

使用 ContentResolver 对 ContentProvider 中的数据进行添加、删除、修改和查询操作：

```
ContentResolver resolver = getContentResolver();
Uri uri = Uri.parse("content://org.hnist.demo.personprovider/person");
ContentValues values = new ContentValues();//添加一条记录
values.put("name", "xiaowang");
values.put("sex", "男");
resolver.insert(uri, values);
//获取person表中所有记录
Cursor cursor = resolver.query(uri, null, null, null, "personid desc");
while(cursor.moveToNext()){
   Log.i("ContentTest","personid="+cursor.getInt(0)+ ",name="+ cursor.getString(1)); }
//把id为1的记录的name字段值更新为zhangsan
ContentValues updateValues = new ContentValues();
updateValues.put("name", "zhangsan");
Uri updateIdUri = ContentUris.withAppendedId(uri, 2);
resolver.update(updateIdUri, updateValues, null, null);
Uri deleteIdUri = ContentUris.withAppendedId(uri, 2); //删除id为2的记录
resolver.delete(deleteIdUri, null, null);
```

Uri 代表了要操作的数据，经常需要解析 Uri，并从 Uri 中获取数据。Android 系统提供了两个用于操作 Uri 的工具类，分别为 UriMatcher 和 ContentUris。

4．UriMatcher 类

在使用 ContentProvider 类操作的时候某一个方法都可能要传递多种 Uri，必须对这些传递的 Uri 进行判断后才可以决定最终的操作形式，为了方便用户的判断，专门提供了 android.content.UriMatcher 类，进行 Uri 的匹配，其常用方法见表 8-13。

表 8-13 UriMatcher 类常用方法

方法	描述
public static final int NO_MATCH	表示一个-1 的整型数据，在实例化对象时使用
public UriMatcher(int code)	实例化 UriMatcher 类对象
public void addURI(String authority, String path, int code)	增加一个指定的 URI 地址
public int match(Uri uri)	与传入的 Uri 进行比较，如果匹配成功，则返回相应的 code，如果匹配失败则返回-1

UriMatcher 类用于匹配 Uri，它的用法如下：

首先把你需要匹配的 Uri 路径全部给列出来，例如：

```
//常量 UriMatcher.NO_MATCH 表示不匹配任何路径的返回码
UriMatcher  sMatcher = new UriMatcher(UriMatcher.NO_MATCH);
//如果 match()方法匹配"content://org.hnist.demo.personprovider/person"
              路径，返回匹配码为 1
sMatcher.addURI("Content://org.hnist.demo.personprovider", "person", 1);
              //添加需要匹配 uri，如果匹配就会返回匹配码 1
//如果 match()方法匹配"content://org.hnist.demo.personprovider/person/5"
              路径，返回匹配码为 2
sMatcher.addURI("content://org.hnist.demo.personprovider","person/#", 2);
              //#号为通配符
switch (sMatcher.match(Uri.parse("content://org.hnist.demo.personprovider/person/10"))) {     case 1
    break;
case 2
    break;
default://不匹配
    break;}
```

注册完需要匹配的 Uri 后，就可以使用 sMatcher.match(uri)方法对输入的 Uri 进行匹配，如果匹配就返回相应的匹配码，匹配码是调用 addURI()方法传入的第三个参数，假设匹配 content://org.hnist.demo.personprovider/person 路径，则返回的匹配码为 1。

5. ContentUris 类

由于所有的数据都要通过 Uri 进行传递，以增加操作为例，当用户执行完增加数据操作后往往需要将增加后的数据 ID 通过 Uri 进行返回，当接收到这个 Uri 的时候就需要从里面取出增加的 ID，Android 中提供了一个 android.content.ContentUris 的辅助工具类，帮助用户来完成这些操作，其常用方法见表 8-14。

表 8-14　ContentUris 类常用方法：

方法	描述
public static long parseId(Uri contentUri)	从指定 Uri 之中取出 ID
public static Uri withAppendedId(Uri contentUri, long id)	在指定的 Uri 之后增加 ID 参数

例如：

```
Uri uri = Uri.parse("content://org.hnist.demo.personprovider/person")
Uri resultUri = ContentUris.withAppendedId(uri, 5);
```

生成后的 Uri 为：content://org.hnist.demo.personprovider/person /5

```
Uri uri = Uri.parse("content://org.hnist.demo.personprovider/person /5")
long personid = ContentUris.parseId(uri);     //获取的结果为：5。
```

8.4.2 创建自己的 ContentProvider

如果希望数据能够共享，要么建立一个自己的 ContentProvider；要么将自己的数据添加到已经存在的 Content Provider。

创建一个自己的 Content Provider，至少要实现以下 3 个方面：

1）需要继承 ContentProvider 并重写以下方法：

```
public class PersonContentProvider extends ContentProvider{
    public boolean onCreate()
    public Uri insert(Uri uri, ContentValues values)
    public int delete(Uri uri, String selection, String[] selectionArgs)
    public int update(Uri uri, ContentValues values, String selection, String[]
            selectionArgs)
    public Cursor query(Uri uri, String[] projection, String selection, String[]
            selectionArgs, String sortOrder)
    public String getType(Uri uri)  }
```

2）实现 SQLiteOpenHelper 类，用于创建和删除 member 表；

```
private static class DatabaseHelper extends SQLiteOpenHelper{
    DatabaseHelper(Context context) {
        super(context, DATABASE_NAME, null, DATABASE_VERSION); }
    public void onCreate(SQLiteDatabase db) {
        //创建用于存储数据的表
        db.execSQL("Create table " + TABLE_NAME + " ( _id INTEGER PRIMARY KEY
                AUTOINCREMENT, USER_NAME TEXT); "); }
    public void onUpgrade(SQLiteDatabase db, int oldVersion, int newVersion) {
        //创建用于删除数据的表
        db.execSQL("DROP TABLE IF EXISTS " + TABLE_NAME);
        onCreate(db); } }
```

3）需要在 AndroidManifest.xml 使用<provider>对该 ContentProvider 进行配置，为了能让其他应用找到该 ContentProvider，ContentProvider 采用了 authorities（主机名/域名）对它进行唯一标识：

```
<provider
    android:name=". MyContentProvider "  //ContentProvider 的名字
    android:authorities=" org.hnist.demo.mycontentprovider"/> //主机名
```

ContentProvider 类的开发比较麻烦，因为涉及的类很多，在这里给出一个实例 exam8-12，限于篇幅，具体的代码这里就不给出了。

8.4.3 操作联系人的 ContentProvider

如果希望数据能够共享，要么建立一个自己的 ContentProvider；要么将自己的数据添加到已经存在的 content provide，如果存在包含着和想公开数据的相同类型数据的 content provider，并且拥有权限，就可以将这些数据写入到这个 content provider 中，从而实现数据共享。

Android 系统提供了多种数据类型的 ContentProvider(声音，视频，图片，联系人等)，它们大都位于 android.provider 包中，例如：

Browser：读取或修改书签，浏览历史或网络搜索；
CallLog：查看或更新通话历史；
Contacts：获取，修改或保存联系人信息；
LiveFolders：由 Content Provider 提供内容的特定文件夹；
MediaStore：访问声音，视频和图片；
Setting：查看和获取蓝牙设置，铃声和其他设置偏好；
SearchRecentSuggestions：该类能为应用程序创建简单的查询建议提供者；

SyncStateContract：用于使用数据数组账号关联数据的ContentProvider约束；
UserDictionary：在可预测文本输入时，提供用户定义的单词给输入法使用；等等。

一般情况下没有必要开发自己的 ContentProvider 类，编程人员只需要操作好这些系统提供的 ContentProvider 就足够了，前提是已获得适当的权限。

android.provider 包中的 Contacts 就是 ContentProvider 对外提供联系人数据及操作的接口。

1. 获取所有联系人信息

Android 系统中的联系人是通过 ContentProvider 来对外提供数据的，其数据库为：/data/data/com.android.providers.contacts/database/contacts2.db，如图 8.22 所示。

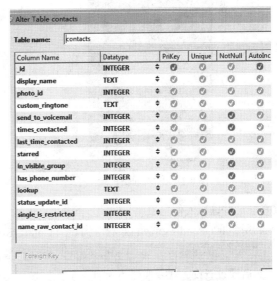

图 8.22　查询所有记录

导出这个数据库，用 SQLite Manager 打开它，发现里面有很多表，如图 8.23 所示，其中主要就是查询"contacts"表，其结构如图 8.24 所示。

图 8.23　contacts2 所有表　　　　　图 8.24　contacts 表结构图

要想进行 ContentProvider 程序访问，需要一个 CONTENT_URI 常量；这些常量在 android.provider.ContactsContract.Contacts 类中有定义，访问联系人的 CONTENT_URI 常量为：ContactsContract.Contacts.CONTENT_URI。

表中的字段较多，这里用到了下面几个字段。

人员 ID：ContactsContract.Contacts._ID；

人员姓名：ContactsContract.Contacts.DISPLAY_NAME；

取得某一个用户的电话信息和保存的内容，则需要以下的几个内容：

手机号码：ContactsContract.CommonDataKinds.Phone.NUMBER；

邮箱地址：ContactsContract.CommonDataKinds.Email.DATA；

所属联系人 ID：ContactsContract.CommonDataKinds.Phone.CONTACT_ID；

首先往手机通信录中添加几条记录，打开模拟器，找到"通信录"，如图 8.25 所示，打开通信录，单击模拟器右边的"MENU"按钮，选择"新建联系人"，如图 8.26 所示，输入姓名、电话、等信息，如图 8.27 所示，输入完毕将显示所有联系人记录，如图 8.28 所示。

图 8.25　打开通信录

图 8.26　新建联系人

图 8.27　输入通信记录

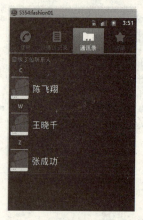

图 8.28　所有联系人记录

要获取这些联系人信息按照下面的步骤进行：

1）利用 getContentResolver()方法得到一个 ContentResolver 实例；例如：

```
ContentResolver cr = this.getContentResolver();  //得到ContentResolver 对象
```

2）然后通过查询方式获取联系所有信息；例如：

```
Cursor cursor = cr.query(ContactsContract.Contacts.CONTENT_URI, null, null,
                null, null);
```

3）循环读取每条记录的信息，逐个读出姓名、电话、email 等信息；例如：

```
for (cursor.moveToFirst(); !cursor.isAfterLast(); cursor.moveToNext()){
Int nameIndex = cursor.getColumnIndex(ContactsContract.Contacts.
    DISPLAY_NAME);
String name = cursor.getString(nameIndex);   //取得联系人名字
sbLog.append("name=" + name + ";");          //取得联系人 ID
String contactId=cursor.getString(cursor.getColumnIndex(Contacts
    Contract.Contacts._ID));                 //根据联系人 ID 查询对应的电话号码
Cursor phoneNumbers = cr.query(ContactsContract.CommonDataKinds.
    Phone.CONTENT_URI,null, ContactsContract.CommonDataKinds.
    Phone.CONTACT_ID + " = "+ contactId, null, null);
```

```
            //根据联系人 ID 查询对应的 email
    Cursor emails = cr.query(ContactsContract.CommonDataKinds.Email.CONTENT_URI,
        null, ContactsContract.CommonDataKinds.Email.CONTACT_ID + " = "
        + contactId, null, null);}
```

4) 显示获取的信息。例如:

```
text.setText(sb.toString());
```

5) 修改 AndroidManifest.xml 权限,在文件中添加如下代码:

```
<uses-permission android:name="android.permission.READ_CONTACTS"/>
```

实例 8-13:显示操作联系人信息实例

新建一个项目,项目的命名为:exam8_13,包名称为:org.hnist.demo,编写程序读取刚才输入的联系人信息,并将姓名、电话、邮箱信息显示出来。

1) 建立布局管理文件 activity_main.xml,代码如下:

```xml
<?xml version="1.0" encoding="utf-8"?>
<LinearLayout xmlns:android="http://schemas.android.com/apk/res/android"
    android:layout_width="fill_parent"
    android:layout_height="fill_parent"
    android:orientation="vertical" >
    <TextView
        android:id="@+id/text"
        android:layout_width="fill_parent"
        android:layout_height="wrap_content" />
    <Button
        android:id="@+id/but"
        android:layout_width="wrap_content"
        android:layout_height="wrap_content"
        android:text="获取联系人信息" />
</LinearLayout>
```

2) 建立 Activity 文件 MainActivity.java,代码如下:

```java
package org.hnist.demo;
import android.app.Activity;
import android.content.ContentResolver;
import android.database.Cursor;
import android.os.Bundle;
import android.provider.Contacts;
import android.provider.ContactsContract;
import android.view.View;
import android.view.View.OnClickListener;
import android.widget.Button;
import android.widget.TextView;
public class MainActivity extends Activity {
    private Button but=null;
    private TextView text=null;
    public void onCreate(Bundle savedInstanceState) {
        super.onCreate(savedInstanceState);
        setContentView(R.layout.activity_main);
```

```
            text=(TextView) this.findViewById(R.id.text);
            but=(Button) this.findViewById(R.id.but);
            this.but.setOnClickListener(new dispOnListener());}
            private class dispOnListener implements OnClickListener{
               public void onClick(View view){
                  StringBuilder sb=getContacts();
                  text.setText(sb.toString()); } }
        private StringBuilder getContacts() {
            StringBuilder  sbLog = new StringBuilder();
            ContentResolver cr = this.getContentResolver(); //得到ContentResolver对象
               //取得通信录信息,主要就是查询"Contacts"表
            Cursor cursor = cr.query(ContactsContract.Contacts.CONTENT_URI, null,
               null, null, null);
//循环读取通信录中的信息
            for(cursor.moveToFirst(); !cursor.isAfterLast(); cursor.moveToNext())
               { //取得联系人名字(显示出来的名字),实际内容在ContactsContract.Contacts中
               int nameIndex = cursor.getColumnIndex(ContactsContract.Contacts.
               DISPLAY_NAME);
                String name = cursor.getString(nameIndex);
                sbLog.append("name=" + name + ";");
                 //取得联系人ID
                String contactId=cursor.getString(cursor.getColumnIndex
                   (ContactsContract.Contacts._ID));
                 //根据联系人ID查询对应的电话号码
                Cursor phoneNumbers = cr.query(ContactsContract.CommonDataKinds.
                   Phone.CONTENT_URI,null, ContactsContract.CommonDataKinds.
                   Phone.CONTACT_ID + " = "+ contactId, null, null);
                //取得电话号码(可能存在多个号码)
                while (phoneNumbers.moveToNext())
                {String strPhoneNumber = phoneNumbers.getString(phoneNumbers.
                   getColumnIndex(ContactsContract.CommonDataKinds.Phone.NUMBER));
                    sbLog.append("Phone=" + strPhoneNumber + ";"); }
                phoneNumbers.close();
                //根据联系人ID查询对应的email
                Cursor emails = cr.query(ContactsContract.CommonDataKinds.
                   Email.CONTENT_URI, null, ContactsContract.CommonDataKinds.
                   Email.CONTACT_ID + " = " + contactId, null, null);
                //取得email(可能存在多个email)
                while (emails.moveToNext())
                {String strEmail = emails.getString(emails.getColumnIndex
                   (ContactsContract.CommonDataKinds.Email.DATA));
                    sbLog.append("Email=" + strEmail + ";");   }
  emails.close();}
            cursor.close();
             return sbLog;      } }
```

3）在 AndroidManifest.xml 文件中添加如下内容:

```
<uses-permission android:name="android.permission.READ_CONTACTS"/>
```

4）保存所有文件,运行该程序,单击按钮,运行结果如图 8.29 所示。

2. 添加联系人信息

添加联系人信息到数据库中的步骤如下:

1）要得到一个 ContentResolver 实例，利用 getContentResolver()方法；例如：

```
ContentResolver resolver = this.getCon-
                    tentResolver();
```

2）定义一个 ContentValues 类，用于封装要添加的数据；

```
ContentValues values = new ContentValues();
```

3）往 raw_contacts 表中插入一条数据，得到主键值；例如：

```
long contactid = ContentUris.parseId(resolver.insert(uri, values));
```

图 8.29　显示所有联系人信息

4）逐个添加姓名、电话、邮箱等信息。例如：

```
uri = Uri.parse("content://com.android.contacts/data");
values.put("raw_contact_id", contactid );
values.put("mimetype", "vnd.android.cursor.item/name");
values.put("data2", "黄腾飞");
resolver.insert(uri, values);          //将信息写入
……
```

联系人相关的 uri：

```
content://com.android.contacts/contacts  操作的数据是联系人信息 Uri
content://com.android.contacts/data/phones  联系人电话 Uri
content://com.android.contacts/data/emails  联系人 Email Uri
```

5）修改 AndroidManifest.xml 权限，在文件中添加如下内容：

```
<uses-permission android:name="android.permission.READ_CONTACTS"/>
<uses-permission android:name="android.permission.WRITE_CONTACTS" />
```

实例 8-14：添加操作联系人信息实例

新建一个项目，项目的命名为：exam8_14，包名称为：org.hnist.demo，在上一个实例的基础上，增加一条记录到操作联系人数据库。

1）建立布局管理文件 activity_main.xml，代码如下：

```xml
<?xml version="1.0" encoding="utf-8"?>
<LinearLayout xmlns:android="http://schemas.android.com/apk/res/android"
    android:layout_width="fill_parent"
    android:layout_height="fill_parent"
    android:orientation="vertical" >
    <TextView
        android:id="@+id/text"
        android:layout_width="fill_parent"
        android:layout_height="wrap_content" />
    <Button
        android:id="@+id/but"
        android:layout_width="wrap_content"
        android:layout_height="wrap_content"
        android:text="获取联系人信息" />
    <Button
        android:id="@+id/insertbut"
        android:layout_width="wrap_content"
        android:layout_height="wrap_content"
```

```
        android:text="添加联系人信息" />
</LinearLayout>
```

2）建立 Activity 文件 MainActivity.java，在上一实例 MainActivity.java 的基础上，增加如下代码：

```
this.insertbut=(Button) this.findViewById(R.id.insertbut);
this.insertbut.setOnClickListener(new insertOnClickListener());
private class insertOnClickListener implements OnClickListener{
    public void onClick(View view){
       try { testAddContacts(); }
catch (Exception e) { e.printStackTrace(); }
        text.setText("添加成功"); } }
public void testAddContacts() throws Exception{
     Uri uri = Uri.parse("content://com.android.contacts/raw_contacts");
     ContentResolver resolver = this.getContentResolver();
     ContentValues values = new ContentValues();
     //往 raw_contacts 表中插入一条数据，得到主键值
     long contactid = ContentUris.parseId(resolver.insert(uri, values));
     //添加姓名
     uri = Uri.parse("content://com.android.contacts/data");
     values.put("raw_contact_id", contactid );
     values.put("mimetype", "vnd.android.cursor.item/name");
     values.put("data2", "黄腾飞");
     resolver.insert(uri, values);
     //添加电话
     values.clear();
     values.put("raw_contact_id", contactid);
     values.put("mimetype", "vnd.android.cursor.item/phone_v2");
     values.put("data2", "2");
     values.put("data1", "13807304561");
     resolver.insert(uri, values);
     //添加 email
     values.clear();
     values.put("raw_contact_id", contactid);
     values.put("mimetype", "vnd.android.cursor.item/email_v2");
     values.put("data2", "2");
     values.put("data1", "htf@qq.com");
     resolver.insert(uri, values); }
```

3）在 AndroidManifest.xml 文件中添加如下内容：

```
<uses-permission android:name="android.
     permission.READ_CONTACTS"/>
<uses-permission android:name="android.
     permission.WRITE_CONTACTS" />
```

图 8.30 添加联系人后的显示结果

保存所有文件，运行该项目，单击"添加"按钮，显示结果如图 8.30 所示。

这里是为了程序简单起见，将信息直接显示在 TextView 中，实际上可以将信息在 ListView 中显示，读者可以适当修改程序来实现。

8.4.4 多媒体信息的 ContentProvider

Android 系统中带有一个图片浏览应用程序和一个音乐播放器，打开图片浏览应用程序可以查看到设备里所有的图片文件，打开音乐播放器可以看到当前设备里所有的 MP3 文件，这个打开的过程很快，这是因为当手机开机或者有 SD 卡插拔等事件发生时，系统会自动扫描 SD 卡和手机内存上的媒体文件，并将相应的信息放到定义好的数据库表格中，这一切都是由于 MediaStore 类的存在，这个类用于存放多媒体信息，它含三个内部类，分别为：MediaStore.Audio(存放音频)、MediaStore.Image(存放图片)、MediaStore.Vedio(存放视频)。

Android 把所有的多媒体数据库接口进行了封装，不用自己创建数据库，直接利用 ContentResolver 去调用封装好的接口就可以进行数据库的操作。

MediaStore 包括的多媒体数据库在/data/data/com.android.providers.media/databases/下，打开 DDMS 视图，在 File Explorer 中能看到这两个数据库，如图 8.31 所示，其中 external.db 存储的是 SD 卡上的媒体信息，由于 SD 卡不一样，所以 external.db 的名字也不完全相同， internal.db 存储的就是手机内存的媒体信息，例如手机出厂时自带的一些铃声等。

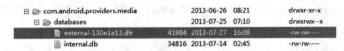

图 8.31 媒体数据库所处位置

导出这两个数据库，利用 SQliteManager 打开它，如图 8.32 所示，其中表 audio_meta：管理 SD 卡中的音频资源；表 video：管理 SD 卡中的视频资源；表 images：管理 SD 卡中的图片资源。后面的实例用到了 audio_meta 表，其结构如图 8.33 所示。

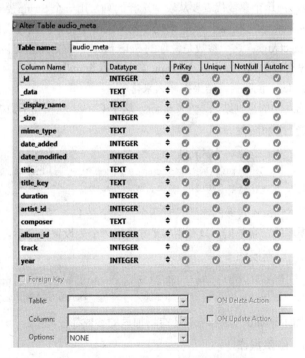

图 8.32 external 数据表   图 8.33 audio_meta 表的结构

要操作这个 ContentProvider，首先要得到一个 ContentResolver 实例，例如：

```
ContentResolver resolver = ctx.getContentResolver();//获得一个ContentResolver实例
```
ContentResolver 实例获得后，就可以进行各种增加、删除、修改、查询操作了。

1. 查询操作

```
Cursor cursor = resolver.query(uri, prjs, selections, selectArgs, order);
```

ContentResolver 的 query 方法接受几个参数，参数意义如下：

Uri：这个 Uri 代表要查询的数据库名称加上表的名称。这个 Uri 一般都直接从 MediaStore 里取得，Android 为多媒体提供的 ContentProvider 的 Uri 如下：

1）MediaStore.Audio.Media.EXTERNAL_CONTENT_URI:存储在手机外部存储器（SD 卡）上的音频文件内容的 ContentProvider 的 Uri；

2）MediaStore.Audio.Media.INTERNAL_CONTENT_URI:存储在手机内部存储器上的音频文件内容的 ContentProvider 的 Uri；

3）MediaStore.Audio.Images.EXTERNAL_CONTENT_URI:存储在手机外部存储器（SD 卡）上的图片文件内容的 ContentProvider 的 Uri；

4）MediaStore.Audio.Images.INTERNAL_CONTENT_URI:存储在手机内部存储器上的图片文件内容的 ContentProvider 的 Uri；

5）MediaStore.Audio.Video.EXTERNAL_CONTENT_URI:存储在手机外部存储器（SD 卡）上的视频文件内容的 ContentProvider 的 Uri；

6）MediaStore.Audio.Video.INTERNAL_CONTENT_URI:存储在手机内部存储器上的视频文件内容的 ContentProvider 的 Uri。

例如：要取所有歌的信息，Uri 的值就为：MediaStore.Audio.Media.EXTERNAL_CONTENT_URI。

Prjs：这个参数代表要从表中选择的列，用一个 String 数组来表示。

Selections：相当于 SQL 语句中的 where 子句，就是代表你的查询条件。

selectArgs：这个参数是说你的 Selections 里有？这里可以以实际值代替这个问号。如果 Selections 这个没有？的话，那么这个 String 数组可以为 null。

Order：说明查询结果按什么来排序。

它返回的查询结果一个 Cursor，这个 Cursor 就相当于数据库查询的中 Result。

2. 增加操作

```
ContentValues values = new ContentValues();
values.put(MediaStore.Audio.Playlists.Members.PLAY_ORDER,0);
resolver.insert(uri, values);
```

这个 insert 传递的参数只有两个，一个是 Uri，另一个是 ContentValues。这个 ContentValues 对应于数据库的一行数据，只要用 put 方法把每个列设置好之后，直接利用 insert 方法去插入即可。

3. 更新操作

```
resolver.update(MediaStore.Audio.Playlists.EXTERNAL_CONTENT_URI,values,
            where, selectionArgs);
```

上面 update 方法和查询还有增加里的参数都很类似。

4. 删除操作

```
resolver.delete(MediaStore.Audio.Playlists.EXTERNAL_CONTENT_URI,where,
            selectionArgs);
```

实例 8-15：多媒体信息的 ContentProvider 操作实例

新建一个项目，项目的命名为：exam8_15，包名称为：org.hnist.demo，操作多媒体信息的 ContentProvider 提供的数据库。

1）先导入两首歌到 SD 卡，打开 DDMS 视图→File Explorer→mnt→sdcard→hnist（不一定是 hnist 文件夹，只要是 sdcard 下的文件夹均可）在这个文件夹下导入两首歌，如图 8.34 所示。

图 8.34　导入歌曲到 SD 卡

2）建立一个布局 songlist.xml 文件，用来列出音乐文件，其代码如下所示：

```
<?xml version="1.0" encoding="utf-8"?>
<TableLayout
android:layout_width="fill_parent"
xmlns:android="http://schemas.android.com/apk/res/android"
android:layout_height="wrap_content">
<TableRow>
    <TextView
        android:id="@+id/title"
        android:textSize="15px"
        android:layout_height="wrap_content"
        android:layout_width="150px"/>
    <TextView
        android:id="@+id/name"
        android:textSize="15px"
        android:layout_height="wrap_content"
        android:layout_width="100px"/>
    <TextView
        android:id="@+id/time"
        android:textSize="15px"
        android:layout_height="wrap_content"
        android:layout_width="50px"/>
</TableRow>
</TableLayout>
```

3）建立布局管理文件 activity_main.xml，代码如下：

```
<?xml version="1.0" encoding="utf-8"?>
<LinearLayout
xmlns:android="http://schemas.android.com/apk/res/android"
android:orientation="vertical"
android:layout_width="fill_parent"
android:layout_height="fill_parent">
<TextView
android:id="@+id/msg"
android:layout_width="fill_parent"
android:layout_height="wrap_content"
android:textSize="20dp"
android:gravity="center"
android:text="所有音乐文件列表" />
<TableLayout
```

```xml
    android:layout_width="fill_parent"
    android:layout_height="wrap_content">
<TableRow>
    <TextView
        android:text="歌曲名"
        android:textSize="15px"
        android:layout_height="wrap_content"
        android:layout_width="150px"/>
    <TextView
        android:textSize="15px"
        android:layout_height="wrap_content"
        android:layout_width="100px"
        android:text="歌 手"/>
    <TextView
        android:text="时 长"
        android:textSize="15px"
        android:layout_height="wrap_content"
        android:layout_width="50px"/>
</TableRow>
</TableLayout>
<ListView
    android:id="@+id/songList"
    android:layout_width="wrap_content"
    android:layout_height="wrap_content" >
</ListView>
</LinearLayout>
```

4）建立 Activity 文件 MainActivity.java，代码如下：

```java
package org.hnist.demo;
import java.util.ArrayList;
import java.util.HashMap;
import java.util.List;
import java.util.Map;
import android.app.Activity;
import android.content.ContentResolver;
import android.database.Cursor;
import android.os.Bundle;
import android.provider.MediaStore;
import android.widget.ListView;
import android.widget.SimpleAdapter;
import android.widget.Toast;
public class MainActivity extends Activity {
private ListView mediaList = null ;              //定义 ListView
String songName;                                 //歌手名
String songTitle;                                //歌曲名
String songtime;                                 //歌曲时间
int AllTime;                                     //歌曲时间
private List<Map<String,String>>list=new ArrayList<Map<String,String>>();
private SimpleAdapter simpleAdapter = null ;
public void onCreate(Bundle savedInstanceState) {
    super.onCreate(savedInstanceState);
    super.setContentView(R.layout.activity_main);
    this.mediaList = (ListView) super.findViewById(R.id.songList) ;
    ContentResolver cr = getContentResolver();//获得一个 ContentResolver 实例
        Cursor c = cr.query(MediaStore.Audio.Media.EXTERNAL_CONTENT_URI, null,
```

第 8 章 Android 数据存储技术

```
                    null, null, MediaStore.Audio.Media.DEFAULT_SORT_ORDER);
                //获得查询信息
        if(c== null) {
            Toast.makeText(this, "没有歌曲信息", Toast.LENGTH_SHORT).show();
        } else {
for (c.moveToFirst(); !c.isAfterLast(); c.moveToNext()) {//循环获取歌曲信息
        //获取歌曲名称
    songTitle=c.getString(c.getColumnIndexOrThrow(MediaStore.Audio.Media.TITLE));
        //获取歌手名
    songName=c.getString(c.getColumnIndexOrThrow(MediaStore.Audio.Media.ARTIST));
        //获取歌曲时间长度
    AllTime = c.getInt(c.getColumnIndexOrThrow(MediaStore.Audio.Media.DURATION));
    songtime=ShowTime(AllTime);//将时间转换为字符串
            Map<String,String>map=new HashMap<String,String>();//定义Map 集合
            map.put("title",songTitle);      //设置显示数据
            map.put("name",songName);        //设置显示数据
            map.put("time",songtime);        //设置显示数据
            this.list.add(map);}             //增加数据
            this.simpleAdapter=new SimpleAdapter(this,//实例化SimpleAdapter
                this.list,                   //增加数据
                R.layout.songinfo,           //使用显示模板 songinfo.xml 文件
                new String[]{"title","name","time"},//定义要显示的Map
                new int[]{R.id.title,R.id.name,R.id.time});
            this.mediaList.setAdapter(this.simpleAdapter);}}
                                            //设置要显示的数据
public String ShowTime(int time) {          //定义 ShowTime()方法
    //将 ms 转换为 s
    time /= 1000;
    //求分
    int minute = time / 60;
    //求秒
    int second = time % 60;
    minute %= 60;
    return String.format("%02d:%02d",
        minute, second);}  }
```

保存所有文件，运行该项目，显示结果如图 8.35 所示。

如果要实现单击歌曲名字就播放这首歌的功能，应该怎么操作呢？

可以设置一个 setOnItemClickListener()方法，触发这个事件会通过 ContentProvider 获取存在数据库中的相关信息，然后播放音乐。

图 8.35 显示所有音乐文件信息

```
this.mediaList.setOnItemClickListener(new OnItemClickListenerImpl());
……
private class OnItemClickListenerImpl  implements OnItemClickListener{
    private MediaPlayer media;
    public void onItemClick(AdapterView<?>parent,View view,int position,long id){
        Map<String,String>map=(Map<String,String>) MainActivity.
                            this.simpleAdapter.getItem(position);
        String  songtitle=map.get("title");
        String  songname=map.get("name");
        String  songtime=map.get("time");
        String  songpath=map.get("path");
            //显示要播放的文件
```

```java
                MainActivity.this.msg.setText(songtitle+"   "+songname+"
                    "+songtime);
                PlaySong(songpath); }}
    private void PlaySong(String songpath) {
        this.media = new MediaPlayer();                    //创建MediaPlayer对象
        ((MediaPlayer) media).reset();                     //恢复到未初始化状态
        try {
            ((MediaPlayer) media).setDataSource(songpath);//获得播放文件
        } catch (IllegalArgumentException e) {
            e.printStackTrace();
        } catch (SecurityException e) {
            e.printStackTrace();
        } catch (IllegalStateException e) {
            e.printStackTrace();
        } catch (IOException e) {
            e.printStackTrace();}
        try {
            ((MediaPlayer) media).prepare();               //准备播放
        } catch (IllegalStateException e) {
            e.printStackTrace();
        } catch (IOException e) {
            e.printStackTrace();}
        (MediaPlayer) media).start();}                     //开始播放文件
```

详细代码见实例 8-16，这里实现了播放音频文件，如果还要进行音量控制和进度控制，建议再建立一个 Java 文件，专门实现某首歌曲的播放，然后用 intent 来实现，限于篇幅，这里不再赘述，感兴趣的读者可以参考实例 7_9 编写代码实现。

前面介绍的几种存储都是将数据存储在本地设备上，除此之外，还有一种存储/读取数据的方式，通过网络来实现数据的存储和读取，将在第 9 章介绍。

本章小结

本章介绍了几种数据的存储形式，着重介绍了如何操作 SQLite 数据库和数据表，以及查询操作。着重介绍了运用系统提供的 ContentProvider 进行数据的共享，操作联系人的 ContentProvider 以及多媒体信息的 ContentProvider。

习题

1. 要在 AndroidManifest.xml 中如何设置才能将一个文件保存到 SD 卡上？
2. 如何才能将信息从 SD 卡读出，给出实现的步骤和关键代码。
3. 如何才能将信息写入 SD 卡保存，给出实现的步骤和关键代码。
4. 查找资料，了解通话记录的 ContentProvider。
5. 参照实例 8-13 实现获取通信记录，写出关键代码。
6. 编程实现实例 8-14 显示的信息在一个 ListView 组件中，写出关键代码。
7. 如果要编程实现实例 8-15 中当歌曲播放时显示播放音量控制条和进度控制条，写出关键代码。

第 9 章　Android 网络通信技术

学习目标：
- 掌握 Android 中的 WebView 组件
- 掌握 Android 中的 HTTP 协议通信技术
- 掌握 Android 中的 Socket 协议通信技术
- 掌握 Android 中的 Web Service 通信技术
- 了解 Android 中的蓝牙通信技术
- 了解 Android 中的 WiFi 通信技术

手机发展到今天，功能已不仅仅是打电话了，手机看新闻、手机炒股、手机缴费、手机银行、手机地图等等，新的应用层出不穷，事实上在 Android 中，掌握了网络通信技术就可以开发出这些网络应用程序。

9.1　Android 网络通信技术基础

手机能够上网是因为手机底层使用了 TCP/IP 协议，可以使手机终端通过无线网络建立 TCP 连接。TCP 协议可以对上层网络提供接口，使上层网络数据的传输建立在"无差别"的网络之上。

所谓无线网络就是采用无线传输媒介（如无线电波、红外线等）的网络，它包括远距离无线连接的全球语音和数据网络以及近距离无线连接的蓝牙技术及射频技术。

Android 基于 Linux 内核，目前，Android 平台提供的与网络相关的包如表 9-1 所示。

表 9-1　Android SDK 中一些与网络有关的包

包	描述
java.net	提供与网络通信相关的类，包括流和数据包 socket、Internet 协议和常见 HTTP 处理。该包是一个多功能网络资源。有经验的 Java 开发人员可以立即使用这个熟悉的包创建应用程序
java.io	虽然没有提供现实网络通信功能，但是仍然非常重要。该包中的类由其他 Java 包中提供的 socket 和链接使用。它们还用于与本地文件的交互
java.nio	包含表示特定数据类型的缓冲区的类。适用于两个基于 Java 语言的端点之间的通信
org.apache.*	表示许多为 HTTP 通信提供精确控制和功能的包。可以将 Apache 视为流行的开源 Web 服务器
android.net	除核心 java.net.*类以外，包含额外的网络访问 socket。该包包括 URI 类，后者频繁用于 Android 应用程序开发，而不仅仅是传统的联网
android.net.http	包含处理 SSL 证书的类

Android 通信可以利用系统提供的 WebView 组件来实现一些网络功能，也可以利用 HTTP 方式通信、Socket 方式通信、Web Service 方式通信、WiFi 方式通信，如果是近距离的通信还可以借助蓝牙通信。

9.1.1　Android 中的 HTTP 协议基础

HTTP（Hyper Text Transfer Protocol，超文本传输协议），是 Web 联网的基础，也是手机联网常用的协议之一，HTTP 协议是建立在 TCP 协议之上的一种应用，用于传送 WWW 方式的数据。HTTP 协议采用了请求/响应模式，是一个属于应用层的面向对象的协议。

HTTP 大致工作流程：客户端向服务器发出 HTTP 请求，服务器接收到客户端的请求后，处理客户端的请求，处理完成后再通过 HTTP 应答返回给客户端。这里的客户端是指 Android 手机端，服务器一般是 HTTP 服务器，HTTP 请求方法有 POST、GET 等方法。

HTTP 协议主要用于 Web 浏览器和 Web 服务器之间的数据交换，当在地址栏中输入 http://host:port/path，其中：http 表示要通过 HTTP 协议来定位网络资源，就相当于通知浏览器使用 HTTP 协议来和 host 所确定的服务器进行通信；host 表示合法的 Internet 主机域名或者 IP 地址；port 指定一个端口号，为空则使用缺省端口 80；path 指定请求资源的 URI。

HTTP 通信编程可以使用 Java 的 java.net.URL 类，也可以使用 Apache 组织提供的 HttpClient 类库，HttpClient 类库已集成到 Android 平台中。

Android 中的 HTTP 协议版本是 HTTP1.1，采用了请求/响应模式：在客户端向服务器发送一个请求并接收以一个响应后，只要不关闭网络连接，就可以继续向服务器发送 HTTP 请求，客户端发送的每次请求都需要服务器回送响应，在请求结束后，会主动释放连接。

HTTP 请求由三部分组成：请求行、消息报头、请求正文。

1. 请求行的格式：Method Request-URI HTTP-Version CRLF

其中 Method 表示请求方法，具体方法如表 9-2 所示，Request-URI 是一个统一资源标识符；HTTP-Version 表示请求的 HTTP 协议版本；CRLF 表示回车和换行。

表 9-2 请求方法 Method 列表

方法	描述
GET	请求获取 Request-URI 所标识的资源
POST	在 Request-URI 所标识的资源后附加新的数据
HEAD	请求获取由 Request-URI 所标识的资源的响应消息报头
PUT	请求服务器存储一个资源，并用 Request-URI 作为其标识
DELETE	请求服务器删除 Request-URI 所标识的资源
TRACE	请求服务器回送收到的请求信息，主要用于测试或诊断
CONNECT	保留将来使用
OPTIONS	请求查询服务器的性能，或者查询与资源相关的选项和需求

例如：

GET 方法：GET /www.hnist.cn/login.html HTTP/1.1 (CRLF)

POST 方法：POST / www.hnist.cn/regedit.jsp HTTP/1.1 (CRLF)

2. HTTP 响应也是由三个部分组成：状态行、消息报头、响应正文

状态行格式：HTTP-Version Status-Code Reason-Phrase CRLF

其中，HTTP-Version 表示服务器 HTTP 协议的版本；Status-Code 表示服务器发回的响应状态代码；Reason-Phrase 表示状态代码的文本描述。

例如：HTTP/1.1 200 OK （CRLF）

状态代码由三位数字组成，第一个数字定义了响应的类别，且有五种可能取值，具体状态码如表 9-3 所示。

常见状态代码、状态描述、说明：

```
200 OK  //客户端请求成功
400 Bad Request  //客户端请求有语法错误，不能被服务器所理解
```

```
401 Unauthorized   //请求未经授权
500 Internal Server Error   //服务器发生不可预期的错误
503 Server Unavailable   //服务器当前不能处理客户端请求,一段时间后,可能恢复正常
```

Android 提供了 HttpURLConnection 类和 HttpClient 类来开发关于 HTTP 的程序。

表 9-3 请求方法 Method 列表

状态代码	描述
1xx	指示信息——请求已接收,继续处理
2xx	成功——表示请求已被成功接收、理解、接受
3xx	重定向——要完成请求必须进行更进一步的操作
4xx	客户端错误——请求有语法错误或请求无法实现
5xx	服务器端错误——服务器未能实现合法的请求

9.1.2 Android 中的 Socket 基础

如果双方建立的是 HTTP 连接,则服务器需要等到客户端发送一次请求后才能将数据传回给客户端,客户端定时向服务器端发送连接请求,不仅可以保持在线,同时也是在"询问"服务器是否有新的数据,如果有就将数据传给客户端。如果是要进行多人联网的游戏,那么 HTTP 就不能很好地满足要求,Socket 通信可以很好地解决这个问题。

Socket(也称为套接字)是一种低级、原始的通信方式,要编写服务器端代码和客户端代码,自己设置端口,自己设置通信协议、验证数据安全和合法性,而且通常还应该是多线程的,开发起来比较麻烦。但是它也有其优点:灵活,不受编程语言、设备、平台和操作系统的限制,通信速度快而高效。

应用层通过传输层进行数据通信时,TCP 会遇到同时为多个应用程序进程提供并发服务的问题。多个 TCP 连接或多个应用程序进程可能需要通过同一个 TCP 协议端口传输数据。为了区别不同的应用程序进程和连接,许多计算机操作系统为应用程序与 TCP/IP 协议交互提供了套接字(Socket)接口。应用层可以和传输层通过 Socket 接口,区分来自不同应用程序进程或网络连接的通信,实现数据传输的并发服务。

Socket 是通信的基石,是支持 TCP/IP 协议的网络通信的基本操作单元。它包含进行网络通信必需的五种信息:连接使用的协议、本地主机的 IP 地址、本地进程的协议端口、远地主机的 IP 地址、远地进程的协议端口。

Socket 有两种主要的操作方式:面向连接和无连接的操作方式。

面向连接的操作使用 TCP 协议,这个模式下的 Socket 必须在发送数据之前与目的地的 Socket 取得连接。一旦连接建立了,Socket 就可以使用一个流接口进行打开、读、写、关闭等操作。所有发送的消息都会在另一端以同样的顺序被接收。它的优点是数据安全性较高,缺点是操作的效率较低。

无连接的操作使用 UDP 协议,这个模式下的 Socket 不需要连接一个目的 Socket,它只是简单发送出数据报,多个数据报的到达顺序可能和出发时的顺序不一样。它的优点是操作的效率较高,缺点是数据安全性较低。

要建立 Socket 连接至少需要一对套接字,其中一个运行于客户端,称为 ClientSocket,另一个运行于服务器端,称为 ServerSocket。

Socket 之间的连接过程分为三个步骤:服务器监听,客户端请求,连接确认。

服务器监听:服务器端套接字并不定位具体的客户端套接字,而是处于等待连接的状态,实时监控网络状态,等待客户端的连接请求。

客户端请求：指客户端的套接字提出连接请求，要连接的目标是服务器端的套接字。为此，客户端的套接字必须首先描述它要连接的服务器的套接字，指出服务器端套接字的地址和端口号，然后就向服务器端套接字提出连接请求。

连接确认：当服务器端套接字监听到或者说接收到客户端套接字的连接请求时，就响应客户端套接字的请求，建立一个新的线程，把服务器端套接字的描述发给客户端，一旦客户端确认了此描述，双方就正式建立连接。而服务器端套接字继续处于监听状态，继续接收其他客户端套接字的连接请求。

创建 Socket 连接时，可以指定使用的传输层协议，Socket 可以支持不同的传输层协议（TCP 或 UDP），当使用 TCP 协议进行连接时，该 Socket 连接就是一个 TCP 连接。

在 Java 中 Socket 相关类都在 java.net 包中，其中主要的类是 Socket 和 ServerSocket。

9.1.3　Android 中的 Web Service 基础

Android 与 Web 服务器以及 Socket 程序通信的操作，这两种程序通信存在平台的制约，如果用户要想搭建一个异步的业务中心操作，上面两种方式可以实现，但是在实际的开发中会使用 Web Service 技术。

Web Services 是一个用于支持网络间不同机器互操作的软件系统，它建立在通用协议如 HTTP、SOAP、UDDI、WSDL 等基础之上，是一种自包含、自描述和模块化的应用程序，它可以在网络中被描述、发布和调用，可以将它看作是基于网络的、分布式的模块化组件。

通常所说的 WebService 就是远程的某个服务器对外公开了某个方法或服务，那么就可以通过编程像调用本地方法一样去调用远程服务器上的这个方法，根本不需要知道远程服务器上的那个方法是用什么语言写的，也不需要知道那个方法是基于什么平台，也就是说 WebService 与平台和语言无关。

Web Services 的优势在于提供了不同应用程序平台之间的互操作，它使得基于组件的开发和 Web 相结合的效果达到最佳。它是基于 HTTP 协议的，调用请求和回应消息都可以穿过防火墙，不需要更改防火墙的设置，这样就避免了使用特殊端口进行通信时无法穿越防火墙的问题，但是这样也存在着安全的隐患。

Web Service 是一个应用组件，它逻辑性地为其他应用程序提供数据与服务。各应用程序通过网络协议和规定的一些标准数据格式来访问 Web Service，通过 Web Service 内部执行得到所需结果。它使得不同计算机语言、不同计算机平台之间的方法调用成为可能，是远程调用和分布式系统的重要实现手段。

在构建和使用 Web Service 时用到以下几个术语：

1）XML：可扩展的标记语言（Xtensible Markup Language 的缩写）是 Web Service 平台中表示数据的基本格式，它主要的优点在于它既与平台无关又与厂商无关，同时还易于建立、易于分析。

2）SOAP：简单对象访问协议（Simple Object Access Protocol 的缩写），是 Web Service 进行消息传递的通信协议，它是一种轻量级的、简单的、基于 XML 的协议，用于在分布式环境中交换格式化和固化信息，所有的 SOAP 消息都使用 XML 编码。

3）WSDL：Web 服务描述语言（Web Service Description Language 的缩写），是一个用来描述如何与 Web 服务通信的 XML 语言。服务的提供者将自己的 Web 服务所有相关内容生成相应的 WSDL 文档，发布给使用者，使用者可以通过这个 WSDL 文档，创建相应的 SOAP 请求消息，通过 HTTP 传递给服务的提供者；Web 服务在完成服务请求后，将 SOAP 返回消息传回请求者，服务请求者再根据 WSDL 文档将 SOAP 返回消息解析成自己能够理解的内容。

4）UDDI：通用描述、发现与集成（Universal Description, Discovery and Integration 的缩写），它是一种独立于平台的，基于 XML 语言用于在互联网上描述商务的协议。

将 Web Service 进行 UDDI 注册发布，UDDI 是一种创建注册表服务的规范，以便大家将自己的 Web Service 进行注册发布供使用者查找。当服务提供者想将自己的 Web Service 向外部发布时，它可以将自己的 Web Service 注册到相应的 UDDI 商用注册网站，因为 WSDL 文件中已经给定 Web Service 的地址 URI，外部可以直接通过 WSDL 提供的 URI 进行相应的 Web Service 调用。UDDI 注册是向世界范围内的公司开放的，不论它们的规模大小。UDDI 并不是一个必需的 Web Service 组件，服务方完全可以不进行 UDDI 的注册。

在 Android 客户端编写一个 Web Service 客户端程序，假设是一个用户登录程序，将用户名和口令以参数的形式传递给远程的 Web Service，由远程 Web Service 处理这个调用，然后再将结果返回给客户端。目前 Android 平台没有提供 Web Service 客户端开发类库，只能借助第三方的 Web Service 客户端开发类库，例如：ksoap 类库。

9.1.4 Android 中的蓝牙基础

蓝牙是目前使用最广泛的无线通信协议之一，实现近距离无线通信。最初由爱立信创建，蓝牙的标准是 IEEE 802.15.1，蓝牙协议工作在无需许可的 ISM(Industrial Scientific Medical)频段的 2.45GHz。最高速度可达 723.1kbps。

蓝牙的优势是：无需驱动程序，小型化无线，低功率、低成本、安全性、稳固，易于使用、即时连接。

Android 平台支持蓝牙网络协议栈，实现蓝牙设备之间数据的无线传输。

蓝牙设备之间的通信主要包括了四个步骤：

1）设置蓝牙设备；
2）寻找局域网内可能或者匹配的设备；
3）连接设备；
4）设备之间的数据传输。

Android 所有关于蓝牙开发的类都在 android.bluetooth 包下，一共有 8 个类，以下是建立蓝牙连接所需要的一些基本类。

BluetoothAdapter 类：代表了一个本地的蓝牙适配器，是所有蓝牙交互的入口点。

BluetoothDevice 类：代表了一个远端的蓝牙设备，使用它请求远端蓝牙设备连接或者获取远端蓝牙设备的名称、地址、种类和绑定状态。

Bluetoothsocket 类：代表了一个蓝牙套接字的接口（类似于 TCP 中的套接字），是应用程序通过输入、输出流与其他蓝牙设备通信的连接点。

Blueboothserversocket 类：打开服务连接来监听可能到来的连接请求（属于 server 端），为了连接两个蓝牙设备，必须有一个设备作为服务器打开一个服务套接字。

9.1.5 Android 中的 Wi-Fi 基础

Wi-Fi 的英文全称为 Wireless Fidelity，无线保真的缩写，又称 802.11b 标准。它是一种可以将个人电脑、手持设备（如 PDA、手机）等终端以无线方式互相连接的技术，即一种无线联网技术，它是通过无线电波来连网。Wi-Fi 为用户提供了无线的宽带互联网访问，能够访问 Wi-Fi 网络的地方被称为热点。Wi-Fi 或 802.11G 在 2.4GHz 频段工作，所支持的速度最高达 54Mbps。

蓝牙和 WiFi 相比，WiFi 是一个更加快速的协议，覆盖范围更大。虽然两者使用相同的频率范围，但是 WiFi 需要更加昂贵的硬件。蓝牙设计被用来在不同的设备之间创建无线连接，而 WiFi 是个无线局域网协议。

Android 开发 Wifi 主要包括以下几个类。
1) ScanResult 类：该类主要是通过 WiFi 硬件的扫描来获取一些周边 WiFi 热点的信息。
2) WifiConfiguration：该类主要用来进行 WiFi 网络的配置，包括安全配置等。
3) WifiInfo：该类主要进行 WiFi 无线连接的描述。
4) WifiManager：该类提供了管理 WiFi 连接的大部分 API。

9.2 WebView 组件介绍

9.2.1 WebView 组件基础知识

Android 手机中内置了一款高性能 webkit 内核浏览器，在 SDK 中封装为一个叫做 WebView 组件。用户可以直接使用 WebView 组件显示网页的内容，或者是将一些指定的 HTML 文件嵌入进来，除了支持各个浏览器的"前进"、"后退"等功能之外，最为强大的是 WebView 还支持 JavaScript 的操作，能与 JavaScript 相互通信。

android.webkit.WebView 类的层次关系如下所示：

```
java.lang.Object
    android.view.View
        android.view.ViewGroup
            android.widget.AbsoluteLayout
                android.webkit.WebView
```

要在 Android 程序中使用 WebView 组件，必须要在程序中使用下面的语句。

```
import android.webkit.WebView;                    //导入 webkit.WebView 类
```

在 Android 中 WebView 组件提供了如表 9-4 所示的常用方法。

表 9-4 WebView 组件的常用方法

方法	描述
public WebView(Context context)	取得 WebView 类的实例化对象
public void addJavascriptInterface(Object obj, String interfaceName)	绑定一个 JavaScript 的对象
public boolean canGoBack()	判断能否实现后退操作
public boolean canGoBackOrForward(int steps)	判断是否可以后退或前进指定步数
public boolean canGoForward()	判断是否可以前进
public boolean canZoomIn()	判断是否可以缩小
public boolean canZoomOut()	判断是否可以放大
public void clearCache(boolean includeDiskFiles)	清空缓存，如果是 false 则只清空 RAM
public void clearFormData()	清空表单的填写记录
public void clearHistory()	清空历史信息
public int getProgress()	得到访问进度
public String getTitle()	取得当前访问页面的标题
public void goBack()	后退一步
public void goBackOrForward(int steps)	后退或前进指定的步数
public void goForward()	前进一步
public void loadData(String data, String mimeType, String encoding)	通过指定的字符串进行页面的加载
public void loadUrl(String url)	读取指定的 URL 地址数据
public void reload()	重新加载页面

方法	描述
public void savePassword(String host, String username, String password)	保存密码
public void setDownloadListener(DownloadListener listener)	对下载文件进行监听
public void setWebChromeClient(WebChromeClient client)	使用 Google Chrome 作为客户端
public void setWebViewClient(WebViewClient client)	使用 WebView 作为客户端
public boolean zoomIn()	是否缩小
public boolean zoomOut()	是否放大
public WebSettings getSettings()	返回 WebSettings 对象

WebView 组件有下面几个比较重要的子类：

1）WebChromeClient 类：会在一些影响浏览器 UI 交互动作发生时被调用，比如 WebView 关闭和隐藏、页面加载进展、js 确认框和警告框、js 加载前、js 操作超时、WebView 获得焦点等等。WebChromeClient 类常用方法列表见表 9-5。

表 9-5　WebChromeClient 类常用方法列表

方法	描述
public void onCloseWindow(WebView window)	窗口关闭操作
public boolean onCreateWindow(WebView view, boolean dialog, boolean userGesture, Message resultMsg)	创建新的 WebView
public boolean onJsAlert(WebView view, String url, String message, JsResult result)	弹出警告框互操作
public boolean onJsBeforeUnload(WebView view, String url, String message, JsResult result)	页面关闭互操作
public boolean onJsConfirm(WebView view, String url, String message, JsResult result)	弹出确认框互操作
public boolean onJsPrompt(WebView view, String url, String message, String defaultValue, JsPromptResult result)	弹出提示框互操作
public boolean onJsTimeout()	计时器已到互操作
public void onProgressChanged(WebView view, int newProgress)	进度改变互操作
public void onReceivedTitle(WebView view, String title)	接收页面标题更改
public void onRequestFocus(WebView view, String title)	WebView 显示焦点

2）WebViewClient 类会在一些影响内容变化的动作发生时被调用，比如表单的错误提交需要重新提交、页面开始加载及加载完成、资源加载中、接收到 HTTP 认证需要处理、页面键盘响应、页面 URL 打开处理等等。WebViewClient 常用方法见表 9-6。

表 9-6　WebViewClient 常用方法

方法	描述
public void doUpdateVisitedHistory(WebView view, String url, boolean isReload)	更新历史记录
public void onFormResubmission (WebView view, Message dontResend, Message resend)	应用程序重新请求网页数据
public void onLoadResource(WebView view, String url)	加载指定地址提供的资源
public void onPageFinished(WebView view,String url)	网页加载完毕
public void onPageStarted(WebView view,String url,Bitmap favicon)	网页开始加载
public void onReceivedError(WebView view, int errorCode,String description,String failingUrl)	报告错误信息
public void onScaleChanged(WebView view,float oldScale, float newScale)	WebView 发生改变
public boolean shouldOverrideUrlLoading(WebView view,String url)	控制新的连接在当前 WebView 中打开

例如：

```
webView.setWebViewClient(webViewClient);        //加载 WebViewClient
```

3）WebSettings 类用来对 WebView 的配置进行配置和管理，比如是否可以进行文件操作、缓存的设置、页面是否支持放大和缩小、字体及文字编码设置、是否允许 js 脚本运行、是否允许图片自动加载、是否允许数据及密码保存等等。

WebSetting 类的常用方法见表 9-7。

表 9-7　WebSetting 类的常用方法

方法	描述
public void　setBuiltInZoomControls(boolean enabled)	设置是否支持缩放控制
public synchronized void setJavaScriptEnabled(boolean flag)	设置是否启动 JavaScript 支持
public void　setSaveFormData(boolean save)	设置是否保存表单数据
public void　setSavePassword(boolean save)	设置是否保存密码
public synchronized void setGeolocationEnabled(boolean flag)	设置是否可以获得地理位置
public void setAllowFileAccess (boolean allow)	启用或禁止 WebView 访问文件数据
public synchronized void setBlockNetworkImage (boolean flag)	是否显示网络图像
public void setCacheMode (int mode)	设置缓冲的模式
public synchronized void setDatabaseEnabled (boolean flag)	设置是否用数据库
public synchronized void setDefaultFontSize (int size)	设置默认字体的大小
public synchronized void setDefaultTextEncodingName(String encoding)	设置在解码时使用的默认编码
public synchronized void setFixedFontFamily (String font)	设置固定使用的字体
public synchronized void setLayoutAlgorithm (WebSettings.LayoutAlgorithm l)	设置布局方式

WebView 在 Android 中的主要作用有如下三个方面：

1）加载网页，直接显示网页；

2）加载 HTML 文件，可以利用 HTML 做界面布局，在手机上实现复杂的显示布局效果；

3）加载 JSP 文件和 JavaScript 相互通信。

9.2.2　使用 WebView 加载网页

实现 WebView 加载网页，按照下述的步骤进行即可：

1）在要 Activity 中实例化 WebView 组件；例如：

```
WebView webView = new WebView(this);            //实例化 WebView 组件
```

或者在布局文件中声明 WebView，然后在 Activity 中实例化 WebView；例如：

```
this.webview = (WebView) super.findViewById(R.id.webview) ;    //取得组件
```

2）调用 WebView 的 loadUrl()方法，设置 WevView 要显示的网页；例如：

```
webView.loadUrl("http://www.hnist.cn");
```

3）调用 Activity 的 setContentView( )方法来显示网页视图；

4）为了让 WebView 支持回退功能，需要覆盖 Activity 类的 onKeyDown()方法，如果不做任何处理，点击系统回退键，整个浏览器会调用 finish()结束，而不是回退到上一页面；

5）需要在 AndroidManifest.xml 文件中添加权限，否则会出现错误。

```
<uses-permission android:name="android.permission.INTERNET" />
```

实例 9-1：使用 WebView 加载网页实例

新建一个项目，项目的命名为：exam9_1，包名称为：org.hnist.demo，利用 WebView 组件加载网页，设置简单的后退与前进功能。

1）修改 activity_main.xml 文件，代码如下：

```
<?xml version="1.0" encoding="utf-8"?>
<LinearLayout
    xmlns:android="http://schemas.android.com/apk/res/android"
```

```xml
    android:orientation="vertical"
    android:layout_width="fill_parent"
    android:layout_height="fill_parent">
    <LinearLayout
        android:orientation="horizontal"
        android:layout_width="fill_parent"
        android:layout_height="wrap_content">
        <Button
            android:id="@+id/open"
            android:layout_width="wrap_content"
            android:layout_height="wrap_content"
            android:text="打开网页" />
        <Button
            android:id="@+id/back"
            android:layout_width="wrap_content"
            android:layout_height="wrap_content"
            android:text="网页后退" />
        <Button
            android:id="@+id/forward"
            android:layout_width="wrap_content"
            android:layout_height="wrap_content"
            android:text="网页前进" />
    </LinearLayout>
    <WebView                                        //加一个 WebView 组件
        android:id="@+id/webview"                   //名为 webview
        android:layout_width="fill_parent"
        android:layout_height="fill_parent"/>
</LinearLayout>
```

2）修改 MainActivity.java 文件，代码如下：

```java
package org.hnist.demo;
import android.app.Activity;
import android.app.AlertDialog;
import android.app.Dialog;
import android.content.DialogInterface;
import android.os.Bundle;
import android.view.View;
import android.view.View.OnClickListener;
import android.webkit.WebView;                     //导入 webkit.WebView 类
import android.widget.Button;
public class MainActivity extends Activity {
private Button open = null ;
private Button back = null ;
private Button forward = null ;
private WebView webview = null ;                   //定义 WebView 组件
private String urlData[] = new String[] { "http://www.hnist.cn",
        "http://law.hnist.cn/", "http://xw.hnist.cn",
        "http://art.hnist.cn/", "http://dyb.hnist.cn/" };   //定义选项数据
@Override
public void onCreate(Bundle savedInstanceState) {
  super.onCreate(savedInstanceState);
```

```java
        super.setContentView(R.layout.activity_main);
        this.open = (Button) super.findViewById(R.id.open) ;
        this.back = (Button) super.findViewById(R.id.back) ;
        this.forward = (Button) super.findViewById(R.id.forward) ;
        this.webview = (WebView) super.findViewById(R.id.webview) ;
                            //取得 WebView 组件
        this.open.setOnClickListener(new OpenOnClickListenerImpl());
                            //单击事件
        this.back.setOnClickListener(new BackOnClickListenerImpl());
                            //单击事件
        this.forward.setOnClickListener(new ForwardOnClickListenerImpl());}
                            //单击事件
private class OpenOnClickListenerImpl implements OnClickListener {
    public void onClick(View view) {
        MainActivity.this.showUrlDailog() ; }   }      //显示对话框
private class BackOnClickListenerImpl implements OnClickListener {
    public void onClick(View view) {
        if (MainActivity.this.webview.canGoBack()) {   //可以后退
            MainActivity.this.webview.goBack() ; }}}   //后退
private class ForwardOnClickListenerImpl implements OnClickListener {
    @Override
    public void onClick(View view) {
        if (MainActivity.this.webview.canGoForward()) {   //可以前进
            MainActivity.this.webview.goForward() ; }}} //前进
private void showUrlDailog(){                              //显示对话框
    Dialog dialog = new AlertDialog.Builder(this)          //实例化 Dialog 对象
        .setTitle("请选择要打开的网站")                      //设置显示标题
        .setNegativeButton("取消",                          //增加取消按钮
            new DialogInterface.OnClickListener() {        //设置操作监听
                public void onClick(DialogInterface dialog,
                    int whichButton) {  }})                //单击事件
        .setItems(this.urlData,                            //设置列表选项
            new DialogInterface.OnClickListener() {
                public void onClick(DialogInterface dialog,
                    int which) {                           //设置显示信息
                    MainActivity.this.webview.loadUrl(
                        urlData[which]);}                  //打开网址
        }).create();                                       //创建 Dialog
    dialog.show();}}
```

3) 修改 AndroidManifest.xml 文件配置权限, 在 AndroidManifest.xml 里添加如下代码:

```xml
<uses-permission android:name="android.permission.INTERNET"/>
```

运行该项目, 弹出如图 9.1 所示界面, 单击"打开网页"按钮, 弹出如图 9.2 所示界面, 选择第三个选项, 弹出如图 9.3 所示界面。

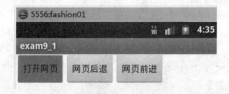

图 9.1 实例 9-1 运行结果

第 9 章 Android 网络通信技术

图 9.2 打开 HTML 文件的运行结果

图 9.3 打开某个链接站点的界面

9.2.3 使用 WebView 加载 HTML 文件

使用 WebView 组件除了可以浏览 Web 站点外，它还可以加载项目中的 HTML 文件，将 HTML 运行的结果显示在手机上，因此可以利用 HTML 进行较为复杂的 Android 手机界面的编写，这主要是通过 android.webkit.WebSetting 类实现的。

按照如下步骤加载 HTML 文件：

1）在要 Activity 中实例化 WebView 组件；例如：

```
WebView webView = new WebView(this);                    //实例化 WebView 组件
```

或者在布局文件中声明 WebView，然后在 Activity 中实例化 WebView；例如：

```
this.webview = (WebView) super.findViewById(R.id.webview);    //取得组件
```

2）开启 WebView 的支持 javascript 的功能；例如：

```
this.webview.getSettings().setJavaScriptEnabled(true);    //启用 javascript
```

3）通过 WebView 的 loadUrl 方法载入相应 html 文件，该 html 文件中包涵了 javascript 方法。例如：

```
this.webview.loadUrl("file:/android_asset/show.html") ;   //读取网页文件
```

4）需要在 AndroidManifest.xml 文件中添加权限，否则会出现错误。

```
<uses-permission android:name="android.permission.INTERNET" />
```

实例 9-2：使用 WebView 加载 HTML 文件实例

新建一个项目，项目的命名为：exam9_2，包名称为：org.hnist.demo，利用 WebView 组件加载项目中的 HTML 文件。

1）建立一个 html 文件，例如：show.html，将它保存在 assets\html 文件夹中，其需要的图片文件也放在 assets 文件夹中，如图 9.4 所示。

该 HTML 文件会出现一个下拉列表，里面含有一些链接。show.html 代码如下：

图 9.4 html 文件放置的位置

```html
<meta http-equiv="Content-Type" content="text/html ; charset=GBK">
<script language="javascript">
function openurl(url) {
    window.location = url ; }     //实现网页跳转
</script>
<center>
<img src="../images/xh.jpg" width="150" height="160">
<h3>请选择您要打开的网站：</h3>
<select name="url" onchange="openurl(this.value)">    //选择指定 url 的值
    <option value="http://www.hnist.cn">湖南理工学院</option>
    <option value="http://xw.hnist.cn">新闻学院</option>
    <option value="http://law.hnist.cn/">政法学院</option>
    <option value="http://art.hnist.cn/">美术学院</option>
    <option value="http://61.187.92.238:8105/">信息学院</option>
</select>
</center>
```

2）修改 activity_main.xml 文件，代码如下：

```xml
<?xml version="1.0" encoding="utf-8"?>
<LinearLayout
xmlns:android="http://schemas.android.com/apk/res/android"
android:orientation="vertical"
android:layout_width="fill_parent"
android:layout_height="fill_parent">
<WebView
    android:id="@+id/webview"
    android:layout_width="fill_parent"
    android:layout_height="fill_parent"/>
</LinearLayout>
```

3）建立 Activity 文件 MainActivity.java，读取 HTML 文件，代码如下：

```java
package org.hnist.demo;
import android.app.Activity;
import android.os.Bundle;
import android.webkit.WebView;
public class MainActivity extends Activity {
private WebView webview = null ;
@Override
public void onCreate(Bundle savedInstanceState) {
    super.onCreate(savedInstanceState);
    super.setContentView(R.layout.activity_main);
    this.webview = (WebView) super.findViewById(R.id.webview) ;//取得组件
    this.webview.getSettings().setJavaScriptEnabled(true) ; //启用 JavaScript
    this.webview.getSettings().setBuiltInZoomControls(true); //控制页面缩放
    this.webview.loadUrl("file:/android_asset/html/show.html");   }}
                                                                //读取网页
```

4）需要在 AndroidManifest.xml 文件中添加权限，否则会出现错误。

```xml
<uses-permission android:name="android.permission.INTERNET" />
```

运行该项目，弹出如图 9.5 所示界面，单击下拉列表，可以选择要打开的网站。

9.2.4 使用 WebView 加载 JSP 文件

类似地可以使用 WebView 组件加载 JSP 文件,将 JSP 程序的运行结果显示在手机上,但是前提是要先在浏览器里正常浏览该 JSP 文件。

按照如下步骤加载 JSP 文件:

1) 要确保该 JSP 文件能在浏览器里正确浏览;

① 下载、安装 Tomcat,安装完成,在浏览器中输入 http://localhost:8080,如果出现如图 9.6 所示的界面,表示安装设置成功。

图 9.5 加载 html 文件运行结果

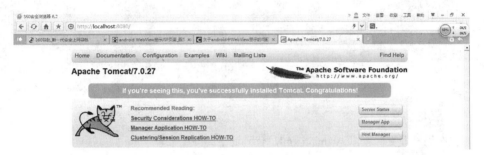

图 9.6 安装 Tomcat 成功

② 找到 Tomcat 的安装目录,里面有个 webapps 文件夹,在里面建立一个文件夹(例如 JSP),将要发布的 JSP 文件(例如:exam1.jsp)复制到该目录,在浏览器中输入 http://localhost:8080/jsp/exam1.jsp,出现如图 9.7 所示的界面。

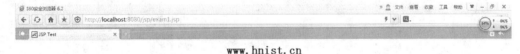

图 9.7 JSP 在 Tomcat 下的运行情况

2) 进入手机编程环境,在 Activity 中实例化 WebView 组件;例如:

```
WebView webView = new WebView(this);          //实例化 WebView 组件
```

或者在布局文件中声明 WebView,然后在 Activity 中实例化 WebView;例如:

```
this.webview = (WebView) super.findViewById(R.id.webview) ;//取得组件
```

3) 要开启 WebView 的支持 JavaScript 的功能;例如:

```
this.webview.getSettings().setJavaScriptEnabled(true) ;//启用 JavaScript
```

4) 通过 WebView 的 loadUrl 方法载入相应 JSP;例如:

```
this.webview.loadUrl("http://10.0.2.2:8080/jsp/exam1.jsp"); //读取 JSP 文件
```

这里的 10.0.2.2 是指本地的计算机的 IP 地址,利用该计算机充当 Tomcat 服务器,当然也可以利用一台在 Internet 实际存在的一个 Tomcat 服务器 IP 地址。

5) 需要在 AndroidManifest.xml 文件中添加权限,否则会出现错误。

```
<uses-permission android:name="android.permission.INTERNET" />
```

实例 9-3：使用 WebView 加载 JSP 文件实例

新建一个项目，项目的命名为：exam9_3，包名称为：org.hnist.demo，利用 WebView 组件加载 JSP 文件。

1）建立一个 JSP 文件，例如：exam1.jsp，将它保存在 webapps\jsp 文件夹中，exam1.jsp 代码如下：

```jsp
<%@ page language="java" import = "java.util.*" pageEncoding = "gb2312" %>
<HTML>
    <HEAD>
     <TITLE>
      JSP 测试
     </TITLE>
    </HEAD>
    <BODY>
<center>
     <% out.println("<h1>www.hnist.cn<br>信息学院</h1>"); %>
</center>
    </BODY>
    </HTML>
```

2）建立布局管理文件 activity_main.xml，代码如下：

```xml
<?xml version="1.0" encoding="utf-8"?>
<LinearLayout
xmlns:android="http://schemas.android.com/apk/res/android"
android:orientation="vertical"
android:layout_width="fill_parent"
android:layout_height="fill_parent">
<WebView
    android:id="@+id/webview"
    android:layout_width="fill_parent"
    android:layout_height="fill_parent"/>
</LinearLayout>
```

3）建立 Activity 文件 MainActivity.java，读取 JSP 文件，代码如下：

```java
package org.hnist.demo;
import android.app.Activity;
import android.os.Bundle;
import android.webkit.WebView;
public class MainActivity extends Activity {
private WebView webview = null ;
@Override
public void onCreate(Bundle savedInstanceState) {
    super.onCreate(savedInstanceState);
    super.setContentView(R.layout.activity_main); //默认布局管理器
    this.webview = (WebView) super.findViewById(R.id.webview) ;  //取得组件
    this.webview.getSettings().setJavaScriptEnabled(true); //启用 JavaScript
    this.webview.getSettings().setBuiltInZoomControls(true); //控制页面缩放
    this.webview.loadUrl("http://10.0.2.2:8080/jsp/exam1.jsp");}}//读取 JSP 文件
```

4）需要在 AndroidManifest.xml 文件中添加权限，否则会出现错误。

```
<uses-permission android:name="android.permission.INTERNET" />
```

保存所有文件，运行该项目，弹出如图 9.8 所示界面。

图 9.8　加载 JSP 文件的运行结果

9.2.5　JavaScript 调用 WebView 中的数据

前面两个例子介绍了 WebView 调用 HTML 文件和 JSP 文件，事实上还可以直接利用 WebView 给 JavaScript 传递参数，或者利用 JavaScript 直接给 WebView 传递参数。

JavaScript 调用 WebView 中的数据按照如下步骤进行：

1）首先，定义一个 WebView 组件，并设置启用 JavaScript。例如：

```
private WebView webview = null ;
this.webview.getSettings().setJavaScriptEnabled(true) ;
```

2）然后，在 WebView 中定义一个内嵌页面中可以触发的事件；必须依靠 WebView 中的方法 public void addJavascriptInterface(Object 操作对象, String 标记名称)来完成。例如：

```
this.webview.addJavascriptInterface(this, "android");
```

这里定义别名为"android"，在操作的对象中设置一个方法，以便在 Javascript 中调用。例如：

```
public String mydata(){
return "好好学习，天天向上"; }
```

3）利用 loadUrl()方法打开网页文件。例如：

```
this.webview.loadUrl("file:/android_asset/html/demo.html");
```

4）在 HTML 文件（或 JSP 文件）中添加 JavaScript 语句，获得 WebView 中的数据。

例如：document.getElementById("msg").innerHTML=window.android.mydata();，其中 getElementById 是指通过元素的 ID 值来获取元素，innerHTML 是指设置或获取某一 ID 的节点里面所包含的 HTML 代码，这里的数据来源于别名为 android 的 mydata()方法。

5）需要在 AndroidManifest.xml 文件中添加权限。

```
<uses-permission android:name="android.permission.INTERNET" />
```

实例 9-4：使用 WebView 给 JSP 文件传递数据实例

新建一个项目，项目的命名为：exam9_4，包名称为：org.hnist.demo，利用 WebView 组件给 JSP 文件传递数据。

1）建立一个 demo.html 文件，放在 asset/html 文件夹中，具体代码如下：

```
<html>
<head>
<meta http-equiv="Content-Type" content="text/html; charset=GBK">
<title>测试页面</title>
```

```
</head>
<body>
   HTML 页面接收来自 android 的数据
   <br>
   <div id="msg">等待参数的传递......</div>
   <input type="submit" value="获得参数"
onclick="document.getElementById('msg').innerHTML=window.android.mydata()"/>
</body>
</html>
```

2）建立布局文件 activity_main.xml，代码如下：

```xml
<?xml version="1.0" encoding="utf-8"?>
<LinearLayout
xmlns:android="http://schemas.android.com/apk/res/android"
android:orientation="vertical"
    android:background="#000000"
    android:layout_width="fill_parent"
android:layout_height="fill_parent">
<WebView
    android:id="@+id/webview"
    android:layout_width="fill_parent"
    android:layout_height="fill_parent"/>
</LinearLayout>
```

3）建立 Activity 文件 MainActivity.java，代码如下：

```java
package org.hnist.demo;
import android.app.Activity;
import android.os.Bundle;
import android.webkit.WebView;
public class MainActivity extends Activity {
private WebView webview = null ;
@Override
public void onCreate(Bundle savedInstanceState) {
    super.onCreate(savedInstanceState);
    super.setContentView(R.layout.activity_main);     //默认布局管理器
    this.webview = (WebView) super.findViewById(R.id.webview);  //取得组件
    this.webview.getSettings().setJavaScriptEnabled(true) ;  //启用 JavaScript
    this.webview.getSettings().setBuiltInZoomControls(true);//控制页面缩放
    this.webview.addJavascriptInterface(this, "android");
    this.webview.loadUrl("file:/android_asset/html/demo.html");}//读取 html 网页
       public String mydata(){
       return "好好学习，天天向上";       }       }
```

4）需要在 AndroidManifest.xml 文件中添加权限，否则会出现错误。

```xml
<uses-permission android:name="android.permission.INTERNET" />
```

保存所有文件，运行该项目，弹出如图 9.9 所示界面，单击按钮，弹出如图 9.10 所示界面，其中"好好学习，天天向上"数据来源于 Android 的 Java 程序。

不知道什么原因这些代码在 Android sdk 2.3.3 下不能顺利执行，什么也不显示，在 Android sdk 2.2 下就可以顺利执行了，执行时要注意更换模拟机的版本。

图 9.9　加载 Html 文件运行结果

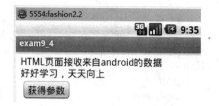
图 9.10　接收来自 Java 的数据

9.2.6　WebView 调用中 JavaScript 的数据

JavaScriptWebView 调用 JavaScript 中的数据按照如下步骤进行：

1）定义一个 WebView 组件，并设置启用 JavaScript。例如：

```
private WebView webview = null ;
  this.webview.getSettings().setJavaScriptEnabled(true) ;
```

2）在 WebView 中定义一个内嵌页面中可以触发的事件；必须依靠 WebView 中的方法 public void addJavascriptInterface(Object 操作对象, String 标记名称)来完成。例如：

```
this.webview.addJavascriptInterface(new runJavaScript(), "myjs");
```

这里定义别名为" myjs "，在操作的对象中设置一个方法，以便调用 JavaScript 中的数据。

3）利用 loadUrl()方法打开网页文件。例如：

```
this.webview.loadUrl("file:/android_asset/html/demo.html");
```

4）在 HTML 文件（或 JSP 文件）中添加 JavaScript 语句，以便将数据可以传递到 WebView 中。

5）需要在 AndroidManifest.xml 文件中添加权限。

```
<uses-permission android:name="android.permission.INTERNET" />
```

实例 9-5：使用 WebView 接收 JSP 文件传来数据实例

新建一个项目，项目的命名为：exam9_5，包名称为：org.hnist.demo，利用 WebView 组件接收来自 JSP 文件传来的数据。

1）建立一个 demo.html 文件，放在 asset/html 文件夹中，具体代码如下：

```
<html xmlns="http://www.w3.org/1999/xhtml ">
<head>
<meta http-equiv="Content-Type" content="text/html; charset=utf-8" />
<script language="javascript" type="text/javascript">
function send2Android(){
 var str = document.getElementById("mess").value;
 window.myjs.runOnAndroidJavaScript(str); }//调用 android 的函数
</script>
</head>
<body>
 <input type="text"  id="mess" />
 <input type="button" value="Send To Android"  onclick="send2Android()"/>
<div id="show"></div>
</body>
</html>
```

2）建立布局文件 activity_main.xml，代码如下：

```
<?xml version="1.0" encoding="utf-8"?>
```

```xml
<LinearLayout
xmlns:android="http://schemas.android.com/apk/res/android"
    android:orientation="vertical"
    android:layout_width="fill_parent"
    android:layout_height="fill_parent"    >
<TextView
   android:id="@+id/show"
   android:layout_width="fill_parent"
   android:layout_height="wrap_content"
   android:text="等待来自javascript的数据……" />
<WebView
   android:id="@+id/wv"
   android:layout_width="fill_parent"
   android:layout_height="66dp" />
</LinearLayout>
```

3）建立 Activity 文件 MainActivity.java，代码如下：

```java
package org.lxh.demo;
import android.app.Activity;
import android.os.Bundle;
import android.os.Handler;
import android.view.View;
import android.view.View.OnClickListener;
import android.webkit.WebView;
import android.widget.Button;
import android.widget.EditText;
import android.widget.TextView;
public class MainActivity extends Activity {
 private EditText txt;
 private WebView wv;
 private Button btn;
 private Handler h = new Handler();
private WebView webview = null ;
    @Override
    public void onCreate(Bundle savedInstanceState) {
      super.onCreate(savedInstanceState);
      setContentView(R.layout.activity_main);
      this.webview = (WebView) super.findViewById(R.id.wv) ;    //取得组件
      this.webview.getSettings().setJavaScriptEnabled(true) ;//启用JavaScript
      this.webview.getSettings().setBuiltInZoomControls(true);//控制页面缩放
      this.webview.addJavascriptInterface(new runJavaScript(), "myjs");
      this.webview.loadUrl("file:/android_asset/html/demo.html ") ; }
                                                                //读取网页
    final class runJavaScript{    //这个 Java 对象是绑定在另一个线程里的，
     public void runOnAndroidJavaScript(final String str){
       h.post(new Runnable(){
    @Override
    public void run() {
     TextView show = (TextView) findViewById(R.id.show);
        show.setText("这是来自 javascript 的数据:"+str); } }); } }}
```

第9章 Android 网络通信技术

运行该项目,弹出如图 9.10 所示界面,单击按钮,弹出如图 9.11 所示界面,其中"Hello!Android!"数据来源于 javascript 程序。

图 9.10 加载 Html 文件运行结果

图 9.11 接收来自 Jsp 文件的数据

9.3 利用 HttpURLConnection 开发 HTTP 程序

9.3.1 HttpURLConnection 基础

在 Android 中除了使用 WebView 组件访问网络以外,还可以用代码方式访问网络,代码方式有时候会显得更加灵活。常见的两个类就是 Android SDK 中的 HttpURLConnection 和 HttpClient 类。

HttpURLConnection 是 Java 的标准类,HttpURLConnection 继承自 URLConnection,包含在包 java.net.*中,它的层次关系如下:

```
java.lang.Object
    java.net.URLConnection
        java.net.HttpURLConnection
```

要在 Android 的 Java 程序中使用 HttpURLConnection 类,必须在程序中使用下面的语句。

```
import java.net.HttpURLConnection;                    //导入 HttpURLConnection 类
```

HttpURLConnection 提供了如表 9-8 所示的常用方法。

表 9-8 HttpURLConnection 常用方法

方法	描述
public int getResponseCode()	获取服务器的响应代码
public String getResponseMessage()	获取服务器的响应消息
public String getResponseMethod()	获取发送请求的方法
public void getOutputStream()	接收数据流
public void getInputStream()	发送数据库
public void flush()	刷新对象输出流
public void close()	关闭流对象
public setUseCaches(boolean newValue)	设置是否使用缓存
public void setRequestMethod(String method)	设置发送请求的方法
public void setDoOutput(boolean newValue);	设置是否向 httpUrlConnection 输出
public void setDoInput(boolean newValue);	设置是否向 httpUrlConnection 输入
public void setRequestProperty("Content-type", String)	设置请求报文头,
public void setConnectTimeout(int timeout)	设置连接超时

HTTP 通信中使用最多的是 GET 请求和 POST 请求，GET 请求可以获得静态页面，也可以把参数放在 URL 字串的后面，传递给服务器；POST 参数不是放在 URL 字串里面，而是放在 HTTP 请求数据中。

还有一个重要的类是 URL，通常用来生成一个指向特定地址的 URL 实例。例如：

```
URL url = new URL("http://www.hnist.cn");  //HttpURLConnection 实例初始化
```

HttpURLConnection 和 URLConnection 都是抽象类，无法直接实例化对象，其对象主要通过 URL 的 openConnection 方法获得，创建一个 HttpURLConnection 连接的代码如下所示：

```
URL url = new URL("http://www.hnist.cn");
HttpURLConnection urlConn = (HttpURLConnection) url.openConnection();
```

openConnection 方法只创建 URLConnection 或者 HttpURLConnection 实例，并不进行真正的连接操作，在连接之前可以对其一些属性进行设置，例如：

```
connection.setDoOutput(true);                //设置输出流
connection.setDoInput(true);                 //设置输入流
connection.setRequestMethod("GET");          //设置请求的方式为 GET
connection.setRequestMethod("POST");         //设置请求的方式为 POST
connection.setUseCaches(false);              //设置 POST 请求方式不能够使用缓存
urlConn.disconnect;                          //关闭 HttpURLConnection 连接
```

在完成 HttpURLConnection 实例的初始化以后，就可以通过 GET 方式或 POST 方式与服务器通信了。

9.3.2 HttpURLConnection 通信：GET 方式

HttpURLConnection 对网络资源的请求在默认情况下是使用 GET 方式，GET 请求可以获取静态页面，也可以把参数放在 URL 字符串后面，传递给服务器。

数据传送和读取会用到 InputStreamReader 读取字节并将其解码为字符，其 read()方法每次读取一个或多个字节。BufferedReader 流能够读取文本行，读取的量比 InputStreamReader 要多，通过向 BufferedReader 传递一个 Reader 对象，来创建一个 BufferedReader 对象，常用的方法如下：

1）BufferedReader(Reader in)：创建一个使用默认大小输入缓冲区的缓冲字符输入流。参数 in：一个 Reader。

2）BufferedReader(Reader in, int sz)：创建一个使用指定大小输入缓冲区的缓冲字符输入流。参数 in：一个 Reader，sz：输入缓冲区的大小。

实例 9-6：HttpURLConnection 实例：获取网络上的文件

新建一个项目，项目的命名为：exam9_6，包名称为：org.hnist.demo，利用 HttpURLConnection 的 GET 方式获取网络上的文件。例如：服务器http://61.187.92.238:8105 上有个 hnist.txt 文件，编程直接将该文件的内容在手机上显示出来。

1）在指定服务器指定位置上建立一个文件 hnist.txt，当然也可以在本机上建立，不过 IP 地址就要改为：10.0.2.2。

2）定义一个布局文件 activity_main.xml，代码如下：

```
<?xml version="1.0" encoding="utf-8"?>
<LinearLayout
xmlns:android="http://schemas.android.com/apk/res/android"
android:orientation="vertical"
android:layout_width="fill_parent"
```

```
android:layout_height="fill_parent"  >
  <TextView
      android:id="@+id/mytxt"
      android:layout_width="wrap_content"
      android:layout_height="wrap_content"/>
</LinearLayout>
```

3）定义一个 Activity 文件 MainActivity.java，代码如下：

```java
package org.hnist.demo;
import java.net.HttpURLConnection;
import java.net.URL;
import android.app.Activity;
import android.os.Bundle;
import android.widget.TextView;
import java.io.InputStreamReader;
import java.io.BufferedReader;
public class MainActivity extends Activity {
private static final String PATH = "http://61.187.92.238:8105/hnist.txt" ;
                                                      //定义路径
    private TextView msg = null ;                     //定义文本显示组件
    @Override
    public void onCreate(Bundle savedInstanceState) {
        super.onCreate(savedInstanceState);
        super.setContentView(R.layout.activity_main);    //调用布局管理器
        this.msg = (TextView) super.findViewById(R.id.mytxt) ; //取得文本组件
        try {
            this.msg.setText(this.getData() ) ;   //在文本框显示获得的文本内容
            } catch (Exception e) {
            //TODO Auto-generated catch block
            e.printStackTrace();} }
public String getData() throws Exception {             //定义 getData()方法
    URL url = new URL(PATH) ;                          //定义 URL
//打开连接
 HttpURLConnection conn = (HttpURLConnection) url.openConnection();
//读取文件到缓冲区，以 GBK 编码形式，否则可能出现乱码
InputStreamReader inputReader=new InputStreamReader(conn.getInputStream(),"GBK");
    //从缓冲区读取内容
BufferedReader bufReader = new BufferedReader(inputReader);
    String line = "";
    String Data = "";
    while ((line = bufReader.readLine()) != null) { Data += line; }//循环
        return Data ;    }}                         //返回获得的数据
```

4）在 AndroidManifest.xml 文件中添加权限，否则会出现错误。

```xml
<uses-permission android:name="android.permission.INTERNET" />
```

运行该项目，结果如图 9.12 所示。

除了可以读取文本文件外，还可以读取图片、MP3 文件、视频文件，不过读取数据的方式可能会不同，建议用字节数组流 ByteArrayOutputStream 来读取。

ByteArrayOutputStream 可以捕获内存缓冲区的数据,转换成字节数组。

ByteArrayOutputStream 类是在创建它的实例时,程序内部创建一个 byte 类型数组的缓冲区,然后利用 ByteArrayOutputStream 和 ByteArrayInputStream 的实例向数组中写入或读出 byte 型数据。

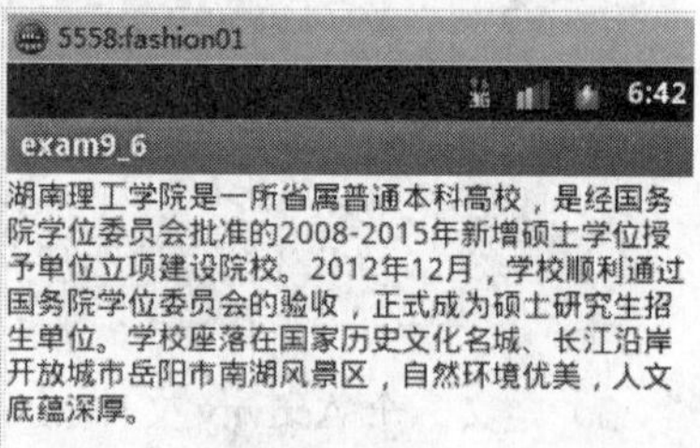

实例 9-7:HttpURLConnection 实例:获取网络上的图片。

新建一个项目,项目的命名为:exam9_7,包名称为:org.hnist.demo,利用 HttpURLConnection 的 GET 方式获取网络上的图片。例如:服务器http://61.187.92.238:8105 上有个 xh.jpg 文件,编程直接将该图片在手机上显示出来。

图 9.12 显示网络上的文本文件

1)在指定服务器指定位置上建立一个图片文件 xh.jpg,当然也可以在本机上建立,不过 IP 地址就要改为:10.0.2.2。

2)定义一个布局文件 activity_main.xml,只需设置一个 ImageView,代码如下:

```xml
<?xml version="1.0" encoding="utf-8"?>
<LinearLayout
xmlns:android="http://schemas.android.com/apk/res/android"
android:orientation="vertical"
android:layout_width="fill_parent"
android:layout_height="fill_parent"  >
    <ImageView
        android:id="@+id/imgt"
        android:layout_width="wrap_content"
        android:layout_height="wrap_content"/>
</LinearLayout>
```

3)定义一个 Activity 文件 MainActivity.java,代码如下:

```java
package org.hnist.demo;
import java.io.ByteArrayOutputStream;
import java.io.InputStream;
import java.net.HttpURLConnection;          //导入 HttpURLConnection 类
import java.net.URL;
import android.app.Activity;
import android.graphics.Bitmap;             //导入 graphics.Bitmap 类
import android.graphics.BitmapFactory;      //导入 graphics.BitmapFactory 类
import android.os.Bundle;
import android.widget.ImageView;
public class MainActivity extends Activity {
private static final String PATH = "http://61.187.92.238:8105/xh.jpg" ;
private ImageView img = null ;              //定义图片显示组件
@Override
public void onCreate(Bundle savedInstanceState) {
    super.onCreate(savedInstanceState);
    super.setContentView(R.layout. activity_main);
    this.img = (ImageView) super.findViewById(R.id.img) ;
    try {
        byte data [] = this.getPic() ;      //接收数据
```

```
            //根据获得的数据生成图形
    Bitmap bm = BitmapFactory.decodeByteArray(data, 0, data.length);
         this.img.setImageBitmap(bm) ;      //在组件中显示图片
      } catch (Exception e) {
         e.printStackTrace();} }
public byte[] getPic() throws Exception {//定义getPic()方法
     ByteArrayOutputStream pic = null ;     //内存输出流
     try {
        URL url = new URL(PATH) ;         //定义URL
        pic = new ByteArrayOutputStream() ;   //定义内存输出流
        byte data [] = new byte[1024] ;       //每次读取1024字节
        //打开连接
HttpURLConnection conn = (HttpURLConnection) url.openConnection();
        InputStream inputstream = conn.getInputStream() ;   //取得输入流
        int len = 0 ;                                       //接收读取长度
        while((len = inputstream.read(data)) != -1) {       //循环读取
            pic.write(data, 0, len) ;  }                    //向内存中保存
        return pic.toByteArray() ;                          //变为字节数组返回
      } catch (Exception e) {
         throw e ;
      } finally {
         if (pic != null) { pic.close(); }}}}              //关闭输出流
```

4）在 AndroidManifest.xml 文件中添加权限，否则会出现错误。

```
<uses-permission android:name="android.permission.INTERNET" />
```

运行该项目，结果如图 9.13 所示。

图 9.13 显示网络上的图片文件

前面介绍的 WebView 组件可以把数据传送到 JavaScript，利用 GET 请求可以把参数放在 URL 字符串后面，也可以把数据传递给服务器的 JavaScript 程序。将上面的代码做适当的修改，例如：

```
//定义一个字符串
String httpUrl = "http://10.0.2.2:8080/jsp/get.jsp?name=aa&password=123456";
URL url = new URL(httpUrl);  //实例化URL
//打开连接
HttpURLConnection urlConn = (HttpURLConnection) url.openConnection();
```

代码表示将变量 name 值为 aa，变量 password 值为 123456，发送到服务器端的 get.jsp 程序，在 get.jsp 中应该有相应的语句接收，例如：

```
String name = request.getParameter("name ") ;
String password = request.getParameter("password") ;
```

实例 9-8：HttpURLConnection 实例：GET 方式传递数据到 JSP 文件

建立一个 Android 项目，项目的名称为 exam9_8，包名称为"org.hnist.demo"利用 HttpURLConnection 的 GET 方式传递数据到指定的 JSP 文件。

1）建立一个 get.jsp 文件，get.jsp 代码如下：

```
<%    //接收发送来的请求参数
String name = request.getParameter("name") ;
String password = request.getParameter("password") ;
if("hnist".equals(name) && "8648870".equals(password)) {
out.println("Receive name is:"+name);
out.println("Receive password is:"+password); %>
Your message are right!   <%}
else {
out.println("Receive name is:"+name);
out.println("Receive password is:"+password);  %>
    Your message are Wrong!
<%}%>
```

在 Tomcat 上发布后，在浏览器上输入地址，显示结果如图 9.14 所示。

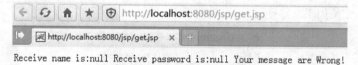

图 9.14　get.jsp 文件浏览结果

表示并没有参数值。

2）定义个布局文件 activity_main.xml，代码如下：

```
<?xml version="1.0" encoding="utf-8"?>
<LinearLayout
xmlns:android="http://schemas.android.com/apk/res/android"
android:orientation="vertical"
android:layout_width="fill_parent"
android:layout_height="fill_parent"  >
    <TextView
        android:id="@+id/msg"
        android:layout_width="wrap_content"
        android:layout_height="wrap_content"/>
</LinearLayout>
```

3）建立 Activity 文件 MainActivity.java 文件，代码如下：

```
package org.hnist.demo;
import java.io.InputStreamReader;            //导入 InputStreamReader 类
import java.net.HttpURLConnection;            //导入 HttpURLConnection 类
```

```java
import java.net.URL;                              //导入URL类
import android.app.Activity;
import android.os.Bundle;
import android.widget.TextView;
public class MainActivity extends Activity {
public void onCreate(Bundle savedInstanceState) {
    super.onCreate(savedInstanceState);
    super.setContentView(R.layout.activity_main);
    TextView msg = (TextView) super.findViewById(R.id.msg);   /
    try {
    //连接地址,同时传递参数
    URL url=new URL("http://10.0.2.2:8080/jsp/get.jsp?name=
        hnist&password=8648870");
    HttpURLConnection conn = (HttpURLConnection) url.openConnection();
                                                            //打开连接
        byte [] data = new byte[200] ;              //开辟空间
        int len = conn.getInputStream().read(data) ; //接收数据
        if(len > 0){
            String temp = new String(data,0,len).trim() ;
            msg.setText(temp) ;              }       //显示接收数据
        conn.getInputStream().close() ;              //关闭输入流
    } catch (Exception e) {
        e.printStackTrace() ;
        msg.setText("服务器连接失败。") ;      }    }}
```

4)在 AndroidManifest.xml 文件中添加权限,否则会出现错误。

```
<uses-permission android:name="android.permission.INTERNET" />
```

运行该项目,结果如图 9.15 所示。

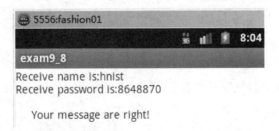

图 9.15 传递参数到 Jsp 文件后,android 接收显示结果

9.3.3 HttpURLConnection 通信:POST 方式

POST 与 GET 的不同在于 POST 的参数不能放在 URL 字符串后面,而是放在 HTTP 请求数据中,这些参数会通过 cookie 或者 session 等其他方式以键值对的形式 key=value 传送到服务器。

按照下面的几个步骤来实现 POST 方式的通信。

1)和 GET 方式一样也需要创建 HttpURLConnection 实例,例如:

```
URL url=new URL("http://10.0.2.2:8080/jsp/post.jsp");//定义一个URL
HttpURLConnection urlConn =(HttpURLConnection) url.openConnection();
```

2)在 POST 方式中,openConnection 方法只创建了 HttpURLConnection 实例,但是并不进行真正

的连接操作，连接之前需要对其一些属性进行设置，比如超过时间等。下面是对 HttpURLConnection 实例的属性设置：

```
urlConn.setDoOutput(true);//设置是否向urlConn输出，设为true，默认为false
urlConn.setDoInput(true);  //设置是否从urlConn读入，默认为true；
//设定传送的内容类型是可序列化的Java对象（如果不设此项，在传送序列化对象时，当Web服务
默认的不是这种类型时可能抛java.io.EOFException出错信息。）
urlConn.setRequestProperty("Content-type","application/x-java-serialized-object");
urlConn.setRequestMethod("POST");       //设置方式为POST，默认为GET方式
urlConn.setUseCaches(false);            //请求不能使用缓存
……                                      //注意，这些配置必须在连接之前完成
```

3）建立连接。

```
urlConn.connect();            //建立HttpURLConnection连接
//getOutputStream()方法会隐含进行连接，开发中不调用connect()也可建立连接
OutputStream outStrm = urlConn.getOutputStream();
```

4）**HttpURLConnection** 写数据与发送数据。

① 定义要上传的数据，传送到服务器，接收返回数据并读出。

```
//定义DataOutputStream对象，传送数据
DataOutputStream out = new DataOutputStream(httpconn.getOutputStream());
String content ="par="+URLEncoder.encode("hello", "gb2312");//定义要上传的参数
out.writeBytes(content);            //将要上传的内容写入流中
out.flush();                        //刷新对象输出流
out.close();                        //关闭输出流对象
//将HTTP请求发送到服务端，获取返回的内容
BufferedReader reader = new BufferedReader(new InputStreamReader
                    (urlConn.getInputStream())) ;
String inputLine = null;
while(((inputLine = reader.readLine()) != null)){//使用循环来读取获得的数据
resultData += inputLine + "\n"; }   //在每一行后面加上一个"\n"来换行
reader.close();                     //关闭输入流对象
urlConn.disconnect();               //关闭HttpURLConnection连接
```

② 或者采用下面的方式进行数据的读写。

```
//构建一个对象输出流对象，以实现输出可序列化的对象
ObjectOutputStream out = new ObjectOutputStream(outStrm);
//向对象输出流写出数据，这些数据将存到内存缓冲区中
out.writeObject(new String("This is a test!"));
out.flush();            //刷新对象输出流，将字节都写入流
//关闭流对象，不能再向对象输出流写入数据，之前写入的数据保存在缓冲区中
out.close();
//调用HttpURLConnection连接对象的getInputStream()函数，将缓冲区中封装好的完整的
    HTTP请求发送到服务端
InputStream inStrm = urlConn.getInputStream(); //将HTTP请求发送到服务端
```

一旦调用 getInputStream()表示本次 HTTP 请求已结束，后面向对象输出流的输出已无意义，即使对象输出流没有调用 close()方法，后面的操作也不会向对象输出流写入任何数据。因此，要重新发送数据到服务器时，就需要重新创建连接、重新设置参数、重新创建流对象、重新写数据。

如果还要将返回的数据读出来，就用上面的方式一样对 inStrm 进行操作。

5）在 AndroidManifest.xml 文件中添加权限，否则会出现错误。

```
<uses-permission android:name="android.permission.INTERNET" />
```

实例 9-9：HttpURLConnection 实例：POST 方式传递数据到 JSP 文件

新建一个项目，项目的命名为：exam9_9，包名称为：org.hnist.demo，利用 HttpURLConnection 的 POST 方式传递数据到指定的 JSP 文件。

1）在服务器上建立一个不需要传递参数的网页 get1.jsp 文件。代码如下：

```
<%    //接收发送来的请求参数
String name = request.getParameter("name") ;
if("hnist".equals(name) ) {
out.println("Receive name is:"+name); %>
    Your message are right!
<%}
else {  out.println("Receive name is:"+name); %>
    Your message are Wrong!
<%}%>
```

在 Tomcat 上发布后，在浏览器上输入地址，显示结果如图 9.16 所示。

图 9.16 get1.jsp 文件浏览结果

2）定义一个布局文件 activity_main.xml，代码如下：

```
<?xml version="1.0" encoding="utf-8"?>
<LinearLayout
xmlns:android="http://schemas.android.com/apk/res/android"
android:orientation="vertical"
android:layout_width="fill_parent"
android:layout_height="fill_parent"  >
    <TextView
        android:id="@+id/msg"
        android:layout_width="wrap_content"
        android:layout_height="wrap_content"/>
</LinearLayout>
```

3）建立 Activity 文件 MainActivity.java 文件，代码如下：

```
package org.lxh.demo;
import java.io.BufferedReader;
import java.io.DataOutputStream;
import java.io.IOException;
import java.io.InputStreamReader;
import java.net.HttpURLConnection;
import java.net.MalformedURLException;
import java.net.URL;
import java.net.URLEncoder;
import android.app.Activity;
import android.os.Bundle;
import android.util.Log;
import android.widget.TextView;
public class MainActivity extends Activity {
    public void onCreate(Bundle savedInstanceState){
```

```java
        super.onCreate(savedInstanceState);
        setContentView(R.layout.activity_main);
        TextView msg = (TextView)this.findViewById(R.id.msg);
        String httpUrl = "http://10.0.2.2:8080/jsp/get1.jsp";
        String resultData = "";//获得的数据
        URL url = null;
        try{
         url = new URL(httpUrl); //构造一个URL对象
        }catch (MalformedURLException e){          }
        if (url != null){
         try{
//使用HttpURLConnection打开连接
HttpURLConnection urlConn = (HttpURLConnection) url.openConnection();
            urlConn.setDoOutput(true); //POST请求需要设置为true
            urlConn.setDoInput(true);
            urlConn.setRequestMethod("POST");//设置以POST方式
            urlConn.setUseCaches(false);//Post请求不能使用缓存
//配置本次连接的Content-type,配置为application/x-www-form-urlencoded
urlConn.setRequestProperty("Content-Type","application/x-www-form-
                       urlencoded");urlConn.connect();
//DataOutputStream流
DataOutputStream out = new DataOutputStream(urlConn.getOutputStream());
//定义要上传的参数
        String content= "name=" + URLEncoder.encode("hnist", "gb2312");
out.writeBytes(content); //将要上传的内容写入流中
        out.flush();//刷新、关闭
        out.close();
        BufferedReader reader = new BufferedReader(new InputStreamReader
                       (urlConn.getInputStream()));//获取数据
        String inputLine = null;
        while(((inputLine = reader.readLine()) != null)){//使用循环来读取数据
         resultData += inputLine + "\n"; }   //在每一行后面加上一个"\n"来换行
        reader.close();
        urlConn.disconnect();//关闭http连接
        if ( resultData != null ){
         msg.setText(resultData);//设置显示取得的内容
        }else{
         msg.setText("读取的内容为NULL");
        }}
        catch (IOException e){          }
        }else{        } }}
```

4）在AndroidManifest.xml文件中添加权限，否则会出现错误。

```
<uses-permission android:name="android.permission.INTERNET" />
```

运行该项目，结果如图9.17所示。

HttpURLConnection要注意的几点：

1）HttpURLConnection的connect()函数，实际上只是建立了一个与服务器的TCP连接，并没有实际发送http请求。

无论是 POST 还是 GET，http 请求实际上直到 HttpURLConnection 的 getInputStream()这个函数里面才正式发送出去。

2）在用 POST 方式发送 URL 请求时，URL 请求参数的设定顺序很关键，对 connection 对象的配置都必须要在 connect()函数执行之前完成。而对 outputStream 的写操作，又必须要在 inputStream 的读操作之前。这些顺

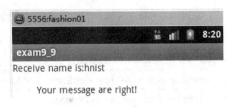

图 9.17　POST 方式传递参数到 JSP 文件

序实际上是由 http 请求的格式决定的。如果 inputStream 读操作在 outputStream 的写操作之前，会抛出例外：

```
java.net.ProtocolException: Cannot write output after reading input.......
```

3）http 请求实际上由两部分组成，一个是 http 头，所有关于此次 http 请求的配置都在 http 头里面定义，一个是正文 content。connect()函数会根据 HttpURLConnection 对象的配置值生成 http 头部信息，因此在调用 connect 函数之前，就必须把所有的配置准备好。

4）在 http 头后面紧跟着的是 http 请求的正文，正文的内容是通过 outputStream 流写入的，实际上 outputStream 不是一个网络流，往里面写入的东西不会立即发送到网络，而是存在于内存缓冲区中，等 outputStream 流关闭时，根据输入的内容生成 http 正文。至此，http 请求的东西已经全部准备就绪。在 getInputStream()函数调用的时候，就会把准备好的 http 请求 正式发送到服务器了，然后返回一个输入流，用于读取服务器对于这次 http 请求的返回信息。由于 http 请求在 getInputStream 的时候已经发送出去了（包括 http 头和正文），因此在 getInputStream()函数之后对 connection 对象进行设置或者写入 outputStream 都没有意义了，执行这些操作会导致异常的发生。

9.4　利用 HttpClient 开发 HTTP 程序

9.4.1　HttpClient 通信基础

在一般情况下，如果只是需要 Web 站点的某个简单页面提交请求并获取服务器响应，HttpURLConnection 是可以轻松实现的，很多情况下，可能需要用户登录而且具有相应的权限才可访问某些页面，这就涉及 Session、Cookie 的处理，使用 HttpURLConnection 来处理就不大合适了。Apache 提供了一个 HttpClient 类来处理这种情况。

可以说，HttpClient 是一个增强版的 HttpURLConnection，HttpURLConnection 可以做的事情 HttpClient 全部可以做，HttpClient 还提供了一些 HttpURLConnection 没有的功能。HttpClient 用于发送 HTTP 请求，接收 HTTP 响应，但不会缓存服务器的响应，不能执行 HTML 页面中嵌入的 JavaScript 代码；也不会对页面内容进行任何解析、处理，它只是关注于如何发送请求、接收响应，以及管理 HTTP 连接，通过它的 API 可以很方便地发出 GET，POST 请求与服务器进行数据的交互。

HttpClient 对 java.net 中的类做了封装和抽象，更适合在 Android 上开发互联网应用。在 HttpClient 类中，提供了下述常用的类及其方法。

1. ClientConnectionManager

ClientConnectionManager 是客户端连接管理器接口，用于对连接到服务器上的客户端进行管理，主要提供了如表 9-9 所示的 6 个抽象方法：

表 9-9 ClientConnectionManager 常用方法

方法	描述
abstract void closeExpiredConnections()	关闭所有无效超时的连接
abstract void closeIdleConnections(long idletime, TimeUnit tunit)	关闭所有空闲的连接
abstract SchemeRegistry getSchemeRegistry()	得到一个注册方案
abstract void releaseConnection(ManagedClientConnection conn, long validDuration, TimeUnit timeUnit)	释放一个连接
abstract ClientConnectionRequest requestConnection(HttpRoute route, Object state)	请求一个新的连接
abstract void shutdown()	关闭管理器并释放资源

2．DefaultHttpClient

DefaultHttpClient 是 HttpClient 的一个直接子类，它是一个默认的 HTTP 客户端，通常可以使用它来创建一个 HTTP 连接。DefaultHttpClient 常用方法见表 9-10。

表 9-10 DefaultHttpClient 常用方法

方法	描述
bublic DefaultHttpClient()	创建 DefaultHttpClient 对象
bublic abstract HttpResponse execute(HttpUriResquest resquest)	根据请求返回 HttpResponse 实例

例如：

```
HttpClient httpclient = new DefaultHttpClient();   //创建一个HTTP 连接
```

3．HttpRequest 接口

HttpGet 和 HttpPost 都是 HttpRequest 接口的直接子类，分别用于向服务器提交 GET 请求和 POST 请求。

HttpGet 类实现了 HttpRequest、HttpUriRequest 接口，常用方法见表 9-11。

表 9-11 HttpGet 常用方法

方法	描述
public HttpGet ()	无参数构造方法用以实例化对象
public HttpGet (URI uri)	通过 URI 对象构造 HttpGet 对象
public HttpGet (String uri)	通过指定的 uri 字符串地址构造实例化 HttpGet 对象

HttpPost 类也实现了 HttpRequest、HttpUriRequest 接口，常用方法见表 9-12。

表 9-12 HttpPost 常用方法

方法	描述
public HttpPost ()	无参数构造方法用以实例化对象
public HttpPost (URI uri)	通过 URI 对象构造 HttpPost 对象
public HttpPost (String uri)	通过指定的 uri 字符串地址构造实例化 HttpPost 对象
public String getMethod()	返回 HTTP 的操作状态，如 get、post。

4．HttpResponse 接口

HttpResponse 是一个 HTTP 连接响应，当执行一个 HTTP 连接后，就会返回一个 HttpResponse，可以通过 HttpResponse 获得一些响应的信息，常用方法见表 9-13。

表 9-13　HttpResponse 常用方法

方法	描述
public abstract HttpResponse execute (HttpUriRequest request)	通过 HttpUriRequest 对象执行返回一个 HttpResponse 对象
public abstract HttpResponse execute (HttpUriRequest request, HttpContext context)	通过 HttpUriRequest 对象和 HttpContext 对象执行返回一个 HttpResponse 对象
public abstract StatusLine getStatusLine ()	获得一个 StatusLine（含有 HTTP 协议版本，服务器发回的响应状态代码，状态码文本描述）接口的实例对象
public abstract Locale getLocale ()	获得一个 Locale 对象
public abstract ProtocolVersion getProtocolVersion ()	获得一个 ProtolVersion 对象它是一个 HTTP 版本的封装类，在这个类里定义了一系列的方法
public abstract String getReasonPhrase ()	状态码的文本描述
public abstract int getStatusCode ()	获得响应状态码
public abstract void setEntity(HttpEntity, entity)	设置一个 HttpEntity 对象
public abstract HttpEntity getEntity ()	获得一个 HttpEntity 对象

5．HttpEntity 接口

HttpEntity 常用方法见表 9-14。

表 9-14　HttpEntity 常用方法

方法	描述
public abstract InputStream getContent()	获得一个输入流对象，可以用这个流来操作文件（例如保存文件到 SD 卡）
public abstract Header getContentEncoding()	获得 Content-Encoding 信息头
public abstract Header getContentType ()	获得 Content-Type 信息头
public String getMethod()	返回 HTTP 的操作状态，如 get、post

通过 EntityUtils 类，它是一个 final 类，一个专门针对于处理 HttpEntity 的帮助类，常用方法见表 9-15。

表 9-15　EntityUtils 常用方法

方法	描述
public static String getContentCharSet (HttpEntity entity)	设置 HttpEntity 对象的 ContentCharset
public static byte[] toByteArray (HttpEntity entity)	将 HttpClient 转换成一个字节数组
public static String toString(HttpEntity entity, String defaultCharset)	通过指定的编码方式取得 HttpEntity 里字符串的内容
public static String toString (HttpEntity entity)	取得 HttpEntity 里字符串的内容 NameValuePair

NameValuePair 接口是一个简单的封闭的键值对，提供了 BasicNameValuePair(String name,String value) 方法指定要传入的参数名称和内容，以及 getName()方法取得参数名称和 getValue 方法取得参数内容。

还有一些类，由于篇幅的原因，这里不一一列出了，可以参考 Android API 学习，了解上述类和方法后，就可以通过下面的三个步骤来访问 HTTP 资源。

1）创建 HttpGet 或 HttpPost 对象，将要请求的 URL 通过构造方法传入 HttpGet 或 HttpPost 对象。

2）使用 DefaultHttpClient 类的 execute 方法发送 HTTP GET 或 HTTP POST 请求，并返回 HttpResponse 对象。

3）通过 HttpResponse 接口的 getEntity 方法返回响应信息，并进行相应的处理。

9.4.2　HttpClient 通信：GET 方式

按照前面描述的三个步骤来实现通过 GET 方式与服务器的通信：

1）创建 HttpGet 对象，将要请求的 URL 通过构造方法传入 HttpGet 对象，例如：

```
//定义http地址
String httpUrl="http://10.0.2.2:8080/jsp/get.jsp?name=hnist&password=8648870";
HttpGet httpRequest = new HttpGet(httpUrl);        //定义HttpGet 连接对象
```

2）使用 DefaultHttpClient 类的 execute 方法发送 HTTP GET 请求，并返回 HttpResponse 对象。

```
HttpClient httpclient = new DefaultHttpClient();    //取得HttpClient 对象
//发送请求，获得HttpResponse
HttpResponse httpResponse = httpclient.execute(httpRequest);
```

3）通过 HttpResponse 接口的 getEntity 方法返回响应信息，并进行相应的处理。

```
if (httpResponse.getStatusLine().getStatusCode() == HttpStatus.SC_OK){
                                                    //请求成功
    String strResult = EntityUtils.toString(httpResponse.getEntity());
                                                    //取得返回的字符串
……
```

4）在 AndroidManifest.xml 文件中添加权限，否则会出现错误。

```
<uses-permission android:name="android.permission.INTERNET" />
```

实例 9-10：HttpClient 通信实例：GET 方式传递数据到 JSP 文件。

建立一个 Android 项目，项目的名称为 exam9_10，包名称为"org.hnist.demo"，利用 HttpClient 的 GET 方式传递数据到指定的 JSP 文件。

1）在服务器上建立一个不需要传递参数的网页 get.jsp 文件。代码如下：

```
<%    //接收发送来的请求参数
String name = request.getParameter("name") ;
String password = request.getParameter("password") ;
if("hnist".equals(name) && "8648870".equals(password)) {
out.println("Receive name is:"+name);
out.println("Receive password is:"+password);%>
Your message are right!
<%}else {
out.println("Receive name is:"+name);
out.println("Receive password is:"+password);   %>
    Your message are Wrong!
<%}%>
```

该文件要在 Tomcat 服务器下发布，显示结果如图 9.18 所示。

Receive name is:null Receive password is:null Your message are Wrong!

图 9.18 POST 方式传递参数到 JSP 文件

表示并没有参数值。

2）定义一个布局文件 activity_main.xml，代码如下：

```
<?xml version="1.0" encoding="utf-8"?>
<LinearLayout
xmlns:android="http://schemas.android.com/apk/res/android"
```

```xml
android:orientation="vertical"
android:layout_width="fill_parent"
android:layout_height="fill_parent"  >
   <TextView
       android:id="@+id/msg"
       android:layout_width="wrap_content"
       android:layout_height="wrap_content"/>
</LinearLayout>
```

3）建立 Activity 文件 MainActivity.java 文件，代码如下：

```java
package org.hnist.demo;
import java.io.IOException;
import java.io.InputStreamReader;
import org.apache.http.HttpResponse;
import org.apache.http.HttpStatus;
import org.apache.http.client.ClientProtocolException;
import org.apache.http.client.HttpClient;
import org.apache.http.client.methods.HttpGet;
import org.apache.http.impl.client.DefaultHttpClient;
import org.apache.http.util.EntityUtils;
import android.app.Activity;
import android.os.Bundle;
import android.widget.TextView;
public class MainActivity extends Activity {
@Override
public void onCreate(Bundle savedInstanceState) {
    super.onCreate(savedInstanceState);
    super.setContentView(R.layout.activity_main);
    TextView msg = (TextView) super.findViewById(R.id.msg);
//定义http地址，设置传递参数
String httpUrl = "http://10.0.2.2:8080/jsp/get.jsp?name=
             hnist&password=8648870";
    HttpGet httpRequest = new HttpGet(httpUrl);          //HttpGet 连接对象
       try{
         HttpClient httpclient = new DefaultHttpClient();//取得HttpClient对象
         //请求HttpClient，取得HttpResponse
         HttpResponse httpResponse = httpclient.execute(httpRequest);
         if (httpResponse.getStatusLine().getStatusCode()==
                     HttpStatus.SC_OK){                  //如果请求成功
//取得返回的字符串
String strResult = EntityUtils.toString(httpResponse.getEntity());
msg.setText(strResult); }                               //显示接收数据
else{
             msg.setText("请求错误!"); }
             }catch (ClientProtocolException e){
             msg.setText(e.getMessage().toString());
             }catch (IOException e){
             msg.setText(e.getMessage().toString());
             }catch (Exception e){
             msg.setText(e.getMessage().toString()); } }}
```

4. 在 AndroidManifest.xml 文件中添加权限，否则会出现错误。

```
<uses-permission android:name="android.permission.INTERNET" />
```

运行该项目，结果如图 9.19 所示。

9.4.3 HttpClient 通信：POST 方式

按照前面描述的三个步骤来实现通过 POST 方式与服务器的通信：

1）创建 HttpPost 对象，将要请求的 URL 通过构造方法传入 HttpPost 对象，例如：

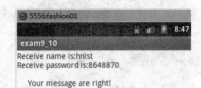

图 9.19 HttpClient GET 传递参数到 JSP 文件

```
String httpUrl="http://10.0.2.2:8080/jsp/get.jsp";    //定义http地址
HttpPost httpRequest = new HttpPost(httpUrl);         //定义HttpPost连接对象
```

需要使用 NameValuePair 来保存客户端传递的参数，可以使用 BasicNameValuePair 来构造一个要被传递的参数，通过 add 方法把这些参数加入到 NameValuePair 中，代码为：

```
//使用NameValuePair来保存要传递的Post参数
List<NameValuePair> params = new ArrayList<NameValuePair>();
params.add(new BasicNameValuePair("par", "Post Style"));  //添加要传递的参数
```

2）使用 DefaultHttpClient 类的 execute 方法发送 HTTP POST 请求，并返回 HttpResponse 对象。

```
//设置字符集，Post需要设置所使用的字符集，否则会出现乱码的现象
HttpEntity httpentity = new UrlEncodedFormEntity(params, "gb2312");
httpRequest.setEntity(httpentity);                     //请求httpRequest
HttpClient httpclient = new DefaultHttpClient();       //取得默认的HttpClient
HttpResponse httpResponse = httpclient.execute(httpRequest);
                                                       //取得HttpResponse
```

3）通过 HttpResponse 接口的 getEntity 方法返回响应信息，并进行相应的处理。

```
if (httpResponse.getStatusLine().getStatusCode() == HttpStatus.SC_OK){
                                                       //请求成功
    String strResult = EntityUtils.toString(httpResponse.getEntity());
                                                       //取得返回的字符串
......
```

4）在 AndroidManifest.xml 文件中添加权限，否则会出现错误。

```
<uses-permission android:name="android.permission.INTERNET" />
```

实例 9-11：HttpClient 通信实例：POST 方式传递数据到 JSP 文件

新建一个项目，项目的命名为：exam9_11，包名称为：org.hnist.demo，利用 HttpClient 的 POST 方式传递数据到指定的 JSP 文件。

1）建立一个 get.jsp 文件，该文件要在 Tomcat 服务器下发布，get.jsp 代码和上面例子中的一致。
2）定义一个布局文件 activity_main.xml，代码和上面例子中的一致。
3）建立 Activity 文件 MainActivity.java，代码如下：

```
package org.hnist.demo;
import java.io.IOException;
import java.io.InputStreamReader;
```

```java
import java.net.URL;
import java.util.ArrayList;
import java.util.List;
import org.apache.http.HttpEntity;
import org.apache.http.HttpResponse;
import org.apache.http.HttpStatus;
import org.apache.http.NameValuePair;
import org.apache.http.client.ClientProtocolException;
import org.apache.http.client.HttpClient;
import org.apache.http.client.entity.UrlEncodedFormEntity;
import org.apache.http.client.methods.HttpGet;
import org.apache.http.client.methods.HttpPost;
import org.apache.http.impl.client.DefaultHttpClient;
import org.apache.http.message.BasicNameValuePair;
import org.apache.http.util.EntityUtils;
import android.app.Activity;
import android.os.Bundle;
import android.widget.TextView;
public class MainActivity extends Activity {
@Override
public void onCreate(Bundle savedInstanceState) {
    super.onCreate(savedInstanceState);
    super.setContentView(R.layout.activity_main);
    TextView msg = (TextView) super.findViewById(R.id.msg);
    String httpUrl = "http://10.0.2.2:8080/jsp/get.jsp"; //http 地址
    HttpPost httpRequest = new HttpPost(httpUrl); //定义 HttpPost 连接对象
    //使用 NameValuePair 来保存要传递的 Post 参数
    List<NameValuePair> params = new ArrayList<NameValuePair>();
    //添加要传递的参数
    params.add(new BasicNameValuePair("name", "hnist"));
    params.add(new BasicNameValuePair("password", "8648870"));
    try{
        //设置字符集，Post 需要设置所使用的字符集
        HttpEntity httpentity = new UrlEncodedFormEntity(params, "gb2312");
        httpRequest.setEntity(httpentity);//请求 httpRequest
        HttpClient httpclient = new DefaultHttpClient(); //取得默认的 HttpClient
        //取得 HttpResponse
        HttpResponse httpResponse = httpclient.execute(httpRequest);
        if(httpResponse.getStatusLine().getStatusCode()==HttpStatus.SC_OK)
            {   //HttpStatus.SC_OK 表示连接成功
                //取得返回的字符串
                String strResult=EntityUtils.toString(httpResponse.getEntity());
                msg.setText(strResult);    //显示取得数据
            }else{
                msg.setText("请求错误!");  }
    }catch (ClientProtocolException e){
        msg.setText(e.getMessage().toString());
    }catch (IOException e){
        msg.setText(e.getMessage().toString());
    }catch (Exception e){
        msg.setText(e.getMessage().toString());   }  }}
```

4. 在 AndroidManifest.xml 文件中添加权限,否则会出现错误。

```
<uses-permission android:name="android.permission.INTERNET" />
```

这么多的数据传递形式到底采用哪个好呢？Android 建议采用 HttpURLConnection 形式传递数据,但是个人觉得 HttpClient 更简单、易于理解。HttpClient 实际上是对 Java 提供方法的一些封装,在 HttpURLConnection 中的输入/输出流操作,在这个接口中被统一封装成了 HttpPost(HttpGet)和 HttpResponse,这样就减少了操作的繁琐性。

GET 方式是通过把参数键值对附加在 URL 后面来传递,在服务器端可以从 "QUERY_STRING" 这个变量中直接读取,效率较高,但缺乏安全性,也无法来处理复杂的数据,长度有限制。主要用于传递简单的参数。

POST 方式：就传输方式讲,参数会被打包在 HTTP 报头中传输,可以是二进制的。从 CONTENT_LENGTH 这个环境变量中读取,便于传送较大一些的数据,同时因为不暴露数据在浏览器的地址栏中,安全性相对较高,但这样的处理效率会受到影响。

在实际运用中,读者应该根据具体情况采用合适的形式来传递参数。

9.4.4 数据的实时更新

前面介绍的只是简单地一次性获取网页数据,而在实际开发中更多的是需要实时获取最新数据,比如实时天气信息,实时交通信息等等。可以通过一个线程来控制视图的更新,要实时从网络获取数据,简单地说就是把获取网络数据的代码写到线程中,不停地进行更新。一般使用 Handler 来实现更新。

实例 9-12：数据实时更新实例

新建一个项目,项目的命名为：exam9_12,包名称为：org.hnist.demo,利用 HttpURLConnection 和 Handler 实现数据的实时更新。例如：创建一个网页来显示系统当前的时间,然后每隔 5 秒系统自动刷新一次视图。

1）创建一个显示当前系统时间的 jsp 网页文件,date.jsp 代码如下：

```
<%@ page language="java" import = "java.util.*" pageEncoding = "gb2312" %>
<%java.text.SimpleDateFormat simpleDateFormat=new java.text.
            SimpleDateFormat ("yyyy-MM-dd HH:mm:ss");
java.util.Date currentTime = new java.util.Date();
String time = simpleDateFormat.format(currentTime).toString();   }
out.println(time); %>
```

在 Tomcat 上发布该文件,显示界面如图 9.20 所示。

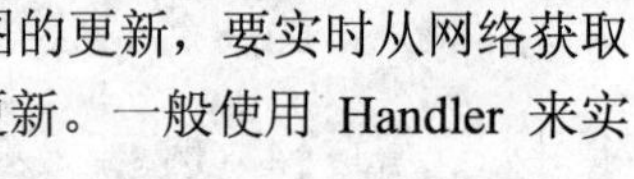

2）定义一个布局文件 activity_main.xml,其中一个文本显示框,用作显示当前系统时间,一个按钮,用作实时刷新,代码如下：

图 9.20 date.jsp 浏览界面

```
<?xml version="1.0" encoding="utf-8"?>
<LinearLayout
    xmlns:android="http://schemas.android.com/apk/res/android"
    android:orientation="vertical"
    android:layout_width="fill_parent"
    android:layout_height="fill_parent"  >
    <TextView
```

```xml
    android:id="@+id/Text"
    android:layout_width="fill_parent"
    android:layout_height="wrap_content"
    android:layout_gravity="center" />
<Button
    android:id="@+id/But"
    android:layout_width="fill_parent"
    android:layout_height="wrap_content"
    android:text="刷新" />
</LinearLayout>
```

3）建立 Activity 文件 MainActivity.java，代码如下：

```java
package org.hnist.demo;
import java.io.BufferedReader;
import java.io.IOException;
import java.io.InputStreamReader;
import java.net.HttpURLConnection;
import java.net.MalformedURLException;
import java.net.URL;
import android.app.Activity;
import android.os.Bundle;
import android.os.Handler;
import android.os.Message;
import android.util.Log;
import android.view.View;
import android.widget.Button;
import android.widget.TextView;
public class MainActivity extends Activity {
    private Button But;
    private TextView msg;
    public void onCreate(Bundle savedInstanceState){
     super.onCreate(savedInstanceState);
     setContentView(R.layout.activity_main);
     msg = (TextView)this.findViewById(R.id.Text);
     But = (Button)this.findViewById(R.id.But);
     But.setOnClickListener(new Button.OnClickListener(){
      public void onClick(View arg0){
        refresh();   }   });    //刷新
     new Thread(mRunnable).start();} //开启线程
    private void refresh(){//刷新网页显示
     String httpUrl = "http://10.0.2.2:8080/jsp/date.jsp";
     String resultData = "";
     URL url = null;
     try{
      url = new URL(httpUrl);//构造一个URL对象
     }catch (MalformedURLException e){      }
     if (url != null){
      try{
       //使用HttpURLConnection打开连接
HttpURLConnection urlConn = (HttpURLConnection) url.openConnection();
       InputStreamReader in = new InputStreamReader(urlConn.
```

```
                    getInputStream(),"GBK");           //得到读取的内容(流)
        BufferedReader buffer = new BufferedReader(in);
                                                //为输出创建BufferedReader
        String inputLine = null;
        while (((inputLine = buffer.readLine()) != null)){//使用循环来读取获得的数据
         resultData += inputLine + "\n"; }        //在每一行后面加上一个"\n"来换行
        in.close();//关闭InputStreamReader
        urlConn.disconnect();//关闭http连接
        if (resultData != null){
         msg.setText(resultData);//设置显示取得的内容
        }else{
         msg.setText("读取的内容为NULL");           }
       }catch (IOException e){            }
      }else{      }    }
private Runnable mRunnable = new Runnable(){
 public void run(){
  while (true){
   try{
    Thread.sleep(5 * 1000);                                //每隔5秒刷新一次
    mHandler.sendMessage(mHandler.obtainMessage());    //发送消息
   }catch (InterruptedException e){ }     }    };
Handler mHandler = new Handler(){
 public void handleMessage(Message msg){
  super.handleMessage(msg);//接收消息
  refresh();}      };}         //刷新
```

4)在 AndroidManifest.xml 文件中添加权限,否则会出现错误。

```
<uses-permission android:name="android.permission.INTERNET" />
```

保存所有文件,运行该项目,结果如图 9.21 所示,显示的时间来自于服务器。

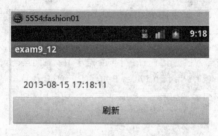

图 9.21　实时更新显示系统时间

9.5　利用 Socket 交换数据

Socket 是 TCP/IP 协议上的一种通信,在通信的两端各建立一个 Socket,从而在通信的两端之间形成网络虚拟链路。一旦建立了虚拟的网络链路,两端的程序就可以通过虚拟链路进行通信。

Socket 通信有两部分,一部分为监听的 Server 端,一部分为主动请求连接的 Client 端。Server 端会一直监听 Socket 中的端口直到有请求为止,当 Client 端对该端口进行连接请求时,Server 端就给予应答并返回一个 Socket 对象,以后 Server 端与 Client 端的数据交换就可以使用这个 Socket 来进行操作了。

Socket 有两种主要的通信方式:基于 TCP 协议和基于 UDP 协议的通信方式。

9.5.1 基于 TCP 协议的 Socket 通信

一个客户端要发起一次通信，首先必须知道运行服务器端的主机 IP 地址，通过指定的端口和服务器建立连接，然后进行通信，通信方式如图 9.22 所示。

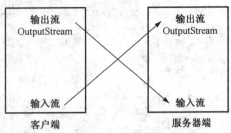

图 9.22 Socket 通信示意图

Socket 类的方法包含在 java.net.Socket 包中，下面通过实例给出常用的几个方法。

客户端有两种方法连接到服务器：

第一种：客户端 Socket 通过构造方法连接服务器。

```
Socket socket = new Socket("211.69.1.17",8000);  //新建 Socket，指定 IP 及端口号
```

第二种：通过 Connect 方法连接服务器。

```
Socket socket = new Socket();  //新建 Socket，未指定 IP 及端口号
socket.connect(new InetSocketAddress("211.69.1.1",80));  //使用默认的连接超时
socket.connect(new InetSocketAddres s("211.69.1.1",80),2000);  //连接超时 2s
socket.setSendBufferSize(8096);     //设置输出流的发送缓冲区大小，默认是 8KB
socket.setReceiveBufferSize(8096);  //设置输入流的接收缓冲区大小，默认是 8KB
socket.setKeepAlive(true);     //防止服务器端无效时，客户端长时间处于连接状态
OutputStream os = socket.getOutputStream();
                      //客户端向服务器端发送数据，获取客户端向服务器端输出流
//判断 Socket 是否处于连接状态
if((socket.isConnected() == true) && (socket.isClosed() == false)){……}
InputStream is = socket.getInputStream();
                      //客户端接收服务器端的响应，读取服务器端向客户端的输入流
byte[] buffer = new byte[is.available()];      //设置缓冲区
is.read(buffer);                               //读取缓冲区
String responseInfo = new String(buffer);      //转换为字符串
Log.i("TEST", responseInfo);                   //日志中输出
socket.close();                                //关闭连接
```

Socket 通信的实现要分别设计服务器端和客户端

1. 创建服务器端的步骤

1) 指定端口实例化一个 ServerSocket，会自动对传入的端口号进行监听；

```
ServerSocket server = new ServerSocket(9090);      //设置监听端口 9090
```

2) 收到连接请求后调用 ServerSocket 的 accept()，返回一个连接的 Socket 对象；

```
Socket client = server.accept();                    //接收客户端请求
```

3) 获取位于该层 Socket 的流以进行读写操作；

```
PrintStream out = new PrintStream(client.getOutputStream()); //获得客户端的输出流
```

4）将数据封装成流；

```
BufferedReader msg = new BufferedReader(new InputStreamReader(client.
                    getInputStream()));        //对收到的数据缓冲区读取
```

5）对 Socket 进行读写；

```
StringBuffer info = new StringBuffer();     //接收客户端发送回来的信息
info.append("I'm server!:");                //回应给客户端的数据
info.append(msg.readLine());                //接收客户端的数据
out.print(info);                            //发送信息到客户端
```

6）关闭打开的流。

```
out.close();                                //关闭输出流
msg.close();                                //关闭输入流
client.close();                             //关闭客户端连接
server.close();                             //关闭服务器端连接
```

2. 创建客户端的步骤

1）通过 IP 地址和端口实例化 Socket，请求连接服务器；

```
Socket client = new Socket("10.0.2.2", 9090);    //指定服务器及端口号
```

注意：10.0.2.2 是本地 PC 的 IP 地址，模拟机的 IP 地址是 127.0.0.1，在没有固定服务器的情况下，可以利用 PC 充当服务器进行实验。

通过上面的步骤后，Server 端和 Client 端就可以连接起来了，要进行数据交换需要用到 IO 流中的 OutputStream 类和 InputStream 类。

OutputStream：当应用程序需要对流进行数据写操作时，可以使用 Socket.getOutputStream()方法返回的数据流进行操作。

InputStream：当应用程序要从流中取出数据时，可以使用 Socket.getInputStream()方法返回的数据流进行操作。

2）获取 Socket 上的流以进行读写；

```
//客户端向服务器端发数据，获取客户端向服务器端的输出流
PrintStream out = new PrintStream(client.getOutputStream());
```

3）把流包装进 BufferReader/PrintWriter 对象；

```
//对返回的数据流进行缓冲区读取
BufferedReader msg = new BufferedReader(new InputStreamReader
                    (client.getInputStream()));
```

4）对 Socket 进行读写；

```
out.println("已经连接上服务器");               //发送数据到服务器
```

5）关闭打开的流。

```
out.close();                                //关闭输出流
msg.close() ;                               //关闭输入流
client.close();                             //关闭连接
```

实例 9-13：基于 TCP 协议的 Socket 实例

新建一个项目，项目的命名为：exam9_13，包名称为：org.hnist.demo，建立一个基于 TCP 协议的 Socket 通信实例。

1）用 Eclipse 在服务器端建立一个项目，server9_13，主程序 Server.java 代码如下：

```java
import java.io.BufferedReader;
import java.io.InputStreamReader;
import java.io.PrintStream;
import java.net.ServerSocket;
import java.net.Socket;
public class Server {
public static void main(String[] args) throws Exception {  //所有异常抛出
    ServerSocket server = new ServerSocket(9090);          //设置监听端口9090
    Socket client = server.accept();                       //接收客户端请求
    //获得客户端的数据流
    PrintStream out = new PrintStream(client.getOutputStream());
    BufferedReader msg = new BufferedReader(new InputStreamReader
    (client.getInputStream()));                            //对收到的数据在缓冲区读取
    StringBuffer info = new StringBuffer();                //接收客户端发送回来的信息
    info.append("I'm server!:");                           //回应给客户端的数据
    info.append(msg.readLine());                           //接收客户端的数据
    out.print(info);                                       //发送信息到客户端
    out.close();                                           //关闭输出流
    msg.close();                                           //关闭输入流
    client.close();                                        //关闭客户端连接
    server.close(); }}                                     //关闭服务器端连接
```

2）建立客户端程序，建立一个布局管理文件 activity_main./xml。

```xml
<?xml version="1.0" encoding="utf-8"?>
<LinearLayout
xmlns:android="http://schemas.android.com/apk/res/android"
android:orientation="vertical"
android:layout_width="fill_parent"
android:layout_height="fill_parent">
<Button
    android:id="@+id/send"
    android:layout_width="fill_parent"
    android:layout_height="wrap_content"
    android:text="连接到服务器" />
<TextView
    android:id="@+id/info"
    android:layout_width="fill_parent"
    android:layout_height="wrap_content"
    android:text="正在连接到服务器..." />
</LinearLayout>
```

3）建立客户端程序，建立一个 Activity 文件 MainActivity.java。

```java
package org.hnist.demo;
import java.io.BufferedReader;
import java.io.InputStreamReader;
import java.io.PrintStream;
import java.net.Socket;
import android.app.Activity;
import android.os.Bundle;
```

```java
import android.view.View;
import android.view.View.OnClickListener;
import android.widget.Button;
import android.widget.TextView;
public class MainActivity extends Activity {
    private Button send = null;
    private TextView info = null;
    @Override
    public void onCreate(Bundle savedInstanceState) {
        super.onCreate(savedInstanceState);
        super.setContentView(R.layout.activity_main);
        this.send = (Button) super.findViewById(R.id.send);
        this.info = (TextView) super.findViewById(R.id.info);
        this.send.setOnClickListener(new SendOnClickListenerImpl());}//设置事件
    private class SendOnClickListenerImpl implements OnClickListener{
        @Override
        public void onClick(View view) {
            try {
                Socket client = new Socket("10.0.2.2", 9090);//指定服务器及端口号
                //客户端向服务器端发送数据,获取客户端向服务器端的输出流
                PrintStream out = new PrintStream(client.getOutputStream());
                BufferedReader msg = new BufferedReader(new InputStreamReader
                    (client.getInputStream()));   //对返回的数据流进行缓冲区读取
                out.println("已经连接上服务器");      //发送数据到服务器
                MainActivity.this.info.setText(msg.readLine());   //设置文本内容
                out.close();                        //关闭输出流
                msg.close() ;                       //关闭输入流
                client.close(); }                   //关闭连接
            catch (Exception e) {
                e.printStackTrace();}}}}
```

4）在 AndroidManifest.xml 中配置相应的权限,在其中添加下面的代码。

```
<uses-permission android:name="android.permission.INTERNET"/>
```

保存所有文件,先运行服务器端的 Server.java,然后运行模拟器上的 MainActivity.java,结果如图 9.23 所示,单击按钮,弹出如图 9.24 所示界面。

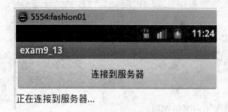 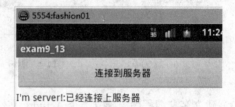

图 9.23　客户端运行界面　　　　　图 9.24　连接到服务器端的客户端界面

9.5.2　基于 UDP 协议的 Socket 通信

使用 TCP 协议这种方式要先和接收方建立连接,然后发送数据,保证数据成功地到达目的地,但是速度慢。而使用 UDP 协议这种方式先把数据打包成数据包,然后直接发送给接收方,不需要建立连接,速度快,但是可能会丢失数据。

使用基于UDP的Socket通信也包含服务器端的设计和客户端的设计。

服务端端的设计：

```
//创建服务端socket，并使之监听9999端口
DatagramSocket socket = new DatagramSocket (9999);
byte data[]=new byte[1024];
//数据包中放的空data来准备接收数据
DatagramPacket packet = new DatagramPacket (data,data.length);
//接收到数据报文，并将报文中的数据复制到指定的DatagramPacket实例中
socket.receive(packet);
String s=packet.getData();   //接收DatagramPacket实例中的数据,转换成字符串
特别的，当数据小于1024时，由于data指定的是1024，所以会出现乱码，解决的办法是：
    String s=new String(packet.getData(),packet.getOffset(),packet.getLength());
```

客户端的设计：

```
DatagramSocket socket = new DatagramSocket (9999);      //创建客户端socket
InetAddress serverAddress=InetAddress.getByName("211.699.1.1");//服务端地址
DatagramPacket packet = new DatagramPacket (data,data.length,
        serverAddress,9999);                 //打包要发送的数据
socket.send(packet );                        //发送DatagramPacket对象
```

一个简单的实例：使用UDP协议完成数据的传送。

1) 服务端的主要代码如下：

```
//创建一个DatagramSocket对象，并指定监听的端口号
DatagramSocket socket = new DatagramSocket(9999);
byte data [] = new byte[1024];
//创建一个空的DatagramPacket对象
DatagramPacket packet =new DatagramPacket(data,data.length);
//使用receive方法接收客户端发来的数据，如客户端没发来数据，就停滞等待
socket.receive(packet);
String result = new String(packet.getData(),packet.getOffset(),packet.getLength());
System.out.println("result--->" + result);  //输出收到的数据
```

2) 客户端主要代码如下：

```
public static void main(String[] args) {
  try {
     //首先创建一个DatagramSocket对象
   DatagramSocket socket = new DatagramSocket(9999);
    //创建一个InetAddree
        InetAddress serverAddress = InetAddress.getByName("211.69.1.1");
    String str = "This is a test! ";        //这是要传输的数据
    byte data [] = str.getBytes();           //把传输内容分解成字节
    DatagramPacket packet= new DatagramPacket(data,data.length,
       serverAddress,9999);
             //创建一个DatagramPacket对象，并指定发送的目的IP地址及端口号
    socket.send(packet);        //调用socket对象的send方法，发送数据
    } catch (Exception e) {
   e.printStackTrace();}}
```

3) 配置AndroidManifest.xml，添加相应的权限，添加如下语句：

```
<uses-permission android:name="android.permission.INTERNET"/>
```

保存所有文件，先运行服务器端，然后运行客户端程序，可以在服务器端发现已经接收到了客服端发来的数据，如图 9.25 所示。

图 9.25　服务器端的显示界面

9.5.3　利用 Socket 实现简易的聊天室

上面的例子中实现了一个客户端和一个服务器的单独通信，并且只能一次通信，在实际中，往往需要在服务器上运行一个程序，它用来接收来自多个客户端的请求，并提供相应服务。这就需要多线程来实现。服务器总是在指定的端口上监听是否有客户请求，一旦监听到客户请求，服务器就会启动一个专门的服务线程来响应该客户的请求，而服务器本身在启动完线程后马上又进入监听状态，等待下一个客户。

如果使用 Socket 通信实现一个简单的聊天室程序，Client1 发信息给 Client2，Client1 的信息先发送到服务器端，服务器端接收到信息后再把 Client1 的信息广播发送给所有的客户端。

服务器端：

1）首先我们要在服务器建立一个 ServerSocket，ServerSocket 对象用于监听来自客户端的 Socket 连接，如果没有连接，它将一直处于等待状态；

```
ServerSocket ss = new ServerSocket(10000);
                        //创建一个 ServerSocket，监听客户端 Socket 的连接请求
while (true){……        //采用不断循环的方式接收来自客户端的请求
```

2）Socket accept()：如果接收到一个客户端 Socket 的连接请求，该方法将返回一个与客户端 Socket 对应的 Socket；

```
Socket s = ss.accept();//接收到客户端 Socket 的请求，服务器端也对应产生一个 Socket
```

3）然后就可以使用 Socket 和客户端进行通信了。

```
……}
```

客户端：

1）创建连接到服务器、指定端口 30000 的 Socket；

```
Socket s = new Socket("192.168.2.214" , 30000);//和服务器建立连接
```

2）然后就可以使用 Socket 和服务器进行通信了。

```
……
```

实例 9-14：基于 TCP 协议的 Socket 实例

使用 Socket 通信实现一个简单的聊天室程序。

1）用 Eclipse 在服务器端建立一个项目，server9_14，主程序 Server.java 的代码如下：

```java
import java.io.BufferedReader;
import java.io.BufferedWriter;
import java.io.IOException;
import java.io.InputStreamReader;
import java.io.OutputStreamWriter;
import java.io.PrintWriter;
import java.net.ServerSocket;
import java.net.Socket;
import java.util.ArrayList;
import java.util.List;
```

```java
public class Server {
 private static final int port = 54321;
 private static List<Socket> clientList = new ArrayList<Socket>();
 private static ServerSocket ss;
 private static Socket s;
 public static void main(String[] args) {
  try {
   ss = new ServerSocket(port);
   while (true) {
    s = ss.accept();
    clientList.add(s);
    Thread mt = new Thread(new ServerThread(s));
    mt.start();   }
  } catch (IOException e) {
   e.printStackTrace();  }  }
 static class ServerThread implements Runnable {
  private BufferedReader in;
  private PrintWriter out;
  private Socket socket;
  private String messages;
  public ServerThread(Socket s) throws IOException {
   this.socket = s;
   in = new BufferedReader(new InputStreamReader(s.getInputStream(),
     "GBK"));
   messages = "user:" + this.socket.getInetAddress()
     + "connction count:" + clientList.size();
   sendmessages();  }
  public void run() {
   try {
    while ((messages = in.readLine()) != null) {
     if (messages.trim().equals("exit")) {
      clientList.remove(socket);
      in.close();
      out.close();
      messages = "user:" + this.socket.getInetAddress()
        + "exit count:" + clientList.size();
      socket.close();
      sendmessages();
      break;
     } else {
      messages = socket.getInetAddress() + ":" + messages;
      sendmessages();     }    }
   } catch (IOException e) {
    e.printStackTrace();   }  }
  private void sendmessages() throws IOException {
   System.out.println(messages);
   for (Socket s : clientList) {
    out = new PrintWriter(new BufferedWriter(
      new OutputStreamWriter(s.getOutputStream(), "GBK")));
    out.println(messages);
    out.flush();    }  }}
```

2) 客户端新建一个项目，项目的命名为：exam9_14，包名称为：org.hnist.demo。
① 建立一个布局管理文件，Activity_main.xml，代码如下：

```xml
<?xml version="1.0" encoding="utf-8"?>
<LinearLayout xmlns:android="http://schemas.android.com/apk/res/android"
    android:orientation="vertical"
    android:layout_width="fill_parent"
    android:layout_height="fill_parent"
    android:background="#000000">
    <EditText
        android:id="@+id/showtext"
        android:layout_width="fill_parent"
        android:layout_height="wrap_content"/>
    <EditText
        android:id="@+id/sendtext"
        android:layout_width="fill_parent"
        android:layout_height="wrap_content"/>
    <Button
        android:text="发送"
        android:id="@+id/send"
        android:layout_width="fill_parent"
        android:layout_height="wrap_content" />
    <Button
        android:text="连接服务器"
        android:id="@+id/connect"
        android:layout_width="fill_parent"
        android:layout_height="wrap_content"/>
</LinearLayout>
```

② 建立一个 Activity 文件 socketClient.java，代码如下：

```java
package org.hnist.demo;
import java.io.BufferedReader;
import java.io.BufferedWriter;
import java.io.IOException;
import java.io.InputStreamReader;
import java.io.OutputStreamWriter;
import java.io.PrintWriter;
import java.net.Socket;
import java.net.UnknownHostException;
import android.app.Activity;
import android.os.Bundle;
import android.os.Handler;
import android.view.View;
import android.view.View.OnClickListener;
import android.widget.Button;
import android.widget.EditText;
public class MainActivity extends Activity {
 private static final String IP = "10.0.2.2";
 private static final int port = 54321;
 private EditText showmsg = null;
 private EditText sendmsg = null;
```

```java
    private Button btnsend, btnconn;
    private Socket s;
    private PrintWriter out = null;
    private BufferedReader in = null;
    private String getmessages = "";
    @Override
    public void onCreate(Bundle savedInstanceState) {
     super.onCreate(savedInstanceState);
     setContentView(R.layout.activity_main);
     showmsg = (EditText) findViewById(R.id.showtext);
     sendmsg = (EditText) findViewById(R.id.sendtext);
     btnconn = (Button) findViewById(R.id.connect);
     btnsend = (Button) findViewById(R.id.send);
     btnconn.setOnClickListener(new OnClickListener() {
      @Override
      public void onClick(View v) {
       try {
        s = new Socket(IP, port);
        in = new BufferedReader(new InputStreamReader(s
          .getInputStream(), "GBK"));
        out = new PrintWriter(
          new BufferedWriter(new OutputStreamWriter(s
            .getOutputStream(), "GBK")), true);
        Thread mThread = new Thread(mRunable);
        mThread.start();
       } catch (UnknownHostException e) {
        e.printStackTrace();
       } catch (IOException e) {
        e.printStackTrace();      }    }  });
    btnsend.setOnClickListener(new OnClickListener() {
     @Override
     public void onClick(View v) {
      String messages = sendmsg.getText().toString().trim();
      try {
       out.println(messages);
       out.flush();
      } catch (Exception e) {
       e.printStackTrace();     }   }); }
   private Runnable mRunable = new Runnable() {
    @Override
    public void run() {
     while (true) {
      try {
       if ((getmessages = in.readLine()) != null) {
        getmessages += "\n";
        mHandler.sendMessage(mHandler.obtainMessage());
       }  } catch (IOException e) {
        e.printStackTrace();   }  }  };
  Handler mHandler = new Handler() {
   public void handleMessage(android.os.Message msg) {
    super.handleMessage(msg);
    showmsg.append(getmessages);  };  };}
```

③ 配置 AndroidManifest.xml，添加相应的权限，添加如下语句：

```
<uses-permission android:name="android.permission.INTERNET"/>
```

保存所有文件，先运行服务器端程序，然后运行客户端程序，如图 9.26 所示，先单击"连接服务器"按钮，然后发送信息，如图 9.27 所示，可以在服务器端发现已经接收到了客服端发来的数据，服务器端显示的信息如图 9.28 所示。

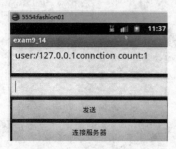
图 9.26　客户端显示信息

图 9.27　客户端发送信息

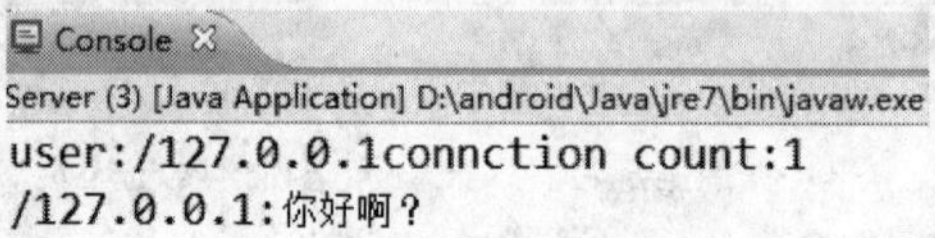

图 9.28　服务器端显示信息

9.6　Web Service 通信

与 HTTP 通信方式相比，HTTP 不能实现远程方法的调用，Web Service 可以实现。Web Service 也有服务器端和客户端，一般采用 XFire 来搭建服务器端。客户端 Android 系统中并没有提供直接与 Web Service 互调用的操作类库，一般通过调用第三方提供的类库才可以完成，比较常用的就是 Ksoap2 类库。

要想使用 XFire 来搭建 Web Service 的服务器端，可以用集成了 XFire 的 MyEclipse（例如：MyEclipse8.5）来实现。

客户端可从http://code.google.com/p/ksoap2-android/downloads/list 下载 KSOAP包，ksoap2-android-assembly-2.4-jar-with-dependencies.jar 文件，通过它来实现客户端与服务器的 WebService 通信。

下面通过一个实例来介绍 Web Service 通信。

一个完整的 Web Service 通信应该有服务器端的建设和客户端的建设两个部分，服务器端的建设比较繁琐，可以借助内置 Tomcat 的 MyEclipse 8.5 来实现，当然也可以有其他的办法，这不是本书讨论的范围，请参考相关资料自行建立。

其实网络上有很多已经建立好了的服务器端，在服务器端提供开发好的方法供外界使用，例如：在 http://www.webxml.com.cn 服务器上对外公开了许多方法，例如，天气情况、航班信息、手机归属地、火车时刻表等，客户端只要获得这些方法的名称和相关参数，就可以编程调用这些方法，没有必要自己再去建立服务器端。

例如：http://webservice.webxml.com.cn/WebServices/WeatherWS.asmx?wsdl可以查看关于"天气"方法的名称和对应参数以及返回值。

例如：

```
getSupportCityString  //获得支持的城市/地区名称和与之对应的 ID
```

输入参数：theRegionCode = 省市、国家 ID 或名称，返回数据：一维字符串数组。

```
getWeather  //获得天气预报数据
```

输入参数：城市/地区 ID 或名称，返回数据：一维字符串数组。

客户端的建设

可从 http://code.google.com/p/ksoap2-android/downloads/list 下载 KSOAP 包，ksoap2-android-assembly-2.4-jar-with-dependencies.jar 文件，通过它来实现客户端与服务器的 WebService 通信。

客户端按照以下几个步骤来调用 WebService 方法。

1）要将它加载到 Android 项目，新建一个项目，右击该项目，选择 Build Path→Configure Build Path，如图 9.29 所示。

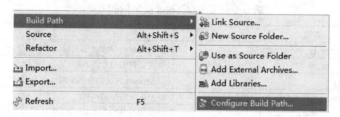

图 9.29 新建项目，添加 KSOAP 包

2）选择 Libraries，如图 9.30 所示，再选择 Add External JARs，选择 KSOAP 包下载的路径，如图 9.31 所示。

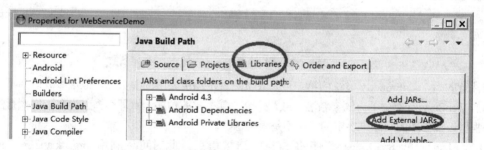

图 9.30 从外部添加 KSOAP 包

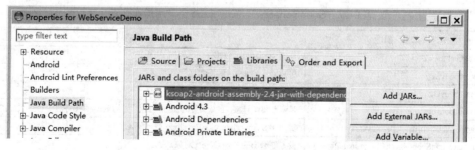

图 9.31 添加 KSOAP 包成功

单击"OK"按钮，加载完成。不加载 KSOAP 包到项目的话，后面的操作将无法进行。

注意：有的时候这样添加不能成功，可以在项目下建立一个文件夹 libs，将 KSOAP 包复制到 libs 文件夹下，右击 KSOAP 包，选择 Build Path→Add to Build Path 来添加。

3）实例化 SoapObject 对象，指定 Web Service 的命名空间（从相关 WSDL 文档中可以查看命名空间），以及调用方法名称；如：

```
private static final String NAMESPACE = "http://WebXml.com.cn/";//命名空间
//定义请求 WSDL 文档 URL
private static String URL="http://www.webxml.com.cn/webservices/
                        weatherwebservice.asmx";
private static final String METHOD_NAME = "getWeatherbyCityName";//调用方法
//定义命名空间+方法名称
private static String SOAP_ACTION= "http://WebXml.com.cn/getWeatherbyCityName";
SoapObject rpc = new SoapObject(NAMESPACE, METHOD_NAME); //实例化 SoapObject 对象
```

SoapObject 类常用的方法见表 9-16。

表 9-16　SoapObject 类常用的方法

方法	描述
public SoapObject(String namespace, String name)	实例化 SoapObject 类对象
public String getName()	取得要调用的方法名称
public String getNamespace()	取得 Soap 对象的命名空间
public Object getProperty(java.lang.String name)	取出指定名称的属性
public SoapObject addProperty(String name, Object value)	设置调用 Web Service 方法时所需要的参数

4）当取得了 SoapObject 的实例化对象之后，就可以通过 SoapObject 类的 addProperty()方法设置调用服务器端提供的方法时所需要的参数，而设置的顺序要与服务器端提供方法的参数顺序符合，如果方法没有参数，可以省略这一步；

```
request.addProperty("参数名称","参数值");
```

例如：

```
rpc.addProperty("theCityName", cityName);      //添加参数
```

要注意的是，addProperty 方法的第 1 个参数虽然表示调用方法的参数名，但该参数值并不一定与服务端的 WebService 类中的方法参数名一致，只要设置参数的顺序一致即可，因为在程序执行的时候也只是根据设置参数的顺序来决定的，而不是根据名称。

5）生成调用 Web Service 程序的 SOAP 请求信息，此时可以利用 org.ksoap2.serialization.SoapSerializationEnvelope 类完成；SoapSerializationEnvelope 类常用操作列表见表 9-17。

表 9-17　SoapSerializationEnvelope 类常用操作列表

方法、常量、属性	描述
public static final int VER11	使用 SOAP 11 版本操作
public Object bodyIn	封装输出的 SoapObject 对象
public Object bodyOut	封装输入的 SoapObject 对象
public boolean dotNet	是否为.NET 连接，此处设置为 false，如果设置为 true，则服务器端无法接收请求参数
public SoapSerializationEnvelope(int version)	实例化 SoapSerializationEnvelope 类对象
public void setOutputSoapObject(Object soapObject)	设置要输出的 SoapObject 对象

设置 SOAP 请求信息（参数部分为 SOAP 协议版本号，与你要调用的 webService 中版本号一致）；例如：

```
//实例化 SoapSerializationEnvelope 类对象 Envelope
SoapSerializationEnvelope envelope=new SoapSerializationEnvelope(SoapEnvelope.VER11);
envelope.bodyOut = rpc;              //封装输入的 SoapObject 对象 rpc
envelope.dotNet = true;              //要访问的服务器是.net 服务器,否则设置为 false
envelope.setOutputSoapObject(rpc);   //设置要输出的 SoapObject 对象
……
```

创建 SoapSerializationEnvelope 对象时需要通过 SoapSerializationEnvelope 类的构造方法设置 SOAP 协议的版本号。该版本号需要根据服务端 Web Service 的版本号设置。在创建 SoapSerializationEnvelope 对象后,设置 SoapSerializationEnvelope 类的 bodyOut 属性,该属性的值就是在第 1 步创建的 SoapObject 对象。

6)创建 org.ksoap2.transport.HttpTransportSE 类对象,定义传输对象,并指明 WSDL 文档 URL,并且利用此对象调用 Web Service 端的操作方法:

```
//定义请求 WSDL 文档 URL
private static String URL="http://www.webxml.com.cn/webservices/
                          weatherwebservice.asmx";
HttpTransportSE ht = new HttpTransportSE(URL);   //指定 WSDL 地址
ht.debug = true;                                  //使用调试
```

HttpTransportSE 类常用操作列表见表 9-18。

表 9-18　HttpTransportSE 类常用操作列表

方法、属性	描述
public boolean debug	是否调试,如果设置为 true 则表示调试
public HttpTransportSE(String url)	实例化 HttpTransportSE 类的对象
public void call(String soapAction, SoapEnvelope envelope) throws IOException, org.xmlpull.v1.XmlPullParserException	调用 Web Service 端的操作方法

7)调用 WebService(其中参数为 1:命名空间+方法名称,2:Envelope 对象):

```
//定义命名空间+方法名称
private static String SOAP_ACTION= "http://WebXml.com.cn/getWeatherbyCityName";
ht.call(SOAP_ACTION, envelope);     //调用 WebService
```

call 方法的第 1 个参数一般为命名空间+方法名称,第 2 个参数就是在第 4 步创建的 SoapSerializationEnvelope 对象。

8)如果现在要接收 Web Service 的返回值,则可以直接通过以下两种方式完成:

```
SoapObject result = (SoapObject) envelope.bodyIn;         //接收返回值
detail = (SoapObject) envelope.getResponse();             //接收返回值
```

9)项目要访问服务器端程序属于网络调用,所以还要修改 AndroidManifest.xml 文件配置网络访问权限,添加如下语句:

```
<uses-permission android:name="android.permission.INTERNET" />
```

实例 9-15:Web Service 通信实例客户端的建设

在客户端新建一个项目,项目的命名为:exam9_15,包名称为:org.hnist.demo。

1)建立布局管理文件 Activity_main.xml,代码如下:

```
<?xml version="1.0" encoding="utf-8"?>
<LinearLayout xmlns:android="http://schemas.android.com/apk/res/android"
    android:orientation="vertical"
```

```xml
        android:layout_width="fill_parent"
        android:layout_height="fill_parent"    >
     <EditText
           android:id="@+id/et"
           android:layout_width="180dp"
           android:layout_height="50dp"
           android:text="请输入城市"
           android:selectAllOnFocus="true"   />
     <Button
           android:id="@+id/search_but"
           android:layout_width="fill_parent"
           android:layout_height="wrap_content"
           android:text="获得天气信息"    />
     <TextView
           android:text=""
           android:id="@+id/TextView01"
           android:layout_width="wrap_content"
           android:layout_height="wrap_content"
           android:textSize="20px"/>
</LinearLayout>
```

2）建立 Activity 文件 MainActivity.java，代码如下：

```java
package org.hnist.demo;
import java.io.UnsupportedEncodingException;
import org.ksoap2.SoapEnvelope;
import org.ksoap2.serialization.SoapObject;
import org.ksoap2.serialization.SoapSerializationEnvelope;
import org.ksoap2.transport.HttpTransportSE;
import android.app.Activity;
import android.os.Bundle;
import android.view.View;
import android.view.View.OnClickListener;
import android.widget.Button;
import android.widget.EditText;
import android.widget.ImageView;
import android.widget.TextView;
public class MainActivity extends Activity {
  private Button search_but;
  private TextView textview1;
  private EditText et;
  public void onCreate(Bundle savedInstanceState) {
   super.onCreate(savedInstanceState);
   setContentView(R.layout.activity_main);
   et=(EditText)findViewById(R.id.et);
   search_but=(Button)findViewById(R.id.search_but);
   search_but.setOnClickListener(listener);     } //注册事件
   private OnClickListener listener=new OnClickListener() {
   public void onClick(View v) {  //事件的实现内容
       String city=et.getText().toString();
```

```java
            getWeather(city);    }   };  //当单击时,调用 getWeather()方法
private static final String NAMESPACE = "http://WebXml.com.cn/";      //命名空间
//定义请求 WSDL 文档 URL
private static String URL="http://www.webxml.com.cn/webservices/
                         weatherwebservice.asmx";
private static final String METHOD_NAME = "getWeatherbyCityName";  //调用方法
//定义命名空间+方法名称
private static String SOAP_ACTION= "http://WebXml.com.cn/getWeatherbyCityName";
   private String weatherToday;
   private SoapObject detail;
   private String weatherNow;
   private String weatherWillBe;
   public void getWeather(String cityName) {  //定义 getWeather()方法
     try {
       textview1 = (TextView) this.findViewById(R.id.TextView01);
       //实例化 SoapObject 对象
       SoapObject rpc = new SoapObject(NAMESPACE, METHOD_NAME);
       rpc.addProperty("theCityName", cityName);  //添加参数
//实例化 SoapSerializationEnvelope 类对象 Envelope
SoapSerializationEnvelope envelope=new SoapSerializationEnvelope(SoapEnvelope.VER11);
       envelope.bodyOut = rpc;         //封装输入的 SoapObject 对象 rpc
       envelope.dotNet = true;         //要访问的服务器是.net 服务器,否则设置为 false
       envelope.setOutputSoapObject(rpc);  //设置要输出的 SoapObject 对象
       HttpTransportSE ht = new HttpTransportSE(URL);  //指定 WSDL 地址
       ht.debug = true;                         //使用调试
       ht.call(SOAP_ACTION, envelope);           //调用 WebService
       detail = (SoapObject) envelope.getResponse();  //获得返回信息
       parseWeather(detail);                    //调用 parseWeather 方法
       return;
     } catch (Exception e) {
       e.printStackTrace();   }  }
     private void parseWeather(SoapObject detail)  //定义 parseWeather 方法
       throws UnsupportedEncodingException {
       textview1 = (TextView) this.findViewById(R.id.TextView01);
       String date = detail.getProperty(6).toString();
       weatherToday = "\n 天气: " + date.split(" ")[1];
       weatherToday = weatherToday + "\n 气温: "
         + detail.getProperty(5).toString();
       weatherToday = weatherToday + "\n 风力: "
         + detail.getProperty(7).toString() + "\n";
       weatherNow = detail.getProperty(8).toString();
       weatherWillBe = detail.getProperty(9).toString();
       textview1.setText(et.getText() + weatherToday);  }  }
```

3)配置 AndroidManifest.xml,添加相应的权限,添加如下语句:

```xml
<uses-permission android:name="android.permission.INTERNET"/>
```

保存所有文件,运行项目,如图 9.32 所示,输入城市名如:岳阳,然后单击按钮,获得天气信息,如图 9.33 所示。

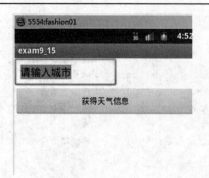

图9.32 运行的初始界面

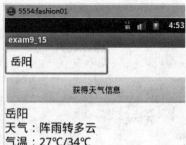

图9.33 获取指定城市的天气信息

9.7 蓝牙通信

9.7.1 蓝牙通信基础

在开发时，蓝牙需要硬件支持，模拟器上不能模拟蓝牙，需要在真正的手机上进行蓝牙功能的测试，如果要实现数据的传输，那么至少需要两台带蓝牙功能的手机。Android 系统的手机一般都带有蓝牙功能，可以很方便地进行程序的调试。

Android 所有关于蓝牙开发的类都在 android.bluetooth 包下，一共有 8 个类，如图 9.34 所示。以下是建立蓝牙连接所需要的一些基本类。

1）BluetoothAdapter 类：代表了一个本地的蓝牙适配器，是所有蓝牙交互的的入口点。利用它你可以发现其他蓝牙设备，查询绑定了的设备，使用已知的 MAC 地址创建BluetoothDevice，建立一个 BluetoothServerSocket（作为服务器端）来监听来自其他设备的连接。

android.bluetooth 类
BluetoothAdapter
BluetoothClass
BluetoothClass.Device
BluetoothClass.Device.Major
BluetoothClass.Service
BluetoothDevice
BluetoothServerSocket
BluetoothSocket

图 9.34 bluetooth 的 8 个子类

BluetoothAdapter 类的常用常量见表 9-19。

表 9-19 BluetoothAdapter 类的常用常量

方法	描述
int STATE_OFF	蓝牙已经关闭
int STATE_ON	蓝牙已经打开
int STATE_TURNING_OFF	蓝牙处于关闭过程中
int STATE_TURNING_ON	蓝牙处于打开过程中
int SCAN_MODE_CONNECTABLE	表明该蓝牙可以扫描其他蓝牙设备
int SCAN_MODE_CONNECTABLE_DISCOVERABLE	表明该蓝牙设备同时可以扫描其他蓝牙设备，并且可以被其他蓝牙设备扫描到
int SCAN_MODE_NONE	该蓝牙不能扫描以及被扫描
ACTION_STATE_CHANGED	蓝牙状态值发生改变
ACTION_SCAN_MODE_CHANGED	蓝牙扫描状态(SCAN_MODE)发生改变
ACTION_DISCOVERY_STARTED	蓝牙扫描过程开始
ACTION_DISCOVERY_FINISHED	蓝牙扫描过程结束
ACTION_LOCAL_NAME_CHANGED	蓝牙设备 Name 发生改变
ACTION_REQUEST_DISCOVERABLE	请求用户选择是否使该蓝牙能被扫描
ACTION_REQUEST_ENABLE	请求用户选择是否打开蓝牙
ACTION_FOUND	蓝牙扫描时，扫描到任一远程蓝牙设备时，会发送此广播

BluetoothAdapter 类的常用方法见表 9-20。

表 9-20 BluetoothAdapter 类的常用方法

方法	描述
public static synchronized BluetoothAdapter getDefaultAdapter()	获得本设备的蓝牙适配器实例,若设备具备蓝牙功能,返回 BluetoothAdapter 实例,否则返回 null 对象
public boolean enable()	打开蓝牙设备
public boolean disable()	关闭蓝牙设备
public boolean startDiscovery()	扫描蓝牙设备
public boolean cancelDiscovery()	取消扫描过程
public boolean isDiscovering()	是否正在处于扫描过程中
public boolean isEnabled ()	是否已经打开蓝牙
public String getName()	获取蓝牙设备 Name
public String getAddress()	获取蓝牙设备的硬件地址
public boolean setName (String name)	设置蓝牙设备的 Name
public Set<BluetoothDevice> getBondedDevices()	获取与本机蓝牙所有绑定的远程蓝牙信息
public static boolean checkBluetoothAddress(String address)	验证蓝牙设备 MAC 地址是否有效
public BluetoothDevice getRemoteDevice (String address)	根据蓝牙地址获取蓝牙设备
public String getState()	获取本地蓝牙适配器当前状态
public BluetoothServerSocket listenUsingRfcommWithServiceRecord (String name, UUID uuid)	创建一个正在监听的 RFCOMM 蓝牙端口

关于 UUID:如果你正试图连接蓝牙串口,那么使用众所周知的 SPP UUID 00001101-0000-1000-8000-00805F9B34FB。如果正试图连接 Android 设备,那么要生成自己的专有 UUID,查询 RFCOMM 通道的服务记录 UUID。

2) BluetoothDevice 类:代表了一个远端的蓝牙设备,使用它请求远端蓝牙设备连接或者获取远端蓝牙设备的名称、地址、种类和绑定状态。(其信息是封装在 bluetoothsocket 中)。

BluetoothDevice 类的常用常量见表 9-21。

表 9-21 BluetoothDevice 类的常用常量

方法	描述
String ACTION_BOND_STATE_CHANGED	指明一个远程设备的连接状态的改变
String ACTION_FOUND	发现远程设备
String ACTION_NAME_CHANGED	指明一个远程设备的名称第一次找到,或者自从最后一次找到该名称开始已经改变
int BOND_BONDED	表明蓝牙已经绑定
int BOND_BONDING	表明蓝牙正在绑定过程中
int BOND_NONE	表明没有绑定

BluetoothDevice 类的常用方法见表 9-22。

表 9-22 BluetoothDevice 类的常用方法

方法	描述
public BluetoothSocketcreateRfcommSocketToServiceRecord (UUID uuid)	根据 UUID 创建并返回一个 BluetoothSocket
public String getAddress()	返回该蓝牙设备的硬件地址
public BluetoothClass getBluetoothClass()	获取远程设备的蓝牙类
public int getBondState()	获取远程设备的连接状态
public String getName()	获取远程设备的蓝牙名称
public String toString()	返回该蓝牙设备的字符串表达式

3）Bluetoothsocket 类：蓝牙通信分为服务器端和客户端，它们之间使用 BluetoothSocket 类的不同方法来获取数据，该类代表了一个蓝牙套接字的接口（类似于 TCP 中的套接字），是应用程序通过输入、输出流与其他蓝牙设备通信的连接点。Bluetoothsocket 类的常用方法见表 9-23。

表 9-23　Bluetoothsocket 类的常用方法

方法	描述
public void close()	与服务器断开关闭
public void connect()	与服务器建立连接
public InputStream getInputStream()	获取输入流
public OutputStream getOutputStream()	获取输出流
public BluetoothDevice getRemoteDevice()	获取 bluetoothSocket 指定连接的那个远程蓝牙设备
public boolean isConnected()	是否与远程蓝牙设备建立连接

4）Blueboothserversocket 类：打开服务连接来监听可能到来的连接请求（属于 Server 端），为了连接两个蓝牙设备，必须有一个设备作为服务器打开一个服务套接字。当远端设备发起连接请求的时候，并且已经连接到的时候，Blueboothserversocket 类将会返回一个 Bluetoothsocket。

Bluetoothsocket 类的常用方法见表 9-24。

表 9-24　Bluetoothsocket 类的常用方法

方法	描述
public void close()	马上关闭端口，并释放所有相关的资源
public BluetoothSocket accept()	返回一个已连接的 BluetoothSocket 类
public BluetoothSocket accept(int timeout)	返回一个指定了过时时间已连接的 BluetoothSocket 类

在服务器端，使用 BluetoothServerSocket 类来创建一个监听服务端口。当一个连接被 BluetoothServerSocket 所接受，它会返回一个新的 BluetoothSocket 来管理该连接。在客户端，使用一个单独的 BluetoothSocket 类去初始化一个外接连接和管理该连接。

通常使用的蓝牙端口是 RFCOMM，它是被 Android API 支持的类型。RFCOMM 是一个面向连接，通过蓝牙模块进行的数据流传输方式，它也被称为串行端口规范（Serial Port Profile，SPP）。

为了创建一个对准备好的新来的连接去进行监听BluetoothServerSocket类，使用 BluetoothAdapter.listenUsingRfcommWithServiceRecord()方法。然后调用accept()方法去监听该连接的请求。会产生一个BluetoothSocket类去管理该连接，如果不再需要接受连接，调用在BluetoothServerSocket类下的close()方法，会马上放弃外界操作并关闭服务器端口。更多的功能可以参考 Android API 进行学习。

9.7.2　蓝牙通信实现

要实现蓝牙操作，一般按照下面几个流程进行：

1. 获取本地蓝牙

```
BluetoothAdapter mAdapter= BluetoothAdapter.getDefaultAdapter();
```

2. 打开、关闭蓝牙

可以在系统设置里开启蓝牙，也可以在应用程序里启动蓝牙功能，有两种方法：

1）直接调用函数 enable()打开蓝牙设备，例如：

```
boolean result = mBluetoothAdapter.enable();
```

2）系统 API 的方式打开蓝牙设备，该方式会弹出一个对话框样式的 Activity 供用户选择是否打开蓝牙设备。

```
        if (!mBluetoothAdapter.isEnabled())  //如果没打开蓝牙功能，就用下面的语句打开
        { Intent intent = new Intent(BluetoothAdapter.ACTION_REQUEST_ENABLE);
          startActivityForResult(intent, REQUEST_OPEN_BT_CODE); }  //以 Dialog 样式
          //显示一个 Activity, 可以在 onActivityResult()方法去处理返回值
```

注意：如果蓝牙已经开启，不会弹出该 Activity 界面。在目前 Android 手机中，是不支持在飞行模式下开启蓝牙的。如果蓝牙已经开启，那么蓝牙的开关状态会随着飞行模式的状态而发生改变。

```
//打开本机的蓝牙发现功能
Intent discoveryIntent=new Intent(BluetoothAdapter .ACTION_REQUEST_DISCOVERABLE);
discoverableIntent.putExtra(BluetoothAdapter.EXTRA_DISCOVERABLE_DURATION, 300);
                         //设置持续时间（最多 300 秒）
mAdapter.disable();      //关闭蓝牙
```

蓝牙功能开启后，就可以查找周边存在的蓝牙设备了。

3．查找设备

使用 BluetoothAdapter 的 startDiscovery()方法来搜索蓝牙设备，startDiscovery()方法是一个异步方法，调用后会立即返回。该方法会进行对其他蓝牙设备的搜索，该方法调用后，搜索过程实际上是在一个 System Service 中进行的，所以可以调用 cancelDiscovery()方法来停止搜索（该方法可以在未执行 discovery 请求时调用）。

请求 Discovery 后，系统开始搜索蓝牙设备，在这个过程中，系统会发送以下三个广播：

ACTION_DISCOVERY_START：开始搜索

ACTION_DISCOVERY_FINISHED：搜索结束

ACTION_FOUND：找到设备，这个 Intent 中包含两个 extra fields：EXTRA_DEVICE 和 EXTRA_CLASS，分别包含 BluetooDevice 和 BluetoothClass。

可以注册相应的 BroadcastReceiver 来接收响应的广播，以便实现某些功能，例如：

```
//创建一个接收 ACTION_FOUND 广播的 BroadcastReceiver
private final BroadcastReceiver mReceiver = new BroadcastReceiver() {
    public void onReceive(Context context, Intent intent) {
        String action = intent.getAction();
        //发现设备
        if (BluetoothDevice.ACTION_FOUND.equals(action)) {
            //从 Intent 中获取设备对象
            BluetoothDevice device = intent.getParcelableExtra
                (BluetoothDevice.EXTRA_DEVICE);
            //将设备名称和地址放入 array adapter, 以便在 ListView 中显示
            mArrayAdapter.add(device.getName() + "\n" + device.getAddress());
        }   };
//注册 BroadcastReceiver
IntentFilter filter = new IntentFilter(BluetoothDevice.ACTION_FOUND);
registerReceiver(mReceiver, filter);  //不要忘了之后解除绑定
```

4．建立连接

两个蓝牙设备之间要进行连接，要有一个服务器端和客户端，这两个设备要在同一个 RFCOMM channel 下且拥有一个连接的 BluetoothSocket，这两个设备才可能建立连接。

服务器设备与客户端设备获取 BluetoothSocket 的途径是不同的。服务器设备是通过 accepted 一个 incoming connection 来获取的，而客户端设备则是通过打开一个到服务器的 RFCOMM channel 来获取的。

1)服务器端的实现:

首先通过调用 listenUsingRfcommWithServiceRecord(String, UUID)方法来获取 bluetoothserversocket 对象。

```
BluetoothServerSocket serverSocket = mAdapter.listenUsingRfcommWith-
                    ServiceRecord(serverSocketName,UUID);
```

其次调用 accept()方法来监听可能到来的连接请求,当监听到以后,返回一个连接上的蓝牙套接字 bluetoothsocket。

```
serverSocket.accept();
```

最后,在监听到一个连接以后,需要调用 close()方法来关闭监听程序。

```
serverSocket. close ();
```

例如:

```
private class AcceptThread extends Thread {
    private final BluetoothServerSocket mmServerSocket;
    public AcceptThread() {
        BluetoothServerSocket tmp = null;
        try {
            tmp=mBluetoothAdapter.listenUsingRfcommWithServiceRecord(NAME,MY_UUID);
        } catch (IOException e) { }
        mmServerSocket = tmp;       }
    public void run() {
        BluetoothSocket socket = null;
        while (true) {
            try {
                socket = mmServerSocket.accept();
            } catch (IOException e) {
                break;            }
            //If a connection was accepted
            if (socket != null) {
                manageConnectedSocket(socket);
                mmServerSocket.close();
                break;           }     }    }
    public void cancel() {
        try {
            mmServerSocket.close();
        } catch (IOException e) { }    } }
```

2)客户端的实现:

通过 bluetoothdevice 对象来获取 bluetoothsocket 并初始化连接,使用 bluetoothdevice 里的方法 createRfcommSocketToServiceRecord(UUID)来获取 bluetoothsocket。

```
BluetoothSocket clienSocket=dcvice.createRfcommSocketToServiceRecord(UUID);
```

调用 connect()方法。如果远端设备接收了该连接,它们将在通信过程中共享 RFFCOMM 信道,并且 connect()方法返回。

```
clienSocket.connect();
```

数据传输完成调用 close()方法来关闭连接。

例如：

```java
private class ConnectThread extends Thread { private final BluetoothSocket mmSocket;
    private final BluetoothDevice mmDevice;
    public ConnectThread(BluetoothDevice device) {
        BluetoothSocket tmp = null;
        mmDevice = device;
        try {
            tmp = device.createRfcommSocketToServiceRecord(MY_UUID);
        } catch (IOException e) { }
        mmSocket = tmp;          }
    public void run() {
        mBluetoothAdapter.cancelDiscovery();
        try {
            mmSocket.connect();
        } catch (IOException connectException) {
            try {
                mmSocket.close();
            } catch (IOException closeException) { }
            return;             }
        manageConnectedSocket(mmSocket);        }
    public void cancel() {
        try {
            mmSocket.close();
        } catch (IOException e) { }     } }
```

5. 数据传递

通过以上操作，就已经建立了 BluetoothSocket 连接了，数据传递无非是通过流的形式。

当设备连接上以后，每个设备都拥有各自的 Bluetoothsocket。就可以实现设备之间数据的共享了。

首先通过调用 getInputStream()和 getOutputStream()方法来获取输入/输出流。然后通过调用 read(byte[]) 和 write(byte[])方法来读取或者写数据。

例如：

```java
private class ConnectedThread extends Thread {
    private final BluetoothSocket mmSocket;
    private final InputStream mmInStream;
    private final OutputStream mmOutStream;
    public ConnectedThread(BluetoothSocket socket) {
        mmSocket = socket;
        InputStream tmpIn = null;
        OutputStream tmpOut = null;
        try {
            tmpIn = socket.getInputStream();
            tmpOut = socket.getOutputStream();
        } catch (IOException e) { }
        mmInStream = tmpIn;
        mmOutStream = tmpOut;     }
    public void run() {
        byte[] buffer = new byte[1024];
        int bytes;
        while (true) {
```

```
        try {
           //从 InputStream 读数据
           bytes = mmInStream.read(buffer);
           mHandler.obtainMessage(MESSAGE_READ, bytes, -1, buffer)
                  .sendToTarget();
        } catch (IOException e) {
           break;              }         }        }
   public void write(byte[] bytes) {
      try {
         mmOutStream.write(bytes);
      } catch (IOException e) { }       }
   public void cancel() {
      try {
         mmSocket.close();
      } catch (IOException e) { }       }  }
```

6. 修改 AndroidManifest.xml 文件配置权限

注意,在使用这些类时要修改 AndroidManifest.xml 文件配置权限,添加如下语句:

```
<uses-permission android:name="android.permission.BLUETOOTH_ADMIN" />
<uses-permission android:name="android.permission.BLUETOOTH" />
```

9.7.3 蓝牙通信实例

Android SDK 里自带了一个蓝牙聊天软件 BluetoothChat,路径如图 9.35 所示(会因为安装时路径不同而不同):

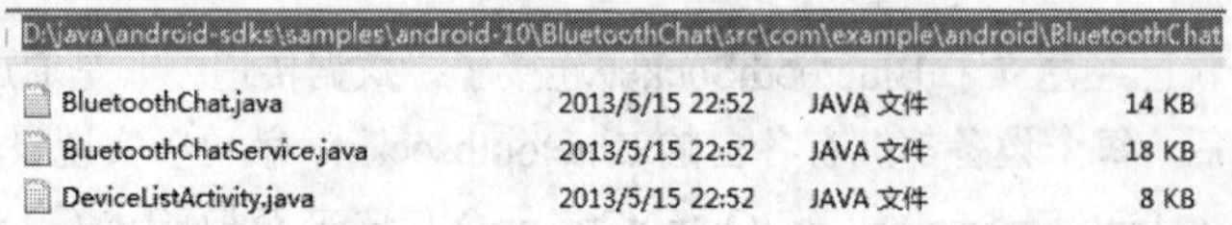

图 9.35 BluetoothChat 程序所处目录

该项目主要有 3 个 Java 文件和 5 个 xml 文件,结构如图 9.36 所示。

图 9.36 BluetoothChat 主程序框架图

限于篇幅原因，具体代码这里不给出了，读者可以将它导入，要注意的是，不能在模拟机上直接运行，要在真实手机上才能运行，具体操作见第 2 章。

9.8 WiFi 通信

在Android中提供了 android.net.wifi 包对 WIFI 进行操作，主要包括以下几个类：

1）ScanResult 类：该类主要是通过 WIFI 硬件的扫描来获取一些周边的 WIFI 热点的信息，包括接入点的地址、接入点的名称、身份认证、频率、信号强度等信息。ScanResult 类常用的变量见表 9-25。

表 9-25　ScanResult 类常用的变量

方法	描述
public String BSSID	WIFI 热点的地址
public String SSID	网络名称
public String capabilities	描述 WIFI 热点的认证，密钥管理以及加密方案等相关信息
public int frequency	客户端与 WIFI 热点通信信道的频率
public int level	主要来判断网络连接的优先数

这里只提供了一个方法，就是将获得信息变成字符串 public String toString()。

2）WifiConfiguration 类：该类主要用来进行 WiFi 网络的配置。WifiConfiguration 类常用的变量见表 9-26。

表 9-26　WifiConfiguration 类常用的变量

方法	描述
public String BSSID	当设置好后，这个网络配置入口只能当是指定 BSSID 的 AP 时候才调用
public String SSID	设置该网络的 SSID
public boolean hiddenSSID	隐藏的 SSID，即该网络不对 SSID 进行广播
public int networkId	客户端与 WIFI 热点通信信道的频率
public int level	这个网络配置入口的 ID
public int priority	配置的优先级
public int status	当前配置状态

这里提供的方法也是 public String toString()。

3）WifiInfo 类：WIFI 已经连接成功以后，可以通过这个类获得一些已经连通的 WIFI 连接的信息获取当前链接的信息，包括接入点、网络连接状态、隐藏的接入点、IP 地址、连接速度、MAC 地址、网络 ID、信号强度等信息。WifiInfo 类常用方法见表 9-27。

表 9-27　WifiInfo 类常用方法

方法	描述
public String getBSSID()	获取 BSSID
public String getSSID()	获得 SSID
public static NetworkInfo.DetailedState getDetailedStateOf(SupplicantState suppState)	获取客户端的连通性
public boolean getHiddenSSID()	获得 SSID 是否被隐藏
public int getIpAddress()	获取 IP 地址
public int getLinkSpeed()	获得连接的速度

方法	描述
public String getMacAddress()	获得 Mac 地址
public int getRssi()	获得 802.11n 网络的信号
public SupplicantState getSupplicanState()	返回具体客户端状态的信息
public String toString()	转换为字符串
public int getnetworkId()	获得通信信道的频率

4）WifiManager 类：该类用来管理 WIFI 连接，这是最重要的一个类。WifiManager 类常用常量见表 9-28。

表 9-28 WifiManager 类常用常量

方法	描述
public static final int WIFI_STATE_DISABLING	0 表示网卡正在关闭
public static final int WIFI_STATE_DISABLED	1 表示网卡不可用
public static final int WIFI_STATE_ENABLING	2 表示网卡正在打开
public static final int WIFI_STATE_ENABLED	3 表示网卡可用
public static final int WIFI_STATE_UNKNOWN	4 表示未知网卡状态

WifiManager 类常用方法见表 9-29。

表 9-29 WifiManager 类常用方法

方法	描述
public int addNetwork(WifiConfiguration config)	通过获取的网络链接状态信息，来添加网络
public static int compareSignalLevel(int rssiA, int rssiB)	对比连接 A 和连接 B
public static int calculateSignalLevel (int rssi , int numLevels)	计算信号的等级
public boolean disableNetwork(int netId)	让一个网络连接失效
public boolean disconnect()	断开连接
public boolean enableNetwork(int netId, Boolean disableOthers)	连接一个连接
public WifiManager.WifiLock createWifiLock (String tag)	创建一个新的 wifi 锁，锁定当前的 wifi 连接
public WifiManager.WifiLock createWifiLock (int lockType, String tag)	创建一个指定类型的 wifi 锁，锁定当前的 wifi 连接
public List<WifiConfiguration> getConfiguredNetworks()	获取网络连接的状态
public WifiInfo getConnectionInfo()	获取当前连接的信息
public List<ScanResult> getScanResulats()	获取扫描测试的结果
public DhcpInfo getDhcpInfo()	获取 DHCP 的信息
public int getWifiState()	获取一个 wifi 接入点是否有效
public boolean isWifiEnabled()	判断一个 wifi 连接是否有效
public boolean pingSupplicant()	ping 一个连接，判断是否能连通
public boolean ressociate()	即便连接没有准备好，是否也要连通
public boolean reconnect()	如果连接准备好了，是否连通
public boolean removeNetwork()	是否移除某一个网络
public boolean saveConfiguration()	是否保留一个配置信息
public boolean setWifiEnabled()	是否让一个连接有效
public boolean startScan()	是否开始扫描
public int updateNetwork(WifiConfiguration config)	更新一个网络连接的信息

更多常量和方法可以参考 android.net.wifi 包的 API，参考网址：
http://developer.android.com/reference/android/net/wifi/package-summary.html

对 WIFI 网卡的基本操作

对 WIFI 网卡进行操作需要通过 WifiManger 对象来进行，首先要获取该对象，方法如下：

```
WifiManger wifiManger = (WifiManger)Context.getSystemService(Service.WIFI_SERVICE);
wifiManger.setWifiEnabled(true);          //打开 WIFI 网卡
wifiManger.setWifiEnablee(false);         //关闭 WIFI 网卡
wifiManger.getWifiState();                //获取网卡的当前的状态
wifiManger.addNetwork();                  //添加一个配置好的网络连接
wifiManger.calculateSignalLevel();        //计算信号的强度
wifiManger.compareSignalLevel();          //比较两个信号的强度
wifiManger.createWifiLock();              //创建一个 WiFi 锁
wifiManger.disconnect();                  //从接入点断开
wifiManger.updateNetwork();               //更新已经配置好的网络
```

Android 应用程序要想对手机的 WIFI 网卡进行操作，需要在 Manifest.xml 中配置相应的权限：

```
<uses-permission                          //修改网络状态的权限
    android:name="android.permission.CHANGE_NETWORK_STATE"></uses-permission>
<uses-permission                          //修改 WIFI 状态的权限
    android:name="android.permission.CHANGE_WIFI_STATE"></uses-permission>
<uses-permission                          //访问网络权限
    android:name="android.permission.ACCESS_NETWORK_STATE"></uses-permission>
<uses-permission                          //访问 WIFI 权限
    android:name="android.permission.ACCESS_WIFI_STATE"></uses-permission>
```

具体实例可以参考 exam9_16 源程序。

本章小结

本章着重介绍了了几种常见网络通信技术：WebView 组件、HTTP 通信技术、Socket 通信技术、Web Services 通信技术、蓝牙和 WiFi 通信技术，其中最简单的就是 WebView 组件，Web Services 通信技术比较复杂，但是运用相当广泛。蓝牙和 WiFi 通信技术要借助硬件设备，在模拟机上不能正常运行。

习题

1. 利用 WebView 组件，如何打开一个指定的网页，给出具体步骤和关键代码。
2. 利用 WebView 组件加载一个 Html 文件和加载一个 Jsp 文件，操作有区别吗？主要区别在哪里？
3. HttpURLConnection 和 HttpClient 利用 POST 方式传递数据有哪些不同？
4. HttpURLConnection 利用 GET 方式传递数据给 JSP 文件的步骤有哪些？写出关键代码。
5. Socket 通信时要建立服务器端和客户端，其中关于端口的部分是如何设计的？分别写出服务器端和客户端关于端口的语句。
6. Web Services 通信时，客户端是通过什么语句传递参数给服务器端的？
7. 在网上查找关于火车时刻表的服务接口程序和参数。
8. 在真实手机上运行实例 9-16 和关于蓝牙的程序 BluetoothChat。

参 考 文 献

[1] Android 中文 API. http://www.android-doc.com/.
[2] Android API 参考资源. http://developer.android.com/reference/packages.html.
[3] 佘志龙等. Android SDK 开发范例大全（第 3 版）[M]. 北京：人民邮电出版社，2011，11.
[4] 李刚. 疯狂 Android 讲义（第 2 版）[M]. 北京：电子工业出版社，2013.3.
[5] 王世江等. Google Android 开发入门指南[M]. 北京：人民邮电出版社，2009，11.
[6] 吴亚峰等. Android 应用案例开发大全[M]. 北京：人民邮电出版社，2011，9.
[7] SatyaKomatineni，杨越译. 精通 Android 3[M]. 北京：人民邮电出版社，2011，11.
[8] 明日科技. Android 从入门到精通[M]. 北京：清华大学出版社，2012，9.
[9] 陈璟，陈平华，李文亮. Android 内核分析[J]. 广东工业大学计算机学院实践与经验，2009:112-115.
[10] 刘卫国，姚昱禹. Android 的架构与应用开发研究[J]. 中南大学：信息科学与工程学院 计算机系统应用 2008-11:110-112.